Nutrigenetics
Applying the Science
of Personal Nutrition

Nutrigenetics
Applying the Science
of Personal Nutrition

Martin Kohlmeier

University of North Carolina, Chapel Hill, NC, USA,
and UNC Nutrition Research Institute, Kannapolis, NC, USA

Gabrielle Z. Kohlmeier, JD, Washington D.C. (Chapter 8)

AMSTERDAM ● BOSTON ● HEIDELBERG ● LONDON ● NEW YORK ● OXFORD
PARIS ● SAN DIEGO ● SAN FRANCISCO ● SINGAPORE ● SYDNEY ● TOKYO
Academic Press is an imprint of Elsevier

Academic Press is an imprint of Elsevier
32 Jamestown Road, London NW1 7BY, UK
225 Wyman Street, Waltham, MA 02451, USA
525 B Street, Suite 1800, San Diego, CA 92101-4495, USA

First edition 2013

Notice
No responsibility is assumed by the publisher for any injury and/or damage to persons or property as a matter of products liability, negligence or otherwise, or from any use or operation of any methods, products, instructions or ideas contained in the material herein. Because of rapid advances in the medical sciences, in particular, independent verification of diagnoses and drug dosages should be made

British Library Cataloguing-in-Publication Data
A catalogue record for this book is available from the British Library

Library of Congress Cataloging-in-Publication Data
A catalog record for this book is available from the Library of Congress

ISBN: 978-0-12-810078-3

For information on all Academic Press publications visit our website at elsevierdirect.com

Typeset by TNQ Books and Journals Pvt Ltd. www.tnq.co.in

Printed and bound in United States of America

13 14 15 16 17   10 9 8 7 6 5 4 3 2 1

The contributions of Olivia Dong are noted with great appreciation.

Foreword

Nothing will change how we think about nutrition more than the new knowledge that is developing in the area of how nutrients interact with genes — the field of nutrigenetics. Before we began to understand the importance of this area, nutrition research results were often confusing. We thought that all people were metabolically similar, and when we studied their response to a nutrient and found responders and non-responders, we attributed this variation to weaknesses in the scientific design of the study. News articles would trumpet a discovery one week, only to have another research study find the opposite result the next week. It was hard for the public to act on this information to improve their health.

Now, we know that people can be metabolically very different because they have differences in genetic and epigenetic coding and we explain why investigators observed that some people responded to a nutrient, while others did not. Using modern genetic methods, it is often possible to accurately predict who the responders will be. Scientists are now developing a comprehensive catalog of gene nutrient interactions that predict requirements and responses to nutrients; today there are hundreds known, but soon tens of thousands will be identified. This catalog will be the foundation for developing individualized recommendations for optimal nutrient intake. In five to ten years, I am certain that doctors, nurses and dietitians will be using this nutrigenetic catalog to change their clinical practice.

Genetics, epigenetics and nutrition are complex scientific areas and people who want to be ready to use nutrigenetics to develop customized nutrition interventions need to master a great deal of new information. Dr. Kohlmeier's book is an excellent place to begin this learning adventure.

Steven H. Zeisel MD, PhD
Kenan Distinguished University Professor of Nutrition and Pediatrics
University of North Carolina at Chapel Hill
Director, UNC Nutrition Research Institute
500 Laureate Way Kannapolis, NC 28081

July 5, 2012

Contents

Contents

Preface

WHY HEALTH CARE PROFESSIONALS NEED TO LEARN ABOUT NUTRIGENETICS

Nutrigenetics investigates inherited differences in nutrient metabolism and explores how to use individual genetic information for making better nutrition choices. The development of the science and practice of nutrigenetics has been going on since the early 1900s. The large and growing number of well-established nutrigenetic conditions leaves little doubt that adjusting personal nutrition patterns to inherited predisposition can greatly improve the health of many people. The question is whether dietitians, nutritionists, physicians, and other health care professionals are prepared to recognize nutrigenetic problems in daily practice. Will they be able to use genetic tests appropriately and provide nutritional guidance that helps the individual they are working with? Nutrigenetics is not a science with easy answers and one standard prescription for each genotype. The interactions of the inherited genome and nutritional factors are very complex and not easily understood. We need competent nutrition professionals who can guide patients and clients effectively through the maze of nutritional and genetic information. Learners will need to work hard to achieve the necessary level of competency, make sense of the molecular differences that define each of us, and fine-tune nutritional interventions for one genetically distinct individual at a time.

The science of nutrigenetics can be traced back to Archibald Garrod's seminal description in 1908 of inborn errors of metabolism, conditions that were defined early on as genetic disruptions of food metabolism. The new biochemical knowledge slowly gave rise to the very successful practice of treating some of these metabolic diseases with nutritional interventions. The discovery by Asbjørn Følling in 1934 that defective metabolism of a dietary amino acid (phenylalanine) causes severe mental impairment was followed in 1953 by Horst Bickel's demonstration that nutritional treatment of the condition is effective. Many countries now routinely screen newborn infants for this metabolic defect (phenylketonuria, PKU) and prevent devastating consequences by starting nutritional treatment of affected children within days after birth. The same is true for a growing number of once untreatable inherited diseases (biotinidase deficiency, maple syrup urine disease, and others) that respond well to early nutritional intervention.

Another group of genetic conditions with altered responses to nutrition are the iron storage diseases. The discovery of disease-causing variants in the *hemochromatosis* gene (*HFE*), *ferroportin, hemojuvelin,* and several other genes involved in iron metabolism have helped with the molecular definition and targeted nutritional treatment of these vicious conditions. Celiac disease (CD) is a similarly insidious genetic condition that is very responsive to nutritional adjustment, in this case the avoidance of gluten-containing foods and products.

The iron storage diseases and CD have brought a new quality to the discussion: the fact that disease-causing gene variants can be relatively common. It is now becoming increasingly difficult to ignore that each of us comes with a few genetic vulnerabilities that can give an opening to full-blown disease in response to nutritional factors. More than 1% of Caucasians will eventually develop celiac disease. About one of 200 Europeans has a strong genetic risk of excessive iron storage due to the presence of at least two copies of iron-retaining gene variants. At least one in ten has a genetic variant that increases risk in combination with excessive iron or alcohol intake. Another very common genetic variant affects folate metabolism. The more than 10% of American women with two copies of the *MTHFR* 677T allele have a potentially increased risk that pregnancy will result in a child with severe disability due to a neural tube defect. Much of this increased risk can be avoided with adequate folate intake from the start of the pregnancy. We now also know that food patterns with high ratios of saturated to unsaturated fat promote obesity in many of the people with a common low-function *APOA2* promoter variant, but usually not in people without this genetic marker. Genetic variation with impact on the efficiency of caffeine metabolism can further illustrate the practical importance of nutrigenetic interactions. Caffeine is a natural ingredient of various traditional beverages. Several cups of coffee or other highly caffeinated beverages increase the risk of a heart attack in the 50% of adults who are slow metabolizers, but the risk for fast metabolizers is not significant. A final example is lactose intolerance, which may have plagued Charles Darwin for much of his adult life. The lack of intestinal lactase and the resulting inability to digest the sugar in milk is the norm in most non-European adult populations and not uncommon even in English people. The mismatch between inherited predisposition and prevalent food culture causes unnecessary discomfort in many adults and keeps gastroenterologists busy attending to patients with irritable bowel syndrome due to unrecognized lactase nonpersistence.

The outlines of this text were originally conceived in preparation for a comprehensive course in nutrigenetics at the University of North Carolina in 1994, probably the first of its kind. Many of the students, then as today, had very limited working knowledge of genetic science. It is a major goal of this text, therefore, to introduce foundational genetic concepts and technical terminology along with the nutrition content. Of course, many key mechanisms and methods were not known at the time and only developed as a result of the explosive evolution of genetics since then. During that time, a lot of new information also emerged about the molecular machinery that processes ingested foods and regulates our nutrient metabolism. This means that the learner will have to examine more molecular details of nutrient metabolism than ever to understand the biological consequences of variants in specific genes. In a discipline that is growing so rapidly, it is particularly important to pause and ask how we have come to know what we seem to know. There are many potential pitfalls that readers of nutrigenetic publications need to appreciate. Furthermore, researchers and practitioners of nutrigenetics must understand what information about a genetic test they need before they start using it in practice. This is of critical importance because the majority of initial nutrigenetic findings has been and will be contradicted by later studies. Students of nutrigenetics have to learn how to balance a healthy enthusiasm for the new paradigms with a keen sense of caution against premature practical use. They also need to become fully aware of the risks that come with inappropriate disclosure of genetic information and learn how to protect the privacy interests of their patients and clients.

The hope is that this text will prepare readers to deal with the inevitable deluge of genetic testing, that they will have learned to consider the science before giving genotype-directed nutrition advice, and that they will be ready to explain the expected benefits and harms from personal nutrition schemes to patients, clients, and the public. Leaving all such professional considerations aside, the biggest insight for many may be that extensive nutrigenetic diversity is part of what defines us as humans. Most of the millions of variants, which make each of us unique, are not flaws, but have helped our various ancestors to survive in starkly different nutritional environments around the world.

CHAPTER 1

Has the Time Come for Genotype-Based Nutrition Decisions?

All nutrition is hereditary; my genes make me do it.

Martin Kohlmeier

Terms of the Trade
- Nutrigenetic: Relating to a nutrition outcome that depends on genetic predisposition.
- OMIM: Online Mendelian Inheritance in Man, a catalogue of all known genes and genetic diseases.

ABSTRACT

Nutrition is personal in more than one sense. Different metabolic and functional set points in each of us determine the kinds and amounts of foods that work best for us. These personal nutrition set points depend as much on our genetic blueprint as on the circumstances of our past and current life. Going against this blueprint can occasionally trigger catastrophic health consequences. More often, any harm from inappropriate intake will be less obvious, but still important over time. What is new is our growing ability to read the genetic blueprint and to predict what detailed nutrition pattern is best for each individual.

1.1 LIFE IS NOT FAIR

1.1.1 We are all similar but not the same

We are extremely complex and finely tuned organisms and our bodies deserve great care to make sure that they will last a while. The owner's manual for American bodies (*Dietary Guidelines for Americans*) [1] provides a lot of great tips. It says, for instance, that 'refined grains should be replaced with whole grains, such that at least half of all grains eaten are whole grains.' This is certainly sound advice for many, but does it really apply to all of us? Another body care manual (*Dietary Reference Intakes*) [2] tells us about folate intake: '[I]t is recommended

Nutrigenetics. http://dx.doi.org/10.1016/B978-0-12-385900-6.00001-0

that all women capable of becoming pregnant consume 400 µg from supplements or fortified foods in addition to intake of food folate from a varied diet.' But then we remember that our bodies are not all the same. We are tall or small, heavy-boned or petite, and different in too many other ways to list. And then there is all this genetic variation. That must surely make a difference. Could we believe that the same food amounts are right for all of us, even if adjusted for body weight and taking into account gender and age? We also have different food preferences. Many would have a big steak for dinner, if they had the choice; but others would much rather have a veggie pizza. Is there something to the notion that our body knows best which of these choices is right for us individually? Some like drinking a few beers at night; others could not imagine doing such a thing. Is this truly a matter of free choice, influenced mainly by moral fiber, culture, habit, and plans for the next day? We are now beginning to recognize genetic factors that drive which foods we should choose and which ones we actually choose. Could it really be our genes that make us eat and drink as we do? Don't expect final answers to these questions, but here you will learn about many of the known facts and concepts.

This section will explore gaps between generalizing assumptions about healthy nutrition and the actual nutritional needs and food preferences of each of us individually. The challenge will be to build a new framework for nutritional guidance that does not assume that everybody is the same. To get started, we will review a few cases to illustrate how ignorance of our genetic blueprint and the resulting uninformed use of unsuitable foods sometimes lead to avoidable, unpleasant consequences. Try to view each of these cases as if it were about you or someone close to you. Then consider how better knowledge might have led to a more favorable outcome or might even have avoided any harm in the first place.

1.1.2 Nomenclatures and databases

Dealing with more than 26,000 human genes is necessarily confusing at times. Names and even functions of different genes often overlap and it is critical to know which one we are talking about. The same is true about the more than eight million common genetic variants. Standardized naming and numbering conventions now cut through the clutter and uncertainty of earlier nomenclatures. Use the right number and detailed online descriptions of the location, function, properties, and sequence variations of a gene are just a few clicks away.

The classification number of enzyme reactions (for instance EC 6.3.4.3) is usually listed after the first mention of an enzyme (EC stands for Enzyme Commission). This number indicates what the enzyme does, not the identity or amino acid sequence of the protein. The EC number can be shared by distinct proteins if they catalyze exactly the same chemical reaction. This number is helpful for locating the reaction and reviewing the substrates, cofactors, and products at the ExPASy website maintained by the Swiss Institute of Bioinformatics (http://enzyme.expasy.org/).

You will also sometimes see the systematic number of the Online Mendelian Inheritance in Man (OMIM) catalogue in parenthesis after a gene name. The purpose of this somewhat pedantic listing is to avoid confusion of one gene with another that may have a similar name or function. You can use the OMIM numbers to access comprehensive descriptions of the genes online at the website of the US National Library of Medicine (http://www.ncbi.nlm.nih.gov/; select OMIM under the databases drop-down menu).

The first mention of a genetic variant is also usually followed in parenthesis by its rs (reference single nucleotide polymorphism [SNP]) number, for instance rs2236225. This number helps to locate the details at the website of the US National Library of Medicine (http://www.ncbi.nlm.nih.gov/, select SNP under the databases drop-down menu).

1.1.3 For the love of fat

A 51-year-old overweight man (body weight, 85 kg; body mass index [BMI], 28.7) has an enlarged liver. Repeated sonograms during the last 2 years have been showing persistent fatty liver (hepatic steatosis). He has stopped drinking alcohol to give his liver a rest, but this has not made any difference. His blood cholesterol level is well controlled by his diet, which is low in meat, eggs, and saturated fat, and high in fruits, vegetables, and whole grains. The concern is that if his condition continues, added inflammation (nonalcoholic fatty liver disease) might eventually cause the development of irreversible liver cirrhosis. Current therapy focuses on weight loss and physical exercise. Are there any other nutritional changes that might help?

Choline (Figure 1.1) is a nutrient that is closely associated with the development of fatty liver. This amine is a key part of phospholipids, which are needed for fat (triglyceride) export from the liver with very-low-density lipoproteins (VLDL). If there is not enough choline, the triglyceride accumulates and fatty liver develops [3]. But not everyone needs the same amount of choline from foods, because the body produces larger amounts in some of us than in others. The carbon for endogenous choline synthesis comes as formate from the mitochondria and has to be bound to tetrahydrofolate (THF) before it can be used (Figure 1.2).

Methylenetetrahydrofolate dehydrogenase 1 (*MTHFD1*; EC 6.3.4.3; OMIM 172460) is the enzyme responsible for linking formic acid to THF. There is a common *MTHFD1* variant, in which the guanine base at position 1958 is replaced by adenine (1958G>A; rs2236225). The enzyme produced from the variant gene channels the formic acid less effectively toward choline synthesis [4]. Higher choline intakes with food have to make up for the diminished rate of

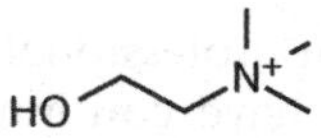

FIGURE 1.1
Choline is needed to make phospholipids for membranes and lipoproteins.

FIGURE 1.2
The carbon for choline synthesis comes from formate. MTHFD1 (methylenetetrahydrofolate dehydrogenase 1) bonds the carbon to tetrahydrofolate and then seven additional steps channel it toward choline synthesis.
ADP, adenosine diphosphate; ATP, adenosine triphosphate; THF, tetrahydrofolate.

endogenous choline synthesis or fatty liver will develop. Knowing that the 51-year-old man described above carries two copies of the *MTHFD1* 1958A variant suggests that he will have a hard time meeting his relatively high daily choline requirement of around 8 mg/kg body weight (i.e., 680 mg) on a diet without eggs and with little meat. Only one in four American men gets as much choline with unrestricted, self-selected food choices. The average intake is 313 mg/day [5]. A large egg contains 125 mg choline, four strips of bacon have 42 mg, and a typical hamburger with four ounces of meat provides 92 mg. Replacing just these three commonly consumed choline-rich foods with grain products and vegetables (which provide little choline) has probably eliminated half of his usual choline intake and the resulting choline deficit may well have triggered the accumulation of fat in his liver. It would be worthwhile for him to try a higher choline intake, either from foods or with a dietary supplement, and see whether it will bring down his liver size.

This is just one example of a common genetic variant that affects the amount of a nutrient we need to consume to avoid diseases or undesirable health conditions. Because requirements are different for everybody, only some will benefit from higher intakes. Others may get too much at those increased levels. As far as choline is concerned, higher than needed intakes have been found [6] to promote the growth of colorectal adenoma (noncancerous polyps of the large intestines, from which cancer sometimes develops), though not of colorectal cancer [7].

1.1.4 The downside of whole grains

A 30-year-old woman had been trying to conceive for over 3 years. She was in good health, but had a long history of bulimia. She had occasional abdominal pain and frequent bowel movements, which were attributed to her irregular eating habits. She says that her eating disorder days are long behind her and that she now maintains a healthy vegetarian diet. This seems to be supported by her now normal BMI of 23. She has moderate microcytic anemia (indicating iron deficiency) and low ferritin concentration in blood (signaling diminished iron stores).

She has been seeking advice from professionals and nonprofessionals without success. Her efforts to conceive have continued to fail. The most common nutritional advice she received was to get her carbohydrates mostly from whole grains and to eat a lot of vegetables and fruits.

Eventually, somebody pays attention and orders an immunoglobulin A (IgA) tissue transglutaminase test (tTG-IgA), a commonly used blood test for celiac disease (CD). It comes back as clearly elevated and the diagnosis of CD is confirmed by the result of a small bowel biopsy. The histology shows flattening of the villi and infiltration of the duodenal epithelia by lymphocytes. She is advised to start a gluten-free diet and does so conscientiously, despite all the necessary adjustments (no more German beer for her!). Eight months later she is pregnant and eventually delivers a healthy child.

At least 1% of all people in North America and Europe have CD. It is slightly less common in other parts of the world. CD is an autoimmune disease in genetically predisposed individuals, mediated by T lymphocytes. They almost always have the tissue compatibility markers HLA DQ-2 and DQ-8 (these genes encode two of the many cell surface protein variants that transplant surgeons try to match between donors and organ recipients). A bewildering number of additional genetic risk factors exist. The disease is triggered by exposure to gluten (gliadin), the storage protein in wheat, rye, and barley. Gluten is the part of wheat flour that makes a nice, crunchy pizza crust. It is what keeps bread, cakes, and other baked goods together and is also used extensively in all kinds of commercial products from puddings to facial creams. More than 5% of women with unexplained infertility have CD [8]. Most people with CD are never diagnosed and the majority do not have the classical intestinal symptoms (diarrhea, abdominal pain). Other autoimmune diseases, such as diabetes type I and thyroiditis, are more common in people with CD than in the general population [9]. Cancers (malignancies of the small intestine, which are otherwise very rare) and kidney disease (IgA-nephropathy and glomerulonephritis) are additional diseases caused by long-term untreated CD [10]. Only early diagnosis and meticulous nutritional adjustment will reduce this extremely varied and grave disease burden.

> At what point would you have recognized early signs and symptoms during this woman's long history?
> Would you have known what diagnostic steps to initiate?
> Do you know which foods are problematic for CD patients?
> In light of the nutrition recommendations mentioned initially, would whole grains have been better for this woman?
> Would you remember to discuss the possibility that her child might have inherited her CD predisposition and will need testing?

1.1.5 A raw deal at the oyster bar

A 56-year-old Florida man eats half a dozen raw oysters as appetizers, has a few beers with his dinner, and feels good. Three days later he suddenly gets high fever (40°C/104 F), nausea, vomiting, and diarrhea. The next day he has severe pain in both legs that keeps getting worse and he decides to go to the emergency room. He becomes increasingly short of breath and his body temperature drops to 35.4°C (95.7°F). His yellow eyes and enlarged liver indicate severe liver disease. A blood culture eventually shows that he has an infection from the Gram-negative marine bacteria *Vibrio vulnificus*. His condition worsens rapidly despite intensive antibiotic treatment, his blood starts clotting throughout the body (disseminated coagulation), and he dies from septicemia (blood poisoning) a few hours later [11].

How could that happen? It was later found that this man carried one copy of the *HFE* gene (OMIM 613609) that encodes a protein in which the normal cysteine in position 282 is replaced by a tyrosine (Cys282Tyr). This gene variant is less effective in suppressing the free iron concentration in blood and lowers the concentration of hepcidin, the antibacterial regulator of iron balance [12]. The single Cys282Tyr allele on its own would have had only a modest effect on iron metabolism. But this man's known long-term alcohol abuse and high iron intake greatly increased the amount of iron in his liver and other tissues, which greatly added to his genetic predisposition. Together, the excessive iron store and the HFE Cys282Tyr variant kept the concentration of free iron in blood abnormally high. The marine bacterium *V. vulnificus*, which is a frequent contaminant of oysters, finds ideal growing conditions in blood with lots of free iron. The bacterium grows readily in blood from otherwise healthy people with hemochromatosis, but not in people without the condition [13]. This explains how this Floridian's dinner companions could eat the same raw oysters without getting sick, while his genetic predisposition put him at risk. About 5% of Americans have inherited hemochromatosis-related gene variants from at least one parent [14] and are potentially at increased risk for food-borne infections. One in about 300 adults has two copies of the Cys282Tyr variant and is at particularly high risk. Long-term alcohol abuse and high iron consumption greatly add to the risk.

Do you understand how iron intake from meats, fortified foods, and dietary supplements affects the iron status of people with these genetic variants? What advice would you give them to minimize their risks?

1.1.6 Another hamburger will kill you

Picture this: an 18-year-old college freshman moves into her dorm. From the start she is troubled by nausea and vomiting. Her counselors are concerned that she might have a severe eating disorder. A few hours after a grill party celebrating the start of the school year, she gets sick again. She thinks it is because of the two hamburgers she ate, because she is not used to eating much meat. Later at night she becomes increasingly incoherent and then passes out. Her roommate thinks that she is intoxicated, consults with other students on the floor, and eventually calls an ambulance. In the emergency room, she can be aroused, but is slow to respond to simple questions and has slightly slurred speech. She is disoriented about time and place and is breathing faster than normal. Blood tests show that she had no alcohol or commonly used recreational drugs. She is admitted for continued monitoring and neurological evaluation, because an epileptic episode without convulsion is suspected. Considering her tachypnea (high breath rate) it is not surprising that she has respiratory alkalosis (high blood pH accompanied by low blood carbon dioxide concentration). Her liver enzymes are slightly elevated, but all other blood chemistries are normal [15]. Further testing shows a greatly elevated blood ammonia concentration. The immediate start of therapy with intravenous sodium benzoate saves her life. Benzoic acid is activated by coenzyme A, which then binds ammonia-containing glycine (Figure 1.3). The resulting hippuric acid can be excreted with urine. This pathway can thus remove ammonia from the body without the need for urea synthesis.

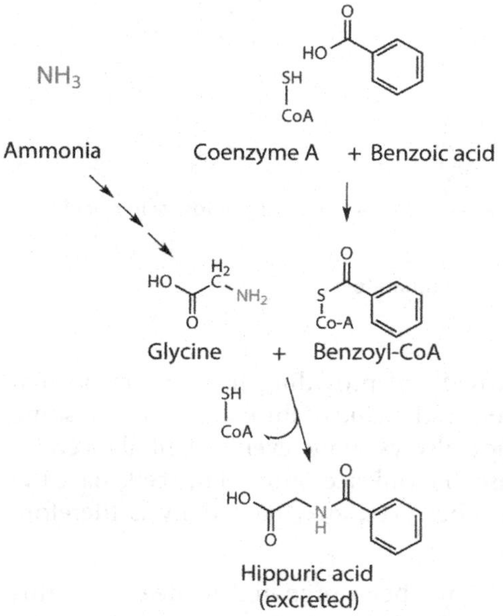

NH_3

Ammonia   Coenzyme A   + Benzoic acid

Glycine   +   Benzoyl-CoA

Hippuric acid
(excreted)

FIGURE 1.3
Benzoate helps to excrete ammonia after condensation to hippuric acid, which is excreted with urine. Alpha-ketoglutarate captures the free ammonia (NH_3) and the resulting glutamate can then transfer its amino group via serine to glycine.

If her roommate had let her sleep off the suspected alcohol excess, the toxic brain edema would have built up rapidly and she would not have made it through the next day.

The suspicion of a genetic urea cycle (Figure 1.4) defect proves correct a few days later, when the report comes back showing abnormally high orotate excretion with urine, an alternative metabolite of carbamoyl phosphate used for the synthesis of the nucleotides uridine triphosphate (UTP) and cytidine triphosphate (CTP). Since the urea cycle defect blocks the use of carbamoyl phosphate for citrulline synthesis, some of the excess is diverted to the synthesis of orotate, which is excreted. Genetic analyses that come in after a few weeks confirm a heterozygous loss-of-function mutation in the *ornithine transcarbamoylase* gene (*OTC*; EC 2.1.3.3; OMIM 311250), which replaces the arginine in position 40 with a histidine (Arg40His). What makes her condition worse is that she has unbalanced X-chromosome inactivation in her liver, which means that even less ammonia is converted into urea. Usually, half of the cells would have the chromosome with the defective gene activated and in the other half the

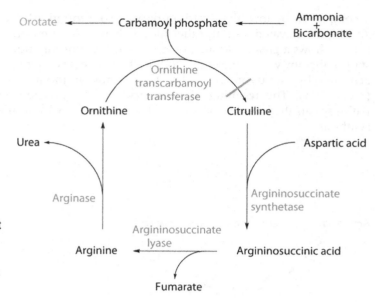

FIGURE 1.4
Ornithine transcarbamoylase is the enzyme that adds carbamyl phosphate to ornithine. This is the second step of the urea cycle, which repackages ammonia as urea for detoxification and easier excretion.

unaffected chromosome would be turned on, providing half of the normal enzyme activity on average. But the random silencing of the extra X chromosome at the four-cell embryo stage does not always work evenly. Unbalanced X-chromosome silencing means in her case that only a quarter of the cells have the unaffected chromosome turned on and her OTC enzyme activity is therefore only about a quarter of normal.

Now ask yourself whether you would have been prepared to deal with this individual. If you were the physician in charge when she was brought to the emergency room, you might have been experienced enough to order early on a blood ammonia test; from then on, the path for follow-up diagnostic procedures and treatment would have been clear. But could you have prevented the incident? It is now thought that about one in 14,000 female newborn infants has low OTC activity and could be at risk of a metabolic crisis due to ammonia accumulation [16]. The vast majority of them will grow up with few or no symptoms until some circumstance boosts ammonia release. There is evidence that without even knowing about it, they can manage their condition quite well for most of the time, particularly by limiting their protein consumption [17]. But dietary changes or medical conditions can trigger increased protein breakdown (catabolism) and ammonia accumulation at any time with potentially catastrophic consequences. A handful of women in every midsized city have precisely this kind of genetic risk, with a life-threatening emergency event waiting to happen. After successfully handling the event, could you counsel her on how to prevent the next episode? Would you start searching for relatives who might share the same genetic risk?

1.2 A GENETIC FRAMEWORK FOR NUTRITION DECISIONS

1.2.1 Personal nutrition

People with a particular genetic variant often differ fundamentally from those without that variant in their response to nutritional challenges. In each of the previous case scenarios, the precipitating cause would pass completely unnoticed and without harm in the vast majority of their family and neighbors.

The recommendations still insist on setting the same nutritional standard for people of the same gender and age. The guidelines just include the little caveat that they apply only to healthy and otherwise typical individuals. But in reality most of us are different from others in respect to some nutrients and foods.

Take the example of milk drinking. Many adults can drink a cup of milk without suffering any discomfort. But the majority of equally healthy and outwardly similar people around the world would get abdominal cramps, bloating, or diarrhea. The main reason for these contrasting experiences is the presence or absence of the lactose-digesting enzyme lactase in the small intestine. Fewer than one in three adults across the world have enough functioning lactase to digest and absorb the sugar (lactose) in milk. The others get abdominal problems from milk drinking because the undigested lactose becomes fodder for intestinal bacteria, which release gas and other unpleasant products. Knowing about this difference, would it make sense to talk about a typical or average response to milk? Should we give the same recommendation for milk consumption to all adults? Of course not.

Here is another example of genetic differences in nutritional requirements. We know that it takes more folate to prevent neural tube defects for some women than for others because their metabolism is tuned differently [18]. We are not talking here about rare inborn errors of metabolism. In the case of folate requirements above the current general recommendations, 10–30% of women fall into that category, depending on ethnicity.

It is not at all unusual to have ten or more percent of healthy individuals with needs that are not met by current general recommendations because of their individual genetic makeup. Even less common gene variants are important for determining intake targets, because a few percent add up by the time we go through the full catalog of important nutrients and foods. The person who does not have at least one nonstandard nutrition requirement is actually very unusual.

On an individual basis, not understanding what the body needs and what it should not be fed can cause significant harm. Following general nutrition recommendations is not enough. Sounds like a massive exaggeration? Just consider the example of our daily bread. What do the *Dietary Guidelines* say about bread consumption? As mentioned earlier, the guidelines reasonably suggest that 'refined grains should be replaced with whole grains, such that at least half of all grains eaten are whole grains.' But maybe there should be at least a footnote pointing out that products containing wheat, rye, or barley are harmful for about 1% of all Americans, who have to avoid them completely. Consuming whole grains will not be in the least helpful for them. We are talking about people who get sick for no other reason than eating regular bread like everybody else. They just have the wrong genes for the food on offer. They are said to have CD, but this disease would not even exist if they lived in a world without gluten in the food supply (such as in the Americas before the Europeans introduced wheat, rye, and barley). There are numerous conditions like this, some rare, some common. It is important that nutrition professionals and other healthcare providers recognize and understand the individual needs of the people they serve. As the necessary diagnostic tests and assessment frameworks become available, shaping advice toward the innate nutrition needs of the individual can become a reality.

1.2.2 Is it Nutrigenetics or Nutrigenomics?

We need to introduce at this point the term *nutrigenetics*, which has been used off and on in previous discussions of this emerging discipline. This contraction of the words nutrition and genetics, very much like pharmacogenetics, makes it much easier to signal the focus of scientific and professional efforts. Nutrigenetics is the science that investigates the combined action of inherited genome and nutrition on health and performance and uses that information in practice. The ultimate aim of practitioners of the science (and art) of nutrigenetics is to use the insight for making better nutrition choices at all levels of decision making, from personal nutrition to international policy.

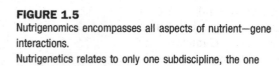

> **KEY POINT**
> Nutrigenetics explores interactions of our inherited genome and nutrition.

Nutrigenomics, in contrast, describes a much broader scope, which includes all nutritional interactions of our genome. This includes, of course, all of our genetic differences; hence, nutrigenetics. But nutrigenomics also refers to all regulatory processes that involve our genome (Figure 1.5). The control of gene transcription by vitamin metabolites (particularly 1,25-dihydroxy-vitamin D

FIGURE 1.5
Nutrigenomics encompasses all aspects of nutrient–gene interactions.
Nutrigenetics relates to only one subdiscipline, the one concerned with inherited predispositions toward nutrition.

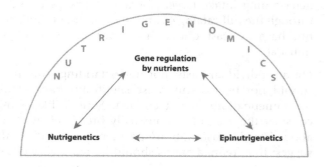

and retinoic acid), iron, and many other nutrients is an important focus of nutrigenomic science. Another area of nutrigenomics explores how methylation and other modifications to our DNA alter its properties and behavior. This important science is called *epigenetics* and has a very strong link to nutrition. *Nutriepigenetics* is not a formally recognized term, but may be used here just once to refer to the nutritional aspects of epigenetics. These two nutrigenomic disciplines intersect and interact in many ways with the third one, nutrigenetics, but they are not the same. Much confusion can be avoided by applying distinct terms for each discipline.

1.2.3 Genetic Nutrition in Practice

A common view on the routine use of genetically informed nutrition practice is that it is still too complex, insufficiently understood, and certainly several decades away from becoming a reality. The simple answer to this opinion is that nutrigenetics has been a reality in clinical practice for decades. For instance, every year we screen millions of newborns for genetic diseases such as phenylketonuria (PKU, OMIM 261600) and biotinidase deficiency (OMIM 253260) because these conditions generally respond well to nutritional therapy [19]. Analysis of the *HFE* variants Cys282Tyr and His63Asp in people with hemochromatosis is also common practice. Table 1.1 lists illustrative examples of long-known genetic conditions with practical nutritional implications.

Table 1.1	Humans have Numerous Genetic Syndromes that Tend to Respond Well to Nutrition Interventions, only a Few of which are Listed here		
Condition	**OMIM**	**Nutritional Factor**	**Worldwide Prevalence (%) [reference]**
L-Gulonolactone oxidase deficiency	240400	(+) Ascorbic acid	100 [20]
Lactose intolerance	223100	(−) Lactose	60−70 [21]
Variant *MTHFR*	607093	(+) Folate	10 [22]
G6PD deficiency	305900	(Ø) Fava beans	5 [23]
Celiac disease	212750	(Ø) Gluten	1 [24]
Hemochromatosis	235200	(−) Iron	0.2 (5−10, mild form)
Congenital sucrase-isomaltase deficiency	222900	(−) Sucrose	0.2 [25]
Trehalose intolerance	612119	(−) Trehalose	0.05 [26]
Phenylketonuria	261600	(−) Phenylalanine (+) Tyrosine	0.01 [27]
OCT deficiency	311250	(−) Protein	0.007 [16]
Hereditary fructose intolerance	229600	(−) Fructose	0.005 [28]
Biotinidase deficiency	253260	(+) Biotin	0.0017 [29]

The plus sign before a listed nutritional factor indicates that people with the condition need more of the factor; a minus sign means that carriers should have less of it; and the Ø symbol signals that this factor needs to be completely avoided. OMIM (Online Mendelian Inheritance in Man) indicates the unified index number in the catalogue of all known genetic diseases.

The harder question is whether and how to expand this established practice. It should not need much argument to offer counseling and functional or genetic testing to close relatives of individuals with clinically relevant variants. This kind of opportunistic screening is suitable for rare conditions like urea cycle defects or hereditary fructose intolerance. Moving this up another notch would be to screen population groups at particularly high risk. The ultimate coverage would be achieved by screening everybody.

But what about the numerous genetic variants with significant health risks that respond to nutritional adjustments (Table 1.2)? For example, 10–30% of women carry *MTHFR* variants that in case of pregnancy increase the risk that their child will have a birth defect [30]. This increased risk is largely avoidable with a higher folate intake than is otherwise needed.

Here is another example. Coagulation factor V (F5; OMIM 172460) plays an important role in blood clotting. About 5% of American women have a genetic variant (prothrombin 20210G>A, Arg506Gln; factor V [F5] Leiden; OMIM 188055; rs6025) that increases their risk of deep vein thrombosis and pulmonary embolism threefold [31]. However, women who take a moderately dosed vitamin E supplement prevent much of that increased risk. Extra vitamin E does not further reduce the already low risk of women without the F5 Leiden variant.

A third example concerns people who cannot fall asleep at night. The adenosine receptor A_{2A} regulates the release of neurotransmitters in the brain. Caffeine can activate this receptor and prevent people from falling asleep. Most sensitive to the sleep-disrupting effects of caffeine are people [32] with the 1083T>C variant of the *adenosine 2A2 receptor* gene (*ADORA2A*; OMIM 102776; rs5751876). Unfortunately, carriers of this variant also tend to be the most avid coffee lovers [33].

With the advent of affordable whole-genome sequencing, it becomes practical to search for both common and rare nutrigenetic variants with predictive algorithms. By the time this book goes into print, thousands of people will have had

Table 1.2	Humans Carry Many Genetic Variants that Increase Health Risks unless Dietary Intake is Adjusted, only a Few of which are Listed Here			
Genotype	**Prevalence (%)**	**Risk**	**Nutrition Factor**	**Reference**
ADORA2A 1083C/T and C/C	69	Poor sleep	(−) Caffeine	[32]
MTHFD1 1958A/G and A/A	65	Fatty liver	(+) Choline	[4]
AGT −6A/A	60	Hypertension	(−) Salt	[34]
ER-beta −13950C/T and C/C	42	Prostate cancer	(+) Isoflavones	[35]
APOA2 −265C/C	15	Obesity	(−) Saturated fat	[36]
MTHFR 677T/T	10	Birth defects	(+) Folate	[22]
Prothrombin 20210A/G and A/A	5	Thrombosis	(+) Vitamin E	[31]

The plus sign before a listed nutritional factor indicates that people with the condition need more of the factor; a minus sign means that carriers should have less of it.

their whole genome sequenced and the cost per whole genome will have dropped to less than US$5000. This is still too expensive for single-purpose, one-time use. But the cost balance changes fundamentally if we are going to use the genetic information lifelong for making all types of medical and life-style decisions. Individual genetic tests are already being used for the selection of type and dose of medications (e.g., the anticoagulant warfarin), preparation for anesthesia (e.g., malignant hyperthermia in response to inhaled anesthetics), and risk assessment (e.g., breast cancer gene testing). None of these tests will have to be ordered individually if the same information can be read instantly from the whole-genome record, right there in the physician's office. The use of whole-genome sequence information for nutritional guidance will be almost an afterthought, just a welcome bonus. This new landscape of widely available personal genetic information compels nutrition and health professionals to develop a comprehensive understanding of the issues from the ground up, starting with the molecular organization and evolutionary dynamic of the human genome, going on to specific nutrigenetic interactions and the evaluation of their impact on health, and concluding with practical implementation and the weighty ethical and legal implications. These are the topics that we want to visit on this journey through the brave new world of nutrigenetics.

SUMMARY AND SEGUE TO THE NEXT CHAPTER

It has become clear that nobody responds the same to everything they consume. Most of the time, we will not even know about a problem, because the outcome is a heart attack, infertility, or another condition that seems to strike randomly and unavoidably. But we do not all have the same risk when we indulge in that extra hamburger or have our daily bread. We have definitely reached the point where our genetic blueprint can reliably predict nutritional triggers of a few undesirable outcomes in a subset of susceptible responders. The number of such well-understood and robustly examined interactions between a specific genetic predisposition and a precipitating nutritional trigger is steadily growing. It is up to us to understand and appropriately use such evidence-based knowledge in everyday practice.

The science of nutritional genetics (nutrigenetics) represents a serious investigation of genetic variants that influence the optimal amount of nutrients for each of us, shape food likes and dislikes, and determine food intolerances.

The following section will revisit the chemistry and architecture of the human genome and the various modes of genetic transmission. It will explore in more detail the nature of genetic variation and how variants influence human nutrition. The section assumes familiarity with foundational concepts of molecular biology and may require some readers to refresh their knowledge with a more basic biochemistry text. But you surely will not want to miss out on our discussion of eternal questions such as Why can people who don't like grapefruit have children who do?, or Does a man's overeating make his future grand-children obese?

References

[1] US Department of Agriculture. US Department of Health and Human Services. Dietary Guidelines for Americans 2010, http://www.health.gov/dietaryguidelines/dga2010/DietaryGuidelines2010.pdf.

[2] Food and Nutrition Board. Institute of Medicine, National Academies. Dietary Reference Intakes, http://www.iom.edu/Activities/Nutrition/SummaryDRIs/~/media/Files/Activity%20Files/Nutrition/DRIs/New%20Material/5DRI%20Values%20SummaryTables%2014.pdf.

[3] Zeisel SH. Choline: critical role during fetal development and dietary requirements in adults. Annu Rev Nutr 2006;26:229–50.

[4] Kohlmeier M, da Costa KA, Fischer LM, Zeisel SH. Genetic variation of folate-mediated one-carbon transfer pathway predicts susceptibility to choline deficiency in humans. Proc Natl Acad Sci U S A 2005;102(44):16025–30.

[5] Cho E, Zeisel SH, Jacques P, Selhub J, Dougherty L, Colditz GA, et al. Dietary choline and betaine assessed by food-frequency questionnaire in relation to plasma total homocysteine concentration in the Framingham Offspring Study. Am J Clin Nutr 2006;83(4):905–11.

[6] Cho E, Willett WC, Colditz GA, Fuchs CS, Wu K, Chan AT, et al. Dietary choline and betaine and the risk of distal colorectal adenoma in women. J Natl Cancer Inst 2007;99(16):1224–31.

[7] Lee JE, Giovannucci E, Fuchs CS, Willett WC, Zeisel SH, Cho E. Choline and betaine intake and the risk of colorectal cancer in men. Cancer Epidemiol Biomarkers Prev 2010;19(3):884–7.

[8] Pellicano R, Astegiano M, Bruno M, Fagoonee S, Rizzetto M. Women and celiac disease: association with unexplained infertility. Minerva Med 2007;98(3):217–19.

[9] Lewis NR, Holmes GK. Risk of morbidity in contemporary celiac disease. Expert Rev Gastro-enterol Hepatol 2010;4(6):767–80.

[10] Jhaveri KD, D'Agati VD, Pursell R, Serur D. Coeliac sprue-associated membranoproliferative glomerulonephritis (MPGN). Nephrol Dial Transplant 2009;24(11):3545–8.

[11] Gerhard GS, Levin KA, Price Goldstein J, Wojnar MM, Chorney MJ, Belchis DA. Vibrio vul-nificus septicemia in a patient with the hemochromatosis HFE C282Y mutation. Arch Pathol Lab Med 2001;125(8):1107–9.

[12] Brissot P, Bardou-Jacquet E, Jouanolle AM, Loreal O. Iron disorders of genetic origin: a changing world. Trends Mol Med. 2011 2011 Dec;17(12):707–13.

[13] Bullen JJ, Spalding PB, Ward CG, Gutteridge JM. Hemochromatosis, iron and septicemia caused by Vibrio vulnificus. Arch Intern Med 1991;151(8):1606–9.

[14] Gerhard GS, Chokshi R, Still CD, Benotti P, Wood GC, Freedman-Weiss M, et al. The influence of iron status and genetic polymorphisms in the HFE gene on the risk for postoperative complications after bariatric surgery: a prospective cohort study in 1,064 patients. Patient Saf Surg 2011;5(1):1.

[15] Pinner JR, Freckmann ML, Kirk EP, Yoshino M. Female heterozygotes for the hypomorphic R40H mutation can have ornithine transcarbamylase deficiency and present in early adoles-cence: a case report and review of the literature. J Med Case Reports 2010;4:361.

[16] Cavicchi C, Malvagia S, la Marca G, Gasperini S, Donati MA, Zammarchi E, et al. Hypoci-trullinemia in expanded newborn screening by LC-MS/MS is not a reliable marker for orni-thine transcarbamylase deficiency. J Pharm Biomed Anal 2009;49(5):1292–5.

[17] Arn PH, Hauser ER, Thomas GH, Herman G, Hess D, Brusilow SW. Hyperammonemia in women with a mutation at the ornithine carbamoyltransferase locus. A cause of postpartum coma. N Engl J Med 1990;322(23):1652–5.

[18] Shaw GM, Lu W, Zhu H, Yang W, Briggs FB, Carmichael SL, et al. 118 SNPs of folate-related genes and risks of spina bifida and conotruncal heart defects. BMC Med Genet. 2009;10:49.

[19] Wolf B. Biotinidase Deficiency: New Directions and Practical Concerns. Curr Treat Options Neurol 2003;5(4):321–8.

[20] Nishikimi M, Yagi K. Molecular basis for the deficiency in humans of gulonolactone oxidase, a key enzyme for ascorbic acid biosynthesis. Am J Clin Nutr 1991;54(6 Suppl):1203S–8S.

[21] Itan Y, Jones BL, Ingram CJ, Swallow DM, Thomas MG. A worldwide correlation of lactase persistence phenotype and genotypes. BMC Evol Biol 2010;10:36.

[22] Wilcken B, Bamforth F, Li Z, Zhu H, Ritvanen A, Renlund M, et al. Geographical and ethnic variation of the 677C>T allele of 5,10 methylenetetrahydrofolate reductase (MTHFR): findings from over 7000 newborns from 16 areas world wide. J Med Genet. 2003;40(8):619–25.

[23] Hedrick PW. Population genetics of malaria resistance in humans. Heredity 2011, Oct;107(4):283–304.

[24] Dube C, Rostom A, Sy R, Cranney A, Saloojee N, Garritty C, et al. The prevalence of celiac disease in average-risk and at-risk Western European populations: a systematic review. Gastroenterology 2005;128(4 Suppl. 1):S57–67.

[25] Peterson ML, Herber R. Intestinal sucrase deficiency. Trans Assoc Am Physicians 1967;80:275–83.

[26] Arola H, Koivula T, Karvonen AL, Jokela H, Ahola T, Isokoski M. Low trehalase activity is associated with abdominal symptoms caused by edible mushrooms. Scand J Gastroenterol 1999;34(9):898–903.

[27] Steinfeld R, Kohlschutter A, Ullrich K, Lukacs Z. Efficiency of long-term tetrahydrobiopterin monotherapy in phenylketonuria. J Inherit Metab Dis 2004;27(4):449–53.

[28] Cross NC, de Franchis R, Sebastio G, Dazzo C, Tolan DR, Gregori C, et al. Molecular analysis of aldolase B genes in hereditary fructose intolerance. Lancet 1990;335(8685):306–9.

[29] Wolf B. Worldwide survey of neonatal screening for biotinidase deficiency. J Inherit Metab Dis 1991;14(6):923–7.

[30] van der Linden IJ, Afman LA, Heil SG, Blom HJ. Genetic variation in genes of folate metabolism and neural-tube defect risk. Proc Nutr Soc. 2006;65(2):204–15.

[31] Glynn RJ, Ridker PM, Goldhaber SZ, Zee RY, Buring JE. Effects of random allocation to vitamin E supplementation on the occurrence of venous thromboembolism: report from the Women's Health Study. Circulation 2007;116(13):1497–503.

[32] Retey JV, Adam M, Khatami R, Luhmann UF, Jung HH, Berger W, et al. A genetic variation in the adenosine A2A receptor gene (ADORA2A) contributes to individual sensitivity to caffeine effects on sleep. Clin Pharmacol Ther 2007;81(5):692–8.

[33] Cornelis MC, El-Sohemy A, Campos H. Genetic polymorphism of the adenosine A2A receptor is associated with habitual caffeine consumption. Am J Clin Nutr 2007;86(1):240–4.

[34] Svetkey LP, Moore TJ, Simons-Morton DG, Appel LJ, Bray GA, Sacks FM, et al. Angiotensinogen genotype and blood pressure response in the Dietary Approaches to Stop Hypertension (DASH) study. J Hypertens 2001;19(11):1949–56.

[35] Hedelin M, Balter KA, Chang ET, Bellocco R, Klint A, Johansson JE, et al. Dietary intake of phytoestrogens, estrogen receptor-beta polymorphisms and the risk of prostate cancer. Prostate 2006;66(14):1512–20.

[36] Corella D, Peloso G, Arnett DK, Demissie S, Cupples LA, Tucker K, et al. APOA2, dietary fat, and body mass index: replication of a gene–diet interaction in 3 independent populations. Arch Intern Med 2009;169(20):1897–906.

CHAPTER 2

How Genetic Transmission Works

Terms of the Trade

- **Alleles:** Alternative DNA sequences at a locus.
- **Autosomal:** Relating to one of the numbered chromosomes.
- **Cis:** Coming from or located on the same DNA strand.
- **Diplotype:** Reference to an individual's two haplotypes.
- **Dominant:** Mode of inheritance where the trait always overrides the other (recessive) trait in the heterozygous state.
- **Epigenetic:** Relating to chemical DNA modifications that influence gene expression.
- **Exon:** A DNA segment within a gene that is translated into protein.
- **Genotype:** Combination of inherited variants at a gene locus.
- **Haplotype:** Combination of alleles on the same strand.
- **Hemizygous:** Carrying one allele when no second copy is present.
- **Heritability:** Estimated percentage to which a trait is inherited.
- **Heteroplasmy:** Presence of mitochondria inherited from a paternal ancestor.
- **Heterozygous:** Carrying different alleles on a chromosome pair.
- **Homozygous:** Carrying the same variant forms (alleles) on a chromosome pair.
- **Intron:** Untranslated DNA segment located between exons.
- **Linkage disequilibrium:** Statistical association of two alleles.
- **Locus (pl. loci):** Specific local sequence on a chromosome.
- **LoD:** Logarithm of odds, often used to estimate the distance between two loci.
- **Memetic inheritance:** Transmission of methods and behaviors from one person to another.
- **Phenotype:** External or biological appearance.
- **Polymorphism:** Variant that occurs in at least 1% of a population.
- **Recessive:** Mode of inheritance where the trait is always overridden by the other (dominant) trait in the heterozygous state.
- **Trans:** Coming from or located on a different DNA strand.
- **Transcription:** Synthesis of messenger RNA from nuclear DNA segments.
- **Translation:** Protein synthesis from messenger RNA templates.

Nutrigenetics. http://dx.doi.org/10.1016/B978-0-12-385900-6.00002-2

ABSTRACT

Humans inherit characteristic features (traits) through nuclear DNA and less often through mitochondrial DNA. A group of features is called the phenotype (presentation form). Traits are transmitted only through the maternal and paternal germline cells that form a new person. Genomic changes in other cells will not reach the next generation. Neighboring DNA sequences on the same chromosome tend to travel together from one generation to the next.

The inheritance patterns of many traits correspond to the transmission of one set of chromosomes from each parent. One form of a trait is dominant if it always overpowers the other form. It is recessive if the other version always wins out. More complex inheritance patterns are less common, but not unusual.

The genome of humans, as of all other organisms, varies in millions of places. These genetic differences are the molecular basis for the different forms of observable traits. The different sequence versions at a particular place (*locus*) in the DNA sequence are called alleles. A group of alleles is called a genotype. All the variants together define who we are and what kind of nutrition is best for us. Common DNA sequence variants are called polymorphisms. It is absolutely critical to understand that polymorphisms must have been transmitted over many generations (because they are so common). They must never be confused with spontaneous mutations, of which 50–100 occur in every individual around conception.

The simplest variant, caused by the exchange of one base for another, is called a single nucleotide polymorphism (SNP). In other instances bases may be deleted or inserted. The length of both deletions and insertions can run into many thousands of bases. Sometimes whole genes are repeated, in some cases many times. Variants of all types can exist without a discernible effect on gene function, but most often they do. Both loss and gain of function are possible, but sometimes the effect is more subtle and depends on metabolic state or other circumstances.

2.1 THE MOLECULAR STAGE

2.1.1 Introduction to the human genome

Genetic science concerns the transmission of inherited characteristics across generations. The rules that govern genetic transmission are the manifestations of the various elements of the human genome. These various parts work together with astounding precision, creating a child from a single newly conceived cell and growing it into a mature adult ready to start the next generation. The following sections will revisit foundational concepts from genome architecture to gene expression, using a few nutritional examples to explore practical implications. Anybody not already familiar with the underlying concepts of molecular biology should consult an introductory text before or while reading this synopsis.

The DNA of the body's tiny population of germline cells (eggs and sperm) is the substrate for all inherited genetic variation. This is the material that is passed on from generation to generation and with it any changes that may have occurred since transmission from the previous generation. The DNA of the vast majority of the other nucleated cells (so-called *somatic* cells) in the body support the day-to-day operations of the body, but any mutations or other damage to them have no relevance for inheritance of traits by the next generation. The distinction between the consequences of structural changes in germline cells and somatic cells is critical and justifies repeated mentioning. The Lamarckian idea that forces acting on tissues and organs beneficially mold them to a particular environment for generations to come is as wrong today as it was 200 years ago, appeals for its revival notwithstanding. But it should be understood that the nutritional environment constantly shapes the genome of both germline cells and somatic cells through epigenetic modification, just not necessarily in a way that is beneficial to the organism. To give a timely example, it has been observed that overfeeding causes parents to transmit through genetic mechanisms the tendency for accumulating excess body fat to their children, possibly across multiple generations [1]. This must not be misunderstood as an effect of over-feeding on the DNA in fat cells or other tissues of the body. Instead, it is only the effects of overfeeding on germline cells that impact the next generation, not DNA modifications in somatic cells. And germline cells from fathers appear to be as much in play as maternal germline cells [2]. Thus, when a man commits to a sustained weight gain program, he might as well consider banking his sperm beforehand to spare his future children an unwelcome inheritance of obesity, diabetes, and hypertension.

> **KEY CONCEPT**
> Only mutations of the DNA in germline cells can be transmitted to the next generation.

The way the nucleotide bases of the human genome are structured and organized in all their complexity is still not known in detail. The following description will be limited to a bare outline of basic principles, risking both reiteration of familiar facts and omission of many important aspects.

The bases in nuclear and mitochondrial DNA (mtDNA) are the two double-ringed purine bases, adenine and guanine, and the single-ringed pyrimidine bases, cytosine and thymine. Ribonucleic acid (RNA) contains uracil in place of thymine. A small percentage of the polynucleotide bases occur in a modified form as 5-methylcytosine (at epigenetically methylated sites) and in even smaller amounts as 5-hydroxymethylcytosine [3], 5-formylcytosine, and 5-car-boxylcytosine [4].

The phosphate group of each DNA base connects to the deoxyribose of the next base. The DNA segment linked through phosphate to the 5′ carbon of the deoxyribose is designated to lie in the 5′ (pronounced five prime) direction and the segment linked to the 3′ deoxyribose carbon is said to lie in 3′ direction (Figure 2.1). DNA-copying enzymes always proceed from 5′ toward 3′. This establishes directionality of the DNA strand and is used to indicate whether

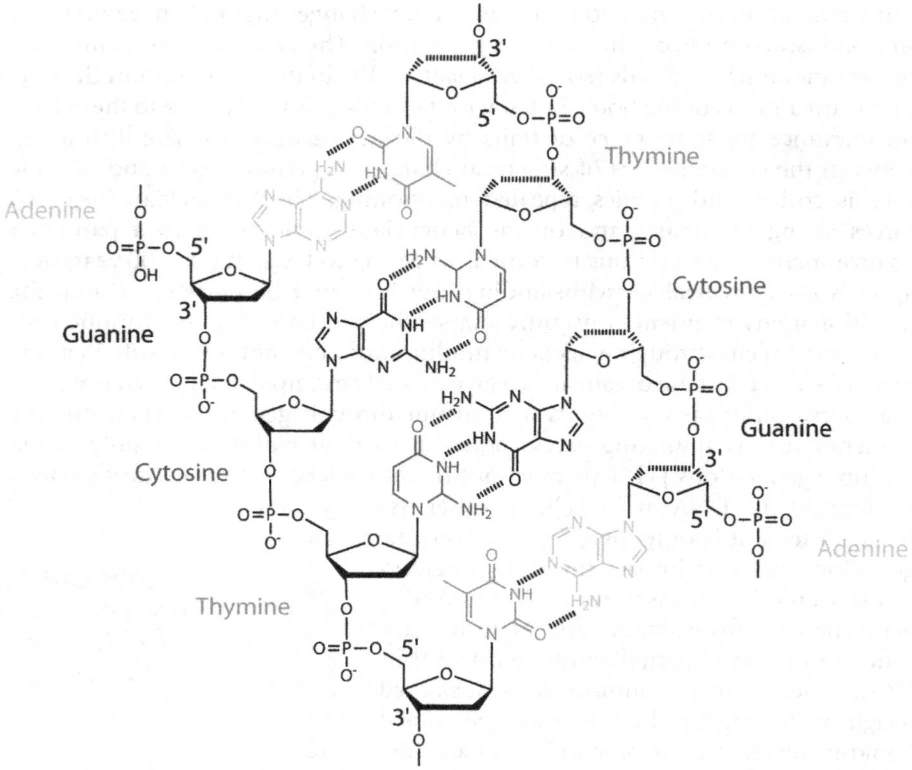

FIGURE 2.1
The 5′ end of DNA is located in the direction where the phosphate links to the 5′ carbon of the deoxyribose sugar. Likewise, the phosphate linked to the 3′ carbon points in the opposite direction.

a particular sequence is located upstream (5′) or downstream (3′) of another position. Human nuclear DNA usually occurs as a double helix with complementary strands winding around each other. Both strands carry genetic information, but the messages are entirely different. This means that it makes a huge difference which of these two strands is read. By convention, one of the strands is designated as the plus strand; the other one is said to be the minus strand. Later we will encounter examples where reading the plus strand from 5′ to 3′ generates one gene product, and reading the complementary minus strand from 5′ to 3′ generates an entirely unrelated product from a different gene.

Mammals transmit both mitochondrial and nuclear DNA to their offspring. Nutrigenetic events and interactions predominantly play in the much larger arena of nuclear DNA with its more than 3.4 billion (10^9) bases [5]. Nonetheless, the mitochondrial genome also modifies responses to the nutritional environment and contains common variants with an impact on nutrition. Each mitochondrion contains a few copies of circular double-stranded DNA, comprising about 16,569 bases encoding 37 genes and additional functional segments [6]. Very importantly, mtDNA encodes 13 of the 83 proteins needed for respiration [7]. Almost all mitochondria, and therefore mtDNA, come from the maternal side. While most mitochondria in an individual share the same lineage, some people have mtDNA from two different lineages [8]. This so-called *heteroplasmy*, an apparent violation of strict matrilineal inheritance, gained some notoriety because an exhumed bone sample from the Russian Czar Nicolas II contained such a mixture of mtDNA, showing both thymine and cytidine at mtDNA position 16169, but a living descendant of the czar's mother did not [9]. This finding seemed to rule out that the bones really were the czar's. Further investigation, however, found the same heteroplasmy in bone samples from the czar's brother. This means that the czar's mother passed a second set of mitochondria (in different admixture ratios) to her sons, but not to her great-great-granddaughter. Large-scale population studies have since demonstrated that 6% of all people have such heteroplasmy, with either point mutations or length variants [10]. It is good to remember that mitochondrial inheritance is not quite as clean and predictable as chromosomal inheritance.

2.1.2 Chromosomes

Nuclear DNA is packaged into chromosomes, wrapped around a scaffold of histone proteins. Most nucleated cells (i.e., most cells in the body with the exception of red blood cells and blood platelets) have 22 identical pairs of distinct chromosomes, numbered by size. Thus, chromosome 1 is the largest and chromosome 22 the smallest (Figure 2.2). These numbered chromosomes are also called *autosomes*. There are also two sex chromosomes; usually a pair of X chromosomes in women and one X chromosome and one Y chromosome in men. These sex chromosomes are also called *allosomes*.

The chromosomes have a central segment, the centromere, which is where the two duplicated chromatids (chromosome copies) are joined during cell division. The two chromosome arms separated by the centromere are labeled by convention as p (for petit, the shorter arm) and q (the longer arm). And while we are at it, let's review how the physical location of a gene is referenced. Stained chromosomes have microscopically visible bands that are numbered from the centromere outwards toward the ends of the arms (Figure 2.3).

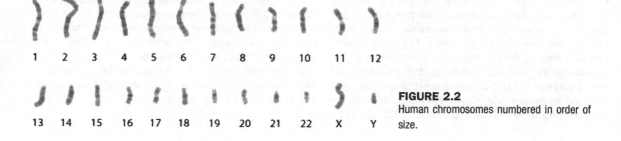

FIGURE 2.2
Human chromosomes numbered in order of size.

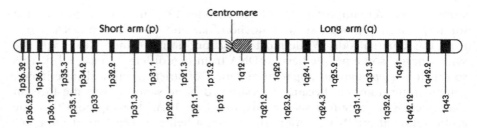

FIGURE 2.3
The bands on stained chromosomes, here for chromosome 1, are landmarks without major functional significance. They are useful reference points for localizing genes and other DNA segments of interest with sequence-specific imaging techniques such as fluorescent in situ hybridization.

The designation of each band on the long arm of chromosome 1 will therefore start with 1q. Both light and dark bands are numbered. For historical reasons some bands are grouped. These chromosome bands have no functional importance, but they are important landmarks for the position of specific DNA segments on a chromosome. A particularly useful technique is fluorescent in situ hybridization (FISH), which can show under the microscope where a small piece of dye-linked DNA binds to a chromosome. Comparing the fluorescent dye binding site to the banding pattern of the chromosome then gives a cytogenetic location, meaning that it indicates the region on the chromosome of a complementary DNA sequence.

The microscopic image of the stained chromosomes gives the karyotype of a healthy person without chromosomal abnormalities. Some individuals have only one X chromosome and no Y chromosome (Turner syndrome); others have two X chromosomes and one Y chromosome (XXY or Klinefelter syndrome). Both conditions cause only moderate physiological and metabolic abnormalities. Other syndromes resulting from an added chromosome (called *trisomy*, since the cells contain three copies of the same chromosome) have much more severe consequences.

How Down syndrome affects nutritional needs

Most common among the severe chromosomal abnormalities is the occurrence of three copies of chromosome 21 (trisomy 21; Down syndrome). Nutritional challenges specific to people with Down syndrome include early feeding and swallowing dysfunction [11], an increased risk of unhealthy weight gain with an emphasis on fat deposits in the abdomen and chest [12], and increased zinc requirement [13]. Celiac disease (CD) is much more common in people with Down syndrome than in the general population, occurring in 5–15% [14–17].

Chromosome 21 carries two genes that are important for folate metabolism, **cystathionine**-*beta-synthase* (*CBS*; EC 4.2.1.22) and the *reduced folate carrier* (*SLC19A1*; OMIM 600424). Because three chromosomes are active instead of the usual two, the resulting overexpression of CBS and SLC19A1 skews one-carbon and folate metabolism [18] and increases folate requirements. Congenital heart defects often accompany Down syndrome, in part due to functional folate deficiency [19].

In very rare cases, part or all of an extra chromosome may come from the same parent because of a chromosome sorting error. An example of such uniparental disomy was identified in an infant with severe malformations of the brain (microcephaly), persistent seizures, and molybdenum cofactor deficiency [20]. As it turned out, a region on the small arm of the maternal chromosome 6 with over 32 million bases, including the defective *MOCS1* (OMIM 603707) gene, had gotten inserted into chromosome 6 inherited from the father. All potentially harmful variants in that region were therefore homozygous in the patient. The protein product of the *MOCS1* gene, molybdenum cofactor biosynthesis protein 1, catalyzes the critical first step of molybdenum activation for its action with sulfite oxidase (EC 1.8.3.1), xanthine oxidase (EC 1.17.3.2), and aldehyde oxidase (EC 1.2.3.1). Unfortunately, this case graphically illustrates the catastrophic consequences of a lack of these molybdenum-dependent enzyme activities.

2.1.3 Genetic recombination

Now is as good a time as ever to introduce one of the most important concepts in human genetics, chromosomal crossover. The eggs and sperms, which might eventually combine and grow into a new person, each contain only a single set of chromosomes. If it was otherwise, there would be just too many chromosomes floating around. The gametes (egg and sperm cells) come about through meiotic division of germline cells with two full chromosome sets. There is first a duplication of the germline cell, followed by two divisions. After the initial duplication, pairs of chromosomes may exchange parts of an arm. This recombination event can be observed directly with imaging techniques [21]. Early investigators realized that the probability of a recombination is roughly proportional to the chromosome length. The distance between two points on a chromosome that have a 1% probability of being separated by a recombination event is 1 centimorgan (cM), named after the early genetics pioneer Thomas Hunt Morgan.

> **KEY CONCEPT**
> Recombination occurs when part of a chromosome is replaced by a segment from the other one of the chromosome pair.

There are typically between one and four recombination events per generation in a chromosome, depending on its length [21]. This corresponds to about one recombination event per every hundred million bases [22]. As a result, each of the chromosomes in the gametes is a patchwork of maternal and paternal segments (Figure 2.4).

The recombination frequency is the basis for various genetic calculations that will be discussed later. It is also the reason why we can make fairly precise estimates about the time elapsed since mutations that helped humans to adapt to their nutritional environments have occurred [23].

Another important aspect of recombination is that it does not always work cleanly and can give rise to all kinds of structural variants, such as deletions,

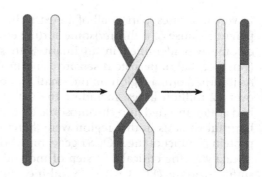

FIGURE 2.4
Recombination during meiosis creates a patchwork chromosome from maternal and paternal chromosome segments.

insertions, duplications, and inversions [24]. Structural variants contribute extensively to the overall genetic diversity, particularly those that persist over many generations and have helped our ancestors to adapt to new environments.

2.1.4 Cell proliferation

The single cell created at conception by fusion of the haploid gametes has to divide numerous times to grow into a viable fetus and those fetal cells have to divide yet further until the adult can have children. The first couple of divisions leave the genome of the cells unchanged. Then one of the two X chromosomes in all cells of female embryos is turned off in a random process that involves methylation of critical cytosine bases [25]. From this handful of starter cells develop the pluripotent stem cells, which form the ectodermal, mesodermal, and endodermal germ layers. Within the first 4 weeks a group of cells separates from all other cell types and over many intermediate stages these precursor cells develop into the mature gametes, eggs (oocytes) in females and sperms (spermatozoa) in males. The term *germline cell* refers to all of those forms, from the earliest precursors to the mature cells ready for conception. It is important to recognize that only germline cell DNA mutations can be transmitted to the next generation. Mutations in all other cells are called *somatic mutations*. Somatic cells with certain mutations can, for instance, grow into a cancer. But none of these mutations can be transmitted to a child conceived by the fusion of two gametes.

The other issue to consider is that a lot can happen to this long line of germline cells from the conception of one generation to conception of the next generation. We know that even identical twins, which start out from the same single cell at conception, differ in a few places because of mutations that occurred during the first few hours of the pregnancy. Any chemical change to germline cell DNA has the potential to be carried over into the next generation, and nutrition is responsible for many of these changes.

2.1.5 Cell-specific gene regulation

Every nucleated cell in the human body has the same DNA sequence, but a muscle cell in the heart looks and behaves very differently from a neuron in the brain. The chemical modifications of nuclear DNA are the main reason for the diversity in appearance and properties of the hundreds of different cell types [26]. Many genes are only expressed in one or a few specialized cells. The main mechanism for turning off genes is the chemical modification of the nuclear DNA, particularly around promoters and other regulatory elements. The methylation of specific cytosine bases silences gene expression. Methylated and otherwise modified cytosines function like on/off switches that direct expression and other genomic events. A related and coordinated mechanism of gene regulation works through the acetylation of the histones around which the DNA double helix is coiled. The program encoded by DNA modifications is partly inherited, but many changes occur in the first few weeks after conception as cells differentiate, and then at a slower pace during pregnancy and early infancy, and even slower later in life. Nutritional and other environmental conditions have a significant influence on the chemical shaping of DNA [27].

Changing the methylation signature of one cell type can transform it into another cell type. It has been shown, for instance, that the change of a few 5-methyl-cytosine switches in the nuclear DNA of an insulin-producing pancreas cell is all it takes to turn it into a glucagon-producing cell [28]. Nutritional and genetic factors together play a major role in both normal and pathological trans-formation of cells through changes of DNA methylation profiles [29]. Folate-rich vegetables and dietary folate supplements can reprogram precancerous cells by increasing the methylation of regulatory DNA segments [30], which may prevent the development of cancer. Similar events affect many other health outcomes, including heart disease, rheumatoid arthritis, and other autoimmune diseases.

Chemical modifications of genomic DNA are called *epigenetic changes*. The branch of genomics science dealing with regulatory modifications of nuclear DNA is called *epigenetics*.

2.1.6 Genome architecture

Less than 2% of the human genome is used as templates for RNA and protein synthesis. Large additional stretches are functional parts of genes, which are the structural units for the production of RNA and proteins. Further portions of the genome have critical regulatory functions for gene activation and inactivation. Such regulatory regions may be many thousands of bases away from the genes on which they act. Genes come in very different sizes, from the enormous *dystrophin* gene (*DMD*; OMIM 300377), spanning more than two million base pairs, to the diminutive *histone H4* gene (*Hist1H4K*; OMIM 602825), stretching to barely 354 base pairs.

It is important to recognize that genes are not neatly arranged one after the other. Instead, the human genome is a veritable jumble of makeshift gene

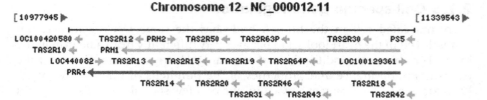

FIGURE 2.5
Genes in the salivary proline-rich protein multigene family cluster and the TAS2 taste receptor cluster overlap extensively on chromosome 12.

assemblies, with wild overlaps. Let's look at the region (Figure 2.5) where clusters of the salivary proline-rich protein multigene family and numerous members of the bitter taste receptors (TAS2R) overlap [31]. This region of not even half a million bases on chromosome 12 accommodates the *PRR4* gene (*proline-rich protein 4*), which takes up over 300,000 bases all on its own, the almost as long *PRH1* (*acid proline-rich protein 2*) gene, and the much smaller *PRH2* (*acid proline-rich protein 2*) genes. Remarkably, the same region also encodes no less than 19 members of the TAS2R family. How is this possible? As the diagram shows, only some parts of these genes, the exons, are actually used for specifying the amino acid sequence of the proteins. The sequences between the exons of a gene are called *introns*. But one gene's intron can be another gene's exon. Most of the *TAS2R* genes, for instance, lie within the introns of *PRH1* and *PRR4*.

Reading one strand of the DNA double helix in the forward (plus) direction will give an entirely different sequence than the complementary strand read in reverse (minus) direction. Specific genes are only read in one direction, which may be forward or reverse. In this region, the *PRH2* gene is read in the opposite direction of *TAS2R* genes.

By the way, *PS5* is a bitter taste receptor pseudogene, a nonfunctioning remnant left behind by evolution in the genomic jumble like a rusty old car wreck dumped in the countryside. Such pseudogenes are common and can cause trouble for genomic analyses because they tend to share large stretches of DNA sequence with related functional genes.

2.1.7 Gene expression

When a genomic sequence needs to be copied for RNA and protein synthesis, the local DNA double helix first has to be partially unwound from its tight coils around histones. Topoisomerases are enzymes that unwind (type I; EC 5.99.1.2) or wind (type II; EC 5.99.1.3) the DNA double helix from its histone spool, just as needed. One of several adenosine triphosphate (ATP)-powered helicases (EC 3.6.-.-) then separates and uncoils the two strands. The unwinding gives other enzyme complexes just enough space to work on a stretch of DNA that is ready for transcription.

DNA-directed RNA polymerase (RNAP; EC 2.7.7.6) produces single-stranded RNA by copying one of the DNA strands in the 5′ to 3′ direction. This RNA strand is then sliced, spliced, and edited.

The selection of the strand to be transcribed depends on specific transcription factors. About one in ten genes encodes a member of this large protein family of transcription factors. Each transcription factor binds to a specific DNA sequence, usually corresponding to one of a range of regulatory elements. The first among these regulatory elements is the promoter, which usually sits a few bases upstream of the start of the transcribed sequence. Transcription starts when a specific set of activating transcription factors binds to the promoter region and attracts DNA-directed RNA polymerase. Some transcription factors have the opposite effect and specifically block transcription.

Additional enhancer and repressor elements on the same DNA strand may act on the promoter. Many of these elements respond specifically to hormones, nutrients, or other molecules that can link, for instance, an excess of a particular nutrient to a regulatory response. Regulatory elements can be located within the transcribed sequence itself or anywhere upstream or downstream, sometimes several thousand bases away. For example, the two critical enhancer regions that control age- and tissue-specific expression of lactase in the small intestine, are located right in the middle of another gene, *MCM6* (OMIM 601806), about 14,000 and 22,000 bases upstream of the *lactase* (*LCT*; EC 3.2.1.108, OMIM 603202) promoter.

It is important to note that in each cell only one X chromosome is available for transcription [32]. A complex sequence of events in the first few days after conception permanently silences one of the X chromosomes in each cell by methylation of most possible cytosine methylation sites [33]. Since inactivation of X chromosomes is a random process, the maternal X chromosome will be inactivated in about half of the cells in a tissue, and the paternal chromosome in the others. The inactivation pattern is often unbalanced, causing significant overrepresentation of one X chromosome copy at the expense of the other. Overrepresentation of a defective allele can cause symptoms that would not occur with balanced X chromosome silencing.

KEY CONCEPT

Epigenetic silencing ensures that only half of the X chromosomes are active in females.

Specialized, large protein complexes, the spliceosomes, remove introns and parts of the 5′ and 3′ ends of the transcript. Some further editing may occur in editosomes. If the gene encodes a protein, the final transcription product will be the messenger RNA (mRNA) template used for protein synthesis (Figure 2.6). The 3′ end of the mRNA anneals to the 5′ start, which stabilizes the mRNA. Synthesis of protein directed by the mRNA strand is called *translation* and occurs in the ribosomes. The same mRNA can potentially serve as template for the

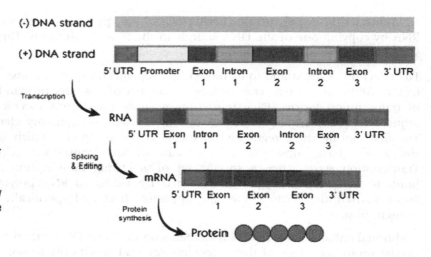

FIGURE 2.6
The initial RNA transcript is processed by removing introns and other unnecessary sequences and sometimes by editing specific bases. The untranslated regions (UTR), located before and after the coding sequences, are targets of regulatory DNA, RNA, and proteins. The mature mRNA is translated into protein in ribosomes.

synthesis of many copies of its protein product. The stability of the mRNA strongly influences, therefore, the final rate of protein synthesis. Many variant genes encode fully functional proteins, but the rate of protein synthesis is so slow that most of the gene function is lost.

2.1.8 Transcript variants

One of the surprising insights from sequencing the human genome was that humanity runs on not much more than 26,000 protein-coding genes. This does not seem to be a lot of genes to keep all of the different body functions going. And to think that over 3% of these precious few genes are already spoken for just for deciphering the odors around us. But obviously there must be enough genes to make it all work, because we are here alive and well. An important reason is that different copies with different functions can be made from the same gene. Depending on the particular cell type, life stage, and other conditions, transcription may start and end at different points within the gene, and may skip part or all of its exons. This provides the flexibility to make very different gene products that meet the needs of the moment and place.

The zinc transporter 2 (*ZnT2* or *SLC39A2*; OMIM 612166), for instance, is expressed most abundantly in tissues with high zinc needs. The tissues lining the milk duct have a greatly increased requirement during lactation because they have to pump zinc into milk. The mammary epithelial cells can make at least two different transcripts: a shorter and a longer version. The short version moves zinc directly across the plasma membrane, and the longer version moves zinc into secretory vesicles [34]. The hormones that stimulate milk production strongly favor expression of the longer transcript. Zinc transfer into milk via secretory vesicles provides much higher capacity than direct transfer across the plasma membrane, helping to move the infant's daily ration of 1–2 mg zinc into milk.

2.2 INHERITANCE

2.2.1 Traits and phenotypes

Traits are distinct characteristics that one can observe more or less directly. One or several traits together represent the phenotype, the directly observable presentation of a state or condition. According to these definitions, the blood pressure response to salt intake could be considered a trait; hypertension can be called a phenotype, because it results from the responses to several other influences. Genetic traits really exist only when there is diversity, such as the presence or absence of the trait, or several different versions of the trait.

> **KEY CONCEPT**
> An inherited trait is the observable consequence of one or multiple DNA variants.

Let's say a middle-aged woman lives on a dairy farm in the Scottish Highlands. She would not expect that any of her family and friends get abdominal pain from drinking a glass of milk. But when offering such a glass of nice cool milk to a couple visiting from the USA, she might be alarmed to hear both declining because they would get severe indigestion. She knows of only one person who had that kind of problem and it turned out to be a serious intestinal condition. They explain to her that where they come from many people cannot drink regular milk and they usually buy treated cow's milk or milk made from soy, almonds, or rice, if they use it at all. For them, the ability to drink cow's milk runs in families. Some can tolerate it and others cannot. Of course, if the US visitors had come to Thailand, their hosts might have offered them a drink with coconut milk, but not mammalian milk, because they would know better than risk digestive problems. To them, the ability to consume large quantities of cow's milk would be a curiosity. But once one starts looking for the ability to drink a cup of milk across diverse populations, it becomes obvious that it is an inherited trait that 25–30% of the adults in the world have, and the others don't. There are many more such traits that most people are blissfully unaware of, because neither they nor anybody else around them have them. Have you ever thought that some healthy people might not be able to drink even small quantities of German beer? Just a hint, it is not the alcohol that is the problem; it is all about gluten intolerance and CD. Or that other people cannot eat peaches, oranges, and pineapples because of an inherited trait (sucrase deficiency)? Read on to find out about such traits in later chapters. And, given the opportunity, we may yet find out that some people have the astounding ability to eat some kind of wild herb, berry, or mushroom that would do grave harm to most other mortals. That might be a worthwhile trait to have when lost in a wilderness and those berries are the only food source in abundant supply.

2.2.2 Inheritance patterns

Many genetic traits are inherited in classical Mendelian fashion. This means that both parents contribute equally because one chromosome of each pair comes from the mother and the other from the father. The transmission to the next generation of a variation on one of the numbered chromosomes (autosomes) is called *autosomal* inheritance. If an inherited allele always prevails over alternative ones, the inheritance pattern is called *autosomal dominant*. If the allele is eclipsed by the alternative, it is called *autosomal recessive*. Let us take the example of adults who cannot drink large amounts of milk because their *lactase* gene in the small intestine is turned off. This is an autosomal recessive trait for all practical purposes. This means that inheritance from only one parent is not enough to make it apparent. If someone is lactose intolerant, the trait has probably come from both parents.

Inheritance of X-linked traits is slightly different, of course. In this case all males inherit one gene dose of the trait from the mother, but none from the father (because he has contributed the Y chromosome instead). A classic example of an X-linked nutrigenetic inheritance pattern can be seen with glucose 6-phosphate (*G6PD*; OMIM 305900) deficiency. If males with G6PD deficiency eat broad beans or take certain medications, some of their red blood cells may break up (hemolyze). Females have the same problem only if both of their X chromosomes are affected, which happens less often. This means that the condition is usually inherited from mothers who are symptom-free carriers; rarely from fathers. Affected fathers can transmit the trait only to daughters, who will not be symptomatic unless they get a second affected X chromosome from their mother.

The inheritance of other traits follows more complex patterns. Genomic imprinting, variable numbers of nucleotide repeats, copy number variants, multilocus traits, and mosaicism are mechanisms that explain deviations from the orthodox Mendelian pattern of inheritance.

Let's look at an example of two independently inherited traits in adults: the ability to drink significant amounts of milk (a cup or more at a time) and the ability to drink a can of sugar-containing soda without adverse effects (Figure 2.7). Milk contains lactose, large quantities of which (for instance the 12 g in one 240 mL glass) cause unpleasant digestive problems unless lactase is active in the small intestine of the drinker. Half of the sugar in soda consists of fructose (for instance 17 g in one 368 mL can), which triggers life-threatening hypoglycemia in people without a functional copy of the *fructoaldolase* gene (*ALDOB*; OMIM 612724) encoding fructose-1,6-bisphosphate aldolase (EC 4.1.2.13). Adverse effects of either lactose or fructose consumption are recessive traits, which means that they occur only when both of the gene copies inherited from the mother and father provide inadequate amounts of functional enzyme. The law of independent assortment holds true in this example. The figure below shows the outcome (phenotype) when children inherit a particular set of alleles from each parent. If, for instance, the father provides the allele for milk intolerance (I), and the mother contributes the allele for milk tolerance (L), the child will have both an "l" and an "L" allele. Because the "l" allele is recessive, the "L" allele wins out and the child will be able to drink milk. Review some parental combinations and ask yourself (without looking) what the outcome (phenotype) will be in the child.

Along those lines, consider the following example. The father is homozygous for the lactose intolerance allele (due to lack of lactase persistence) and has a problem with drinking milk (Figure 2.8). The mother is homozygous for the fructose-intolerance allele (due to defective aldolase B) and gets low blood sugar (hypoglycemia) after drinking sugar-sweetened soda. Now, because both alleles are recessive, any children resulting from this union will be able to drink both milk and sugar-sweetened soda.

Mendelian principles of heredity

Law of segregation
An individual inherits one allele from the mother and another from the father.

Law of independent assortment
Traits are inherited independently of each other.

In the year 1866, Gregor Mendel presented in lectures and published proceedings his insightful formulation of the two basic principles of inheritance, which established the scientific study of genetics. It took several decades until the significance of his concepts was recognized, but they then guided the rapid growth of genetic research.

FIGURE 2.7

Inheritance of two independent traits: The ability to safely drink sugar-containing soda (indicated by the soda can) and the ability to drink milk (indicated by the milk carton) without abdominal discomfort are linked to different chromosomes and inherited independently of each other. The table shows the genotypes resulting from specific combinations of parental chromosomes. Since both traits are recessive, only children with two *intolerance* alleles have a problem.

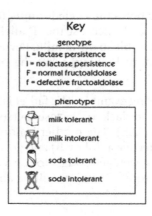

Key

genotype
L = lactase persistence
l = no lactase persistence
F = normal fructoaldolase
f = defective fructoaldolase

phenotype
milk tolerant
milk intolerant
soda tolerant
soda intolerant

Maternal chromosomes

Children's possible
Genotypes and phenotypes

On the other hand, if both parents were heterozygous for both traits, there would be a 25% chance that a child would not be able to drink milk and a 25% chance that she could not tolerate the soda. Multiplication of the individual likelihoods ($0.25 \times 0.25 = 0.0625$) indicates that one child in 16 from such a heterozygous couple will not be able to tolerate either milk or soda.

It is important to understand that traits are inherited independently only if they derive from genes on different chromosomes or at different ends of the same chromosome. Genes located near each other are often inherited together as explained in more detail below.

Analysis of a trait in multigenerational families is the most direct way to determine the mode of inheritance. An individual's presentation (phenotype) will then depend on how critical the single-strand DNA variant is for the trait. If a trait becomes apparent even when the responsible variant is present only on one of the two DNA strands in a chromosome, it is called a *dominant trait*. If it takes two copies to show, it is called a *recessive trait*. Rare metabolic diseases (inborn errors of metabolism) such as phenylketonuria (PKU) or abetalipoproteinemia are usually recessive. This means that a single functional DNA copy

FIGURE 2.8

This is an example of two traits that are transmitted independently. Each parent is homozygous for a different recessive trait, one for the inability to drink a cup of milk and the other one for the inability to drink sugar-sweetened soda without causing discomfort.

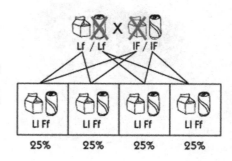

can do the job and the trait becomes apparent only when both copies are altered.

2.2.3 The concept of genotypes

Without much fanfare, we have now slipped from talking about phenotypes to talking about genotypes, from traits to genes. The physical organization of a gene on a DNA strand has been described above. Genes are for all practical purposes the basic units of trait inheritance. If the DNA sequence varies at a certain position between people, the different versions are called *alleles*. Note that the term *allele* is used similarly in classical genetics to describe alternative versions of a trait. The position of a DNA sequence on the chromosome is called a *locus* (plural *loci*) from the Latin word for place. If there are two sequence versions, the genetic is biallelic; with several versions it is multiallelic.

> **KEY CONCEPT**
> An allele is one of two or more alternative versions of a genetic trait or short DNA sequence.

A genotype describes DNA sequences at one or more positions (loci) in the DNA sequences of both the maternal and the paternal chromosomes. For instance, the homozygous genotype for the European type of lactase persistence would be given as *LAC* -13910T/T, indicating that both the maternal and the paternal chromosomes have a T (thymine) in position -13910.

The difference is primarily a question of focus, going from outward appearance with all its underlying complexities (phenotype) to the very specific and directly measurable molecular unit of inheritance (genotype). It is the difference between referring to the ability to drink milk and a specific DNA sequence responsible for persistent lactase expression, which in turn helps adults to digest milk.

The Danish geneticist Wilhelm Johannsen was an early advocate (at around 1900) of the existence of a particularized genotype [35] and he was the one who established the use of the terms *genotype* and *gene*. The stepwise (quantum) mode of inheritance seems so obvious today because we have grown up with the concept of genes. Think about this: the exchange of one base for another one cannot be a continuous transition, because a sequence change has either occurred or not. But at that time it was not known that the genome is written like a text, with the nucleotide bases as the letters. A common view was that traits are like paint colors that can be mixed in any ratio. But this is not how it works, because letters cannot be mixed, only replaced, added, or deleted. And most of the time only a very small number of letters (bases) in a region can change without upsetting the finely tuned function of the sequence. If the change from a new mutation is really bad, the newly conceived child may not survive long enough or function well enough to have offspring. If the altered genome functions even slightly worse than the current model, the mutation probably will stay rare or die out eventually.

Several decades of hard work and many brilliant insights during the first half of the twentieth century were eventually consolidated into our current understanding of how inheritance works, well before these theories were conclusively vindicated by direct DNA analyses. Today we know that physical traits are inherited through nuclear and mitochondrial DNA, and also that variation of DNA segments outside the protein-encoding genes gives rise to complex patterns of inheritance.

2.2.4 Relatives

It is easy to understand that genes are inherited from both the mother and father, because we get a full set of chromosomes from each parent. From the two copies at each locus, the child will get either one or the other, with a 50% probability of each. There is also a 50% probability that a sibling will have inherited the same allele from this parent, if the variation is biallelic. The percentage of allelic variants that two family members share is called the *coefficient of relationship*.

The best way to explore how closely someone is related to another person is to draw up a family tree. One can then count the number of generations from the client to a common ancestor plus the number back to the other person (Figure 2.9). The coefficient of relationship of someone with a grandparent would thus be $0.5 \times 0.5 = 25\%$. For an aunt or uncle that would be two steps to the common ancestor, the grandparent, and then one step back down, making it three steps in total and a $0.5 \times 0.5 \times 0.5 = 12.5\%$ probability. For a half-sibling at any level just add one generational step. By the way, geneticists use the term *avuncle* to indicate the relationship to either an aunt or uncle. The coefficient of relationship is only correct to the extent that none of the couples producing an offspring are related (consanguineous). In populations with a long history of

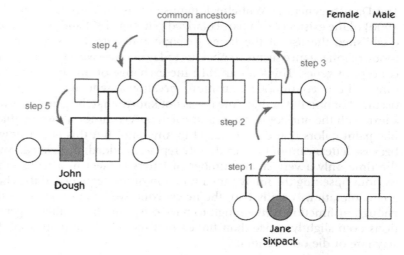

FIGURE 2.9
The coefficient of relatedness is the probability that we share a particular DNA sequence variant with a family member because of our shared inheritance. To estimate the coefficient of relationship between Jane Sixpack and John Dough, draw first a family tree. Then count the steps it takes from Jane to the shared ancestor and back to John. Finally, multiply by 0.5 as many times as there are steps. In this case we get a coefficient of $0.5 \times 0.5 \times 0.5 \times 0.5 \times 0.5 = 0.03125$.

in-group marriage, two people may be much more closely related to each other than they know, because of multiple shared ancestors, possibly in the distant past.

2.2.5 A simple culinary experiment

Next time you have at least two adult generations of your biological family at the breakfast table, offer each of them a glass of unsweetened white grapefruit juice (not one of the much milder pink varieties). Then track the extent to which your family reports tasting the bitter flavor of the fruit and disliking the juice. Draw your findings into a family tree and consider whether the individual preferences are inherited.

Here is an explanation of what is going on behind the scenes: The two variants of the gene for the bitter taste receptor 19 (*TAS2R19*; OMIM 613961) encode either a cysteine or an arginine in position 299 (rs10772420) of the receptor protein. The 299Cys version is much more sensitive to quinine and other bitter compounds than the more common form 299Arg [36]. Individuals with one or two copies of 299Cys usually find white grapefruit too bitter for their liking, while people without this variant have no problem with the bitter taste [37]. Since one copy is enough to make people taste and dislike the bitter grapefruit flavor, the trait is dominant. Grapefruit juice preference is not strictly tied to the *TAS2R19* genotype, but is close enough to illustrate the point. If both of your parents like white grapefruit juice, then you will probably also like it. If neither of them like it, the situation is less clear, because they could be either heterozygous or homozygous for the Cys299 variant. Still, more likely than not, you will be no fan of grapefruit juice (see Figure 2.10). If you do like white grapefruit juice and neither of your parents does, both of them where probably heterozygous and each contributed their Arg299 variant copy to your grapefruit-loving palate. As surprising as it may seem, this kind of inheritance pattern explains how someone can like grapefruit even if neither parent likes it.

A final concept needs to be introduced here: *heritability*. This term indicates the estimated percentage to which a trait is inherited. The remainder is attributable to environmental influences. In the example of grapefruit choice, much of the preference will be attributable to the inherited taste receptor allele. But another important influence is probably childhood experience. Some children may be taught to tolerate the bitter taste as the way food should taste. If such conditioning had a significant influence on current taste perception, then heritability of grapefruit choice would be considerably less than 100%.

2.2.6 Inheritance through mitochondrial DNA

The mitochondria carry with them a full complement of DNA, polymerase, and transcription regulators. Most importantly, mitochondria come almost exclusively from the maternal side, because the mitochondria of sperm are

FIGURE 2.10
The ability to taste the bitter flavor in grapefruit is transmitted from the parents to the children as a recessive Mendelian trait related to a single nucleotide polymorphism affecting the amino acid at position 299 of the bitter taste receptor 19. Note how it is possible that two people with a dislike of grapefruit can have an offspring who likes this bitter fruit. This can happen when both parents are heterozygous and both contribute their g allele to create an offspring with the grapefruit-liking gg genotype. Not a very likely event, but it does occur.

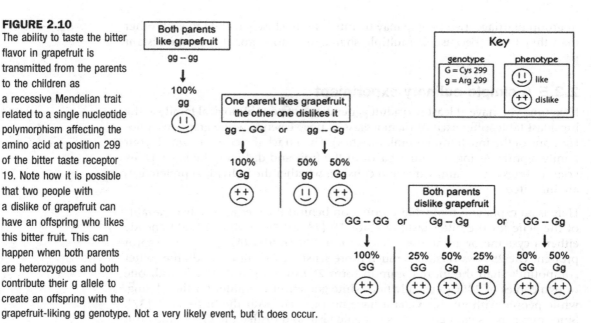

expelled after conception. It is a rare event if a paternal mitochondrion stays on [8]. This special way in which mitochondria are handed down to the next generation determines patterns of inheritance. It means that conditions and diseases related to variants in mtDNA are inherited almost exclusively through the mother.

The circular DNA of mitochondria contains less than 17,000 bases, but encodes key actors in oxidative phosphorylation [7]. It should not be surprising, therefore, that an increasing number of investigations link mtDNA variants to insulin resistance, diabetes, body composition, and metabolic syndrome [38]. One example is the common single nucleotide variant 16189T>C of mtDNA, which appears to increase the risk of obesity, diabetes, and coronary heart disease in the many people who carry it [39,40].

The maternal mode of inheritance was studied in an extended kindred of Caucasians (Figure 2.11) with a mtDNA variant that changes the thymine next to the isoleucine anticodon to a cytosine [41]. The signature finding, which pointed to a mtDNA mutation, was that mothers transmitted low magnesium concentration in blood (hypomagnesemia) to three-quarters of their children, but none of the biological fathers did. Almost half of the family members in the maternal lineage had the hypomagnesemia trait, but none in the paternal lineage.

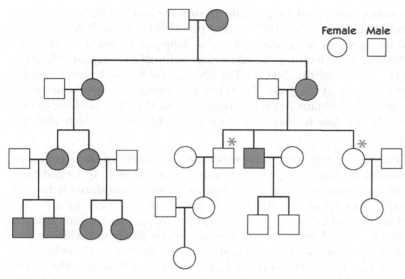

FIGURE 2.11
All members of a large kindred with abnormally low magnesium concentration in the blood (filled symbols) have inherited it exclusively through the maternal lineage [41]. The reason for this uncommon mode of inheritance is that the condition is caused by a mutation in mitochondrial DNA. Note that transmission is not as clear-cut as it would be with the much more common inheritance through nuclear DNA. The reason is that occasionally a small number of paternal mitochondria slip through the exclusion mechanism (heteroplasmy) and thereby make up for a maternal defect. This seems to have happened for the individuals marked with an asterisk, which should have the trait by maternal inheritance, but don't.

Mitochondrial inheritance always should be considered when a trait appears to be inherited only from mothers (but as we all know, it is very rare that it is all the mother's fault).

2.2.7 Epigenetic inheritance

Our genetic material is not a simple assembly of the four standard nucleotide bases on a string. Many of the cytosine bases are methylated at carbon 5. Such chemical changes to DNA are called *epigenetic changes*. Methylation of cytosine residues in promoter regions and other regulatory elements of a gene occurs by a tightly regulated process and can block its expression [42].

Some epigenetic DNA modifications are transmitted from parents to children, often in a way specific to the parental gender. The transmitted DNA is said to be *imprinted*. Some of these epigenetic signatures reflect previous exposure of the parents, and probably even of the grandparents, to nutritional and other environmental influences [43].

KEY CONCEPT
The process of imprinting controls the transmission of certain critical methylation patterns from parent to offspring.

Mutation of one of the genes that control normal imprinting is the cause for inherited disruptions of normal imprinting, such as Angelman and Prader-Willi syndromes, which cause severe harm to health. These syndromes demonstrate the impact of defective control of DNA imprinting on a wide range of molecular, metabolic, and physiological functions. The typical features in Prader-Willi syndrome (PWS), for example, include overeating and rapid weight gain during the toddler years, as well as obesity later in life. A loss of SNORD116 is responsible for most PWS cases [44]. SNORD116 functions as noncoding RNA that helps to organize other small pieces of RNA in the nucleolus, probably by controlling how they are spliced. It apparently takes a lot of this SNORD116 RNA to do its work, since healthy people have 29 copies of the gene, like pearls on a string. The gene is normally expressed mainly in the brain, but not in patients with PWS. We still have to find out how exactly the loss of this gene causes the imprinting defects and the observed symptoms.

As pointed out above, epigenetic DNA modification profiles of some genes are inherited, usually without change across untold generations [45]. In addition, the epigenetic profiles at some loci depend on the parents' nutritional states and to some extent also on nutritional factors after birth and later in life. This influence of nutritional factors on heritable epigenetic profiles is a rare instance where dietary exposure changes inherited traits. There is now good evidence that both starvation and overfeeding during pregnancy influence the likelihood of chronic disease later in life. This concept goes back to observations of the Dutch famine during German occupation. Germany occupied the Netherlands during World War II and triggered a severe famine by cutting off food supplies. This caused severe food shortages and starvation during the winter of 1944–1945. Typical daily energy intakes were considerably below 1000 calories. During this specific time, a large number of women carried pregnancies to term under starvation conditions. A later comparison of health risks in people born before, during and after this hunger winter showed that they were more susceptible to obesity, diabetes, and hypertension, most likely as the result of persistent epigenetic programming [46]. Epigenetic DNA modifications in response to the parents' nutritional status before and during pregnancy are the focus of active research [47]. Several specific gene targets have been suggested that could be involved in this effect [48]. Numerous variations in imprinting are likely to impact health, with the potential for both beneficial and undesirable consequences. Cataloging and understanding such inherited variations in DNA imprinting profiles is still in its early stages.

To be very clear, it is not only the mother's lifestyle that influences obesity risk and other outcomes in the offspring. Overeating changes the epigenetic profile of germline DNA of men, and some of these modifications appear to be transmitted to the children and by them in turn to later generations [49]. One cohort from a remote and very stable Swedish community was tracked from childhood to the end of the life of the grandchildren. Researchers investigated records from the last 200 years. Historical harvest data were used to determine who among the grandparent generation experienced at least one plentiful harvest near puberty (at ages 9–12 years), calling them *overfed*. Those with at least one poor harvest year during that critical time period were called *underfed*. It turned out that the grandchildren of men from the overfed group did not live as long as grandchildren from men growing up during normal harvest times [50]. And compared to grandchildren of underfed men, their lifespan was 32 years shorter! This pattern of transmission to grandchildren, which strongly suggests an imprinting (epigenetic) mechanism, can be recreated in experimental animal models [51].

Now, how does that diet and fitness program sound, not just to impress a future spouse but also for the sake of the children and grandchildren?

2.2.8 The Hardy-Weinberg equilibrium

The combination of alleles at a locus gives us genotypes. If the combination is truly random in a large enough population sample with no significant outward or inward migration and which is not subject to selection pressure, the genotype frequencies will reflect the probabilities of chance combinations.

Polynomials of the form $(p_1 + p_2 + \ldots)^2$ describe the genotype distributions where p_1, p_2, etc. are the frequencies of the possible alleles at a particular locus. Observed genotype frequencies are closest to the frequencies calculated with one of these polynomials if Mendelian inheritance applies and none of the previously mentioned factors interfere with the random combination of alternative alleles in a population sample. These polynomials describing the normal Mendelian distribution of genotypes have been named after the pioneer geneticists Godfrey Harold Hardy and Wilhelm Weinberg.

Let us examine the simplest example of a locus on an autosomal chromosome with two alternative alleles: A with frequency p and a with frequency q. By definition, the two frequencies add up to 1, because only those two versions occur. If each parent contributes one of their two allele copies with equal probability, there will be four possible combinations: A/A, A/a, a/A, and a/a. The ideal genotype frequencies are then p^2 for A/A, pq for A/a, pq for a/A, and q^2 for a/a.

It can easily be seen that this follows the polynomial $p^2 + 2pq + q^2$. With a greater number of alleles, the number of genotypes and calculation of genotype frequencies become more complex, of course.

> **REMEMBER THIS**
> The Hardy-Weinberg polynomial for biallelic variants is $p^2 + 2pq + q^2$.

If the observed distribution is close to the theoretical one, the population is said to be in Hardy-Weinberg equilibrium [52]. In practice, the observed numbers can only be an approximation of the theoretical distribution because the combinations inherently carry a random error. Additional deviations from the expected normal distribution will come from factors such as analytical errors, high-frequency deletions, selection bias, increased prenatal losses, later mortality with one genotype, frequent inbreeding, a mitochondrial mode of inheritance, or epigenetic transmission.

> **KEY CONCEPT**
> Hardy-Weinberg equilibrium exists when the genotypes in a large, stable population combine randomly following Mendelian rules and are not subject to selection pressure.

2.2.9 Memetic transmission

Throughout history, human populations have dealt with the threat of food scarcity and nutrient deficiency by inventing methods for getting access to new food sources and for preparing their foods in a more nutritious way. The instructions for these new methods evolved over time and were handed down from one generation to another. The technologies have also spread to unrelated group members and to other populations. Such transmitted instructions have been dubbed memes, because their transfer from one person to another demonstrates some striking parallels to the transmission of genes. There also has been the meme that a meme is a virus of the mind.

KEY CONCEPT
A meme is anything that we can learn and then teach to others. This text is full of memes.

Nutritional patterns of many populations are strongly influenced by traditional norms (memes) with a strong influence on their nutritional environment. The adaptation of cooking practices to combat niacin deficiency is a classic example. Human metabolism cannot always make enough niacin from tryptophan to make up for a niacin-poor food supply. Some people learned to overcome that limitation early on. The outer layer of corn contains niacin precursors that can be turned into niacin with alkali pretreatment. The Aztecs and other pre-Columbian Americans dehulled corn and cooked it with alkali-rich ashes. The process is still called *nixtamalization* from the Nahuatl words *nextli* (ashes) and *tamalli* (corn gruel).

There are numerous other examples of prehistoric memetic transmission of food technologies and behaviors shaping human nutrition, such as cooking methods, milking of animals, brewing of alcoholic beverages, and cultivation of food plants.

Today's memes come in untold shapes and forms and enter our minds all the time. They are transmitted by people (family, friends, neighbors, teachers, campaigners, preachers), text, movies, radio, television, and of course the internet. Recipes and food fads are contemporary examples of nutrition memes. When we talk about a new fad diet going viral, we explicitly acknowledge the infectious (memetic) nature of this information.

2.3 MOLECULAR EFFECTS OF GENETIC VARIATION

2.3.1 Polymorphisms

Genetic variants that occur in at least 1% of sizeable populations are defined as *polymorphisms*. The two or more forms of variant DNA segments are called *alleles*. The position of a variant in the DNA sequence of a chromosome is the locus. If there are two versions, the genetic variant is biallelic; with several versions it is multiallelic. You will notice that there is never a monoallelic variant where everybody has the same version.

A few caveats are in order at this point. First, one should strongly resist the unfortunate habit to refer to all polymorphisms as SNPs (single nucleotide polymorphisms). The acronym certainly has a snappy sound, but that does not make it helpful. The many types of polymorphisms that are not properly SNPs include insertions, deletions, and copy number variants. The improper use of the term makes us forget that commonly used analytical methods detect single base changes well, but not the other equally common forms of variation. This should become clear in the descriptions of the various types of variants below. Second, there is a common misconception that polymorphisms are mutations. This is incorrect by definition, because it would be very unlikely for a particular variation to spontaneously occur precisely at the same locus and with precisely the same change in millions of people. All variants that are common enough to be called polymorphism have been inherited over many hundreds or thousands of generations, sometimes spanning millions of years and multiple distinct species! Third, it is established practice to refer to the most common allele as the wild type. One needs to recognize that this convention has no functional implications. It does not mean that this allele produces the most active protein or that it is the ancestral form. For example, the G allele at the 1958 locus of *methylene tetrahydrofolate dehydrogenase* (*MTHFD1*; OMIM 172460) is less common than the A allele, but it confers higher enzyme activity [53] and it is the original form shared with all other primates. One should also recognize that some gene products have more than one action. An enzyme may act for instance on different substrates and an allelic variant may then have reduced activity with one substrate but increased activity toward another one. Similarly, hormonal and other mediators often act on diverse targets and variation may have very different effects on these.

So far, more than 15 million polymorphisms have been identified [54]. Many of them are catalogued with a unique rs (reference SNP) number that helps looking them up in databases and simplifies the identification of the polymorphism in reports or assay orders. Try, for instance, the Short Genetic Variation database, dbSNP, of the National Center for Biotechnology Information (NCBI) National Library of Medicine at http://www.ncbi.nlm.nih.gov/. DNA sequence differences (genotype) sometimes lead to differences in appearance (phenotype), but more often will not do so. Indeed, most common genetic variants in humans are not clearly linked to a difference in appearance or function [55]. Of course, initially overlooked phenotypic differences may still be recognized in later investigations. Rare, sporadic mutations are much more likely to cause a dramatic difference, most often loss of function. It is easy to see that such harmful mutations are much less likely to be passed on to multiple generations than harmless or useful ones.

The molecular character of a variant may sometimes suggest its biological significance. In most cases, however, this can only be the starting point for functional investigations.

2.3.2 Genetic linkage

Genetic variants draw interest most often when they are associated with an important trait. Almost regardless of how strong that association is, one should consider whether linkage to a different nearby variant is the cause. What is linkage you ask? Consider this: recombination events during the generation of gametes (eggs and sperms) randomly mix very large blocks of chromosome pairs. Neighboring DNA segments usually stay together over many generations until a recombination event separates them and moves one to the twin chromosome. The odds that two DNA segments on a chromosome get separated in a single generation are proportional to their distance. There is therefore a less than 0.0001% chance that two points (loci) in a thousand base-long DNA segment will become separated from one generation to the next, because there is on average only about one recombination event per 125 million bases [21]. Because the odds can differ so greatly, it makes sense to express them as the exponent to base ten. The customary measure to estimate the distance between two loci is therefore the logarithm of odds (LoD). Ten thousand generations after a new variant x appears in the promoter a thousand bases upstream of a previously existing polymorphic allele a in exon 1 of a gene, more than 98% of the descendants with promoter allele x will still have allele a in exon 1.

Definition of Linkage Disequilibrium

Linkage disequilibrium is a measure of how much the association of the frequencies of two variants (which may be on the same chromosome or not) deviates from randomness. The precise definition is given by the equation

$$D_{AB} = p_{AB} - p_A\, p_B$$

for the variant pairs A/a and B/b on a particular chromosome copy. In this equation D_{AB} is the linkage equilibrium coefficient and the various p terms indicate frequencies in a population. The term p_{AB} is the frequency of chromosomal copies that carry both A and B. The term $p_A\, p_B$ is the multiplication product of the individual frequencies of variant A (p_A) and variant B (p_B).

Take the example of the variant pair C and T (rs1076991) at position -105 of the *MTHFD1* (OMIM 172460) promoter and the variant pair G and A (rs2236225) at position 1958, encoding either arginine or glutamate in position 653 of the mature protein. Together they determine the 10-formyl tetrahydrofolate synthase activity of this trifunctional enzyme complex [56]. And this enzyme activity also happens to determine how much choline its carrier needs [53] and what a woman's risk is to have a child with a neural tube defect, if she became pregnant [56].

Let's assume we found the following frequencies in a population:

- 105C/1958G 0.53 (both alleles on the same chromosome copy)
- 105C 0.49 (on either chromosome copy)
  1958G 0.68 (on either chromosome copy)

We can then calculate

$$D_{-105C1958} = 0.53 - 0.49 \times 0.68 = 0.2$$

indicating that the association between these two variations is so low that it is close to random. In other words, if we determined only the allele at position 1958, we could not predict what the allele at position -105 was. We will have to determine it analytically.

Linkage is thus defined as the tendency of neighboring alleles to occur together on the same chromosome. When a variant in the coding region of a gene is associated with an interesting trait, it is quite possible that the true causative variant is located a few hundred or thousand bases away.

2.3.3 Haplotypes

Linkage provides a good context for exploring the concept of *haplotype* [57]. This term relates to a long shared DNA sequence on the same chromosome copy, containing the same variants from the beginning to the end of that sequence. Haplotypes are an important concept for understanding how several variants in combination affect the same gene. Many of our genes contain dozens of functionally important variants.

KEY CONCEPT
A haplotype is a segment on a chromosome strand that has been inherited unchanged from a distant ancestor.

Haplotypes tend to be transmitted in their entirety over hundreds of generations. For example, the *lactase* gene (OMIM 603202) and all its major regulatory elements extend over about 70,000 bases on chromosome 2. The likelihood that a sequence of this length is broken up by a recombination event is less than one in a thousand per generation. This means that there is near certainty that the gene we get from a parent matches exactly, base for base, the sequence of that parent and also the sequence of the grandparent from where it originated. If we know which of the two haplotypes we inherited from a parent, we know that we carry all the same DNA variants in the same order exactly as that parent. Let's assume that we know the full lactase gene sequences of our parents' four lactase copies. We will then need to analyze (or have analyzed) only small stretches of our two lactase gene copies to know the full sequences. We will then also know from which parent each of our two lactase gene copies came. We don't even need to know the full sequence. Analyzing just a few gene variants will achieve the same in most instances.

Today's standard genetic tests cannot determine directly whether two variants lie on the same or on different DNA strands without full sequence analysis. To answer this important question, the phase (telling us whether a set of alleles is on the same strand or not) has to be determined by direct molecular analysis, computational haplotyping, or with family studies. If variants are on the same DNA strand they are said to be in the *cis* position (from the Latin for *on this side*). If they are on different strands, they are in the *trans* position (Latin for *on the other side*).

For example, when one of two variants renders the product of a gene inactive and the other one blocks expression of the gene, it is absolutely essential to know the phase of the two variants. If both variants are on the same strand, the other strand can still produce fully functional protein at 50% or more of the normal rate; the carrier may be healthy and fully functional under most circumstances. If the variants are on different strands, on the other hand, the function of the gene will be completely disrupted.

Apolipoprotein E haplotypes

Differences in the amino acids at protein position 112 and 158 determine which apolipoprotein E (*APOE*; OMIM 107741) isoforms someone carries. Both positions contain either arginine or cysteine, encoded by the DNA polymorphisms 334C>T (rs419358) and 472C>T (rs7412), respectively [58]. In addition, the common promoter variant -219T (rs405509) slightly reduces protein expression compared to the more common variant -219G [59].

The combination of 112Arg and 158Arg constitutes the ancestral [60] Apo-E isoform ε4, which is linked to low vitamin K concentration [61], increased risk of late-onset Alzheimer's dementia, [62] and the promotion of heart disease [63]. Reduced expression of isoform ε4 in carriers of the -219T variant appears to add to the already increased risk [64].

If someone is heterozygous at all three polymorphic sites, four different haplotype combinations are conceivable. However, the haplotype 334C-472T has been observed only in a single African pedigree and can be excluded. This leaves the likely haplotypes 334T-472T (*apoE*ε2*) and 334C-472C (*apoE*ε4*). The next challenge then is to determine whether the variant -219T slows expression of either ε2 or ε4. Consultation of tables with previously observed haplotype frequencies suggests that strands with the ε2 isoform usually carry the -219G variant, and strands with the ε4 isoform tend to have the -219T variant. Such imputation would not work, however, if one of the isoforms was ε3, because this isoform is associated with the -219T allele almost as often as with the -219G allele.

Available software packages use the estimation-maximization (EM) algorithm [65], composite haplotype method [66], and other approaches for these calculations. Computational resolution of haplotypes, if it is possible at all, leaves a small error rate, particularly when done for individuals from populations with unknown haplotypes and haplotype frequencies. If the individual haplotype needs to be known with certainty, each of the two strands has to be sequenced or analyzed using another molecular method.

PHASE, http://depts.washington.edu/uwc4c/express-licenses/assets/phase/

fastPHASE, http://depts.washington.edu/uwc4c/express-licenses/assets/fastphase/

HaploBlock, http://bioinfo.cs.technion.ac.il/haploblock/

	DNA			Protein	
	-219	334	472	112	158
Low E2	T	T	T	Cys	Cys
High E2	G	T	T	Cys	Cys
Low E3	T	T	C	Cys	Arg
High E3	G	T	C	Cys	Arg
Low E4	T	C	C	Arg	Arg
High E4	G	C	C	Arg	Arg

2.3.4 Nonsynonymous variants

The simplest change in a gene sequence is the replacement of one base with a different one. This type of variant (and only this) is called a *SNP*. It is usually described with its location, the base in the most common form, an arrow or > character, and the base of the variant (for instance as *MTHFR* 677C>T in the example below). The location is usually counted from the expression start site and a negative sign indicates that the location is upstream of the start site. When a base change occurs within an exon, the new codon triplet may encode a different amino acid. Such a DNA sequence change that leads to an alteration in the amino acid sequence of the encoded protein is conventionally called a *missense variant*. If the variant introduces a premature stop codon or otherwise disrupts normal expression of the gene product, it is called a *nonsense variant*. Amino acid changes will often abolish some or all of the protein's activity, but a gain of function is also possible. When a gene encoding an enzyme has several common variants that differ in their amino acid sequence, the individual forms are called *allozymes*. A star followed by a number usually indicates the specific form. Allozyme *1 is the most common form in this notation. Replacement in 5,10-methylenetetrahydrofolate reductase (MTHFR; EC 1.5.1.20) of the alanine in position 222 to valine in the 677C>T variant (allozyme *4, Figure 2.12) cuts the normal activity of this enzyme in half [67]. But sometimes there is no change at all, such as with the variant with alanine instead of glycine in position 429 (allozyme *6). And in gain-of-function variants the altered protein is actually more active than the original, as with the variant with arginine instead of glutamate in position 422 (allozyme *5).

It is important to understand that the sequence modification may change functional properties that are not measured. The MTHFR variant containing valine in position 222, for instance, is more sensitive to loss of the FAD cofactor and thermal inactivation than the unchanged enzyme [68], which is not apparent when enzyme activity is tested only under standard conditions. Alternatively, the difference may be apparent with one substrate but not with another [69]. In other cases there may be a change in response to a metabolic activator or inhibitor. Sometimes the combination of two amino acid changes modifies function in a different way than either of the changes alone. As just mentioned, the MTHFR variant Gly429Ala (allozyme *6) retains normal activity, and the variant Arg519His (allozyme *7) may even have slightly increased activity. But a protein that contains both variants (allozyme *16) has less than half of the normal activity [67].

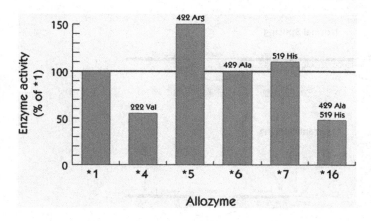

FIGURE 2.12
Change in activity of recombinant constructs with different amino acid sequences (allozymes) relative to the most common MTHFR (5-mTHF) allozyme *1.
Arg, arginine; His, histidine; Val, valine.

2.3.5 Synonymous variants

Not all base changes in exonic DNA sequences change the expected sequence of the gene product, because different base triplets can encode the same amino acid. Many of these synonymous variants in exons have no biological effect at all. That is why they are frequently called *silent variants*. This is not really a helpful term, however, since in many instances synonymous alterations of a DNA sequence change the amount or quality of the synthesized protein. This can be because the speed of DNA transcription changes or because the stability of the resulting mRNA differs [70,71], possibly due to altered folding [72]. It should be noted that both transcription rate and mRNA stability can increase or decrease with a sequence change. Both scenarios occur fairly commonly in polymorphic variants.

Molecular case study

Systematic genetic screening of patients with familial hypercholesterolemia [73] detects a novel C>A nucleotide variant at position 1216 in exon 9 of the *low-density lipoprotein receptor* gene (*LDLR*; OMIM 606945). One might assume that this synonymous variant cannot be responsible for hypercholesterolemia, because the predicted amino acid sequence of LDLR is conserved (Arg385Arg). But the base change creates a sequence in exon 9 that accepts exon 8 more effectively for splicing than the original site 31 bases upstream. This aberrant splicing skips these first 31 bases in exon 9. As can be seen in the sequence diagram below, the skipping moves the reading frame one base forward (this is called a *frameshift*) and encodes different amino acids than in the undulated sequence. The frameshift continues into exon 10 until the codon for amino acid 391 is reached. The stop codon at this position then terminates translation and triggers release of the incomplete protein chain. In addition, the abnormal mRNA decays much faster than the unchanged mRNA and little of the altered protein is produced.

A synonymous mutation in exon 9 causes the splicing apparatus to skip the 31 bases before it, resulting in a shortened and frame-shifted mRNA encoding nonsense after the mutation until it prematurely terminates translation because the frame-shifted mRNA presents a stop codon.

Moral of the story: never assume that synonymous variants are silent! They may disrupt normal splicing, change the transcription rate, or alter mRNA stability.

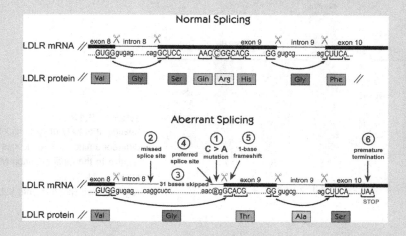

2.3.6 Variants in untranslated regions

For a long time it was thought that only 1–2% of the human genome is important and that most of the rest was just jumbled fragments, inactive pseudogenes, and other unimportant garbage. It is now becoming increasingly clear that many regions outside the translated gene sequence have direct roles in controlling gene expression.

Transcription of a gene starts many bases before the segments encoding the final product and continues for hundreds of bases beyond what is needed to encode a protein or functional RNA. These excess pieces are lopped off during mRNA editing and are not translated into protein. The untranslated region in the 5′ direction (5′ UTR) folds in specific ways until a helicase adjusts the secondary structure and allows ribosomes and regulatory elements to bind [74]. The untranslated region at the 3′ end of a gene (3′ UTR) is equally important for controlling expression. This long slack region bends back and links up to the 5′ end, forming a circle. What happens at the 3′ UTR often determines the amount of mRNA that is made by controlling the rate of its translation and degradation. Both regulatory proteins and small RNA segments interact with specific sequences in the 3′ UTR, modifying the amount of protein that is produced [75]. This makes it obvious that variants at either end of the mRNA can significantly alter the amount of expressed protein.

For example, a common polymorphism of the *dihydrofolate reductase* gene (*DHFR*; OMIM 126060) has either a cytosine or a thymine base at position 829, which is located deep in the 3′ UTR following the end of the sequence encoding the enzyme. The mRNA version with cytosine is translated into protein less efficiently than the thymine version. How can this happen? In the cytosine version, miR-24 binds to the 3′ UTR, interfering with circularizing of the mRNA and accelerating its degradation. With thymine in position 829, miR-24 cannot bind [76]. What we have here is a region that responds to a regulatory microRNA, and the 829T variant has lost its responsiveness to miR-24-induced repression at this site. MicroRNAs are usually 22 bases in length and control as much as 30% of all genes [75], often by binding somewhere in the 3′ UTR.

2.3.7 Repeats

Repetition of short sequences is the trademark of a fairly common type of genetic variant. These repeats may consist of just a few or several dozen replications. The patterns in short tandem repeats (STR) may consist of a handful of nucleotides, but may sometimes run to as many as 50. Alternative terms for these STR are microsatellites or simple sequence repeats (SSR). They are usually located in noncoding sequences. Their impact depends as much on their location as on their pattern.

The *ALOX5* gene (OMIM 152390) promoter region, for instance, contains a variable number of a Sp1-binding motif, which consists of the hexanucleotide

GGGCGG. Each of these variants affects gene expression differently, though not in the same direction in all cell types [77]. High arachidonic acid intake (with meat and eggs) will cause thickening of artery walls in carriers of the variant with four repeats, but not in people with five repeats [78]. The reason is that the higher enzyme activity in the four-repeat variant generates more inflammatory eicosanoids, such as leukotriene B4, which cause thickening and narrowing of arteries. The enzyme in people with the five-repeat variant seems to be much less active and high arachidonic acid consumption seems to leave the arteries of these individuals largely untouched.

Another example is the polymorphism with seven to nine repeats of the trinucleotide ATT (rs3832406) in the gene *MTHFD1L* [79]. The number of repeats influences splicing efficiency despite the fact that they lie well within intron 7, 11 bases before the splice site. This difference in splicing efficiency actually has consequences: women with only seven ATT repeats have a greater risk of having a child with a neural tube defect than women with more repeats.

2.3.8 Copy number variants

Sometimes multiple copies of a functional gene can be present on the same DNA strand [80]. Copy number variants may extend over hundreds of thousands of bases, or just a few hundred. More than 11,000 variants have been identified, affecting more than a thousand genes [81]. For instance, copy number variants occur in most genes for glycolytic enzymes and thus should affect enzyme activity levels [82].

KEY CONCEPT

Some people have extra copies of a gene, which usually increases its function.

The inheritance patterns of such copy number variants may appear complex at first, but upon closer examination still follow classical Mendelian genetics. Children inherit one of their mother's DNA strands with however many copies she has on that strand plus one of their father's DNA strands with however many copies present on that one.

2.3.9 Inversion polymorphisms

Another type of variant that is not readily detected with many conventional methods occurs when the arrangement of gene segments is inverted. Most of the rearranged segments are between hundreds of thousands and several million bases long. Hundreds of such variants have been confirmed that occur in more than 20% of the investigated individuals [83]. The functional consequences of most of these inversion polymorphisms remain to be explored. The most severe disruption of gene function will occur if the rearrangement splits a gene. About half of all cases of severe hemophilia

A (a bleeding disorder due to factor VIII deficiency) are caused by an inversion leading to factor VIII gene disruption [84]. Such disruption with complete loss of function is unlikely to occur in polymorphisms because by definition such common variants have been handed down and favored for many generations; otherwise they would not have become as common as they are. However, the disruption may affect regulatory elements that are separated from their target gene and may be favorable under the right circumstances.

2.3.10 Deletions and hemizygosity

A base insertion or deletion will shift the reading frame if it does not occur in a multiple of three. Such frameshifts lead to nonsense transcripts with unstable mRNA and little protein is produced [85]. Frameshift variants also may generate a stop codon that terminates transcription prematurely. In other cases, the frameshift may obliterate a stop codon and generate an overlong protein.

A particular situation arises when only one parent has the gene to begin with and a variant copy from the other parent cannot compensate for the absence. Such representation of a heritable trait by only one chromosome is called *hemizygosity*. Of course, males are usually hemizygous for genes on the X chromosome, because they have only a single copy. Several nutritionally relevant genes are located on the X chromosome. An important example is *ornithine transcarbamoylase* (OTC; EC 2.1.3.3). When it does not work efficiently, ammonia removal as urea is slowed and ammonia will accumulate to potentially lethal levels. It is no wonder then that males with defective OTC usually die in early childhood, while females may only learn about a single copy defect when a change to a high-protein diet, fasting, or surgery triggers a life-threatening episode of high ammonia concentration [86].

Sometimes deletions of thousands or millions of bases (called *microdeletions*, because they are not visible under a light microscope) can be the reason that a gene is hemizygous. This concept is illustrated by the case of a child with mild PKU. The cause was determined to be biopterin deficiency due to a change in the 6-pyruvoyl-tetrahydropterin synthase (EC 4.2.3.12) amino acid sequence encoded by the *PTPS* gene (OMIM 612719) [87]. The same change was present in the mother, but the father had a large deletion on one of his chromosome 11 copies spanning several million bases, including the PTPS gene. Standard genetic screening would show a normal PTPS gene sequence, but not the fact that he had only a single copy of it.

For many analytical methods, high-frequency deletions will suggest a deviation from the Hardy-Weinberg equilibrium with a low proportion of heterozygotes. The reason is that only one strand will produce a signal when the other strand is deleted.

High-frequency deletions exist, for instance, in the bitter taste receptor TAS2R43, where they cause considerable variability in the taste threshold for caffeine. More than half of a small German population sample had at least one of their two alleles deleted, and more than a quarter did not have any functional copy [88]. Looking for extended regions of (apparent) homozygosity is another way of identifying deletions. Using this measure, more than a quarter of all people in one genomic survey had deletions of up to 25 million bases [89].

2.3.11 Epimutations

It is now thought that an imprinting change, which is called an *epimutation*, can improperly silence a gene on its own, in the absence of other DNA mutations, and thereby impact biological function and health [90]. One emerging example is the inherited silencing of the mismatch repair gene *MLH1* (OMIM 120436), which greatly increases the risk of colorectal cancer [91]. Aflatoxin from molds growing on stored foods [92], endocrine disruptors, and other compounds can cause such inherited changes. If such an epimutation is inherited, it will affect the cells in a particular tissue equally. If it occurs during early embryogenesis, the abnormal epigenetic profile will impact all the cells directly descended from the epimutated cell. This will lead to a mosaic-like distribution of affected and unaffected cells (mosaicism). If the epimutation happens later in life, only the descendants of the originally epimutated cell will be affected. This is largely analogous to the inheritance patterns of regular DNA mutations.

Practice questions

If trehalose intolerance is a recessive trait and your mother's brother and your father's sister have this trait, but neither of your parents or grandparents have it, what is the likelihood that your sister also has this condition?

What are likely explanations for the occurrence in a small ethnic group of 1% heterozygous and no homozygous carriers of a particular *G6PD* variant?

If one in ten Swedes is lactose intolerant and the genotype distribution conforms to a Hardy-Weinberg distribution, what percentage of Swedish men do you expect to carry two copies of the lactase persistence allele?

How would you assess the impact of double heterozygosity with one allele of a low-activity *DHFR* promoter variant and one allele of a low-activity synonymous variant on the ability of the enzyme to reactivate dihydrofolate?

Assuming that you are an adult and lactose intolerant, what is the likelihood that your mother is also lactose intolerant?

Why is the insertion of eight bases more likely than the deletion of nine bases to abolish the function of a gene?

SUMMARY AND SEGUE TO THE NEXT CHAPTER

This section revisited the biological basis of inheritance, including the architecture of the human genome as well as mechanisms of transcription and translation. We also reviewed common modes of genetic transmission, which can usually be followed by simply counting the alleles located on the maternal and paternal chromosome copies. Occasionally we find a more complex pattern of inheritance, which may be explained by circumstances such as inheritance of different numbers of copies from each parent, the absence of one parental gene copy, transmission via epigenetic mechanisms, or in very rare cases inheritance through maternal mtDNA.

We will now move on to examine where genetic variation with impact on nutrition is coming from. The next chapter will explore the origins of some genetic variants that influence which foods are best for us. We will touch on vital issues such as the nature of Paleolithic diets, what persuaded most Chinese not to drink alcohol, and how 800 years ago most Easter Islanders came to have the haptoglobin variant Hp1.

References

[1] Dunn GA, Morgan CP, Bale TL. Sex-specificity in transgenerational epigenetic programming. Horm Behav 2011 Mar;59(3):290−5.

[2] Yazbek SN, Spiezio SH, Nadeau JH, Buchner DA. Ancestral paternal genotype controls body weight and food intake for multiple generations. Hum Mol Genet 2010 Nov 1;19(21): 4134−44.

[3] Globisch D, Munzel M, Muller M, Michalakis S, Wagner M, Koch S, et al. Tissue distribution of 5-hydroxymethylcytosine and search for active demethylation intermediates. PLoS One 2010;5(12):e15367.

[4] Ito S, Shen L, Dai Q, Wu SC, Collins LB, Swenberg JA, et al. Tet proteins can convert 5-methylcytosine to 5-formylcytosine and 5-carboxylcytosine. Science 2011 Jul 21;33(6047): 1300−3.

[5] Lander ES, Linton LM, Birren B, Nusbaum C, Zody MC, Baldwin J, et al. Initial sequencing and analysis of the human genome. Nature 2001;409(6822):860−921.

[6] Taylor RW, Taylor GA, Durham SE, Turnbull DM. The determination of complete human mitochondrial DNA sequences in single cells: implications for the study of somatic mitochondrial DNA point mutations. Nucleic Acids Res 2001 Aug 1;29(15):E74−4.

[7] Eynon N, Moran M, Birk R, Lucia A. The champions' mitochondria: is it genetically determined? A review on mitochondrial DNA and elite athletic performance. Physiol Genomics 2011 Jul 14;43(13):789−98.

[8] Reiner JE, Kishore RB, Levin BC, Albanetti T, Boire N, Knipe A, et al. Detection of heteroplasmic mitochondrial DNA in single mitochondria. PLoS One 2010;5(12):e14359.

[9] Ivanov PL, Wadhams MJ, Roby RK, Holland MM, Weedn VW, Parsons TJ. Mitochondrial DNA sequence heteroplasmy in the Grand Duke of Russia Georgij Romanov establishes the authenticity of the remains of Tsar Nicholas II. Nat Genet 1996 Apr;12(4):417−20.

[10] Irwin JA, Saunier JL, Niederstatter H, Strouss KM, Sturk KA, Diegoli TM, et al. Investigation of heteroplasmy in the human mitochondrial DNA control region: a synthesis of observations from more than 5000 global population samples. J Mol Evol 2009 May;68(5): 516−27.

[11] Cooper-Brown L, Copeland S, Dailey S, Downey D, Petersen MC, Stimson C, et al. Feeding and swallowing dysfunction in genetic syndromes. Dev Disabil Res Rev 2008;14(2):147−57.

[12] Gonzalez-Aguero A, Ara I, Moreno LA, Vicente-Rodriguez G, Casajus JA. Fat and lean masses in youths with Down syndrome: Gender differences. Res Dev Disabil 2011 Sep-Oct;32(5): 1685–93.

[13] Marreiro Ddo N, de Sousa AF, Nogueira Ndo N, Oliveira FE. Effect of zinc supplementation on thyroid hormone metabolism of adolescents with Down syndrome. Biol Trace Elem Res 2009;129(1–3):20–7.

[14] Book L, Hart A, Black J, Feolo M, Zone JJ, Neuhausen SL. Prevalence and clinical characteristics of celiac disease in Downs syndrome in a US study. Am J Med Genet 2001 Jan 1;98(1): 70–4.

[15] Shamaly H, Hartman C, Pollack S, Hujerat M, Katz R, Gideoni O, et al. Tissue transglutaminase antibodies are a useful serological marker for the diagnosis of celiac disease in patients with Down syndrome. J Pediatr Gastroenterol Nutr 2007 May;44(5):583–6.

[16] Cerqueira RM, Rocha CM, Fernandes CD, Correia MR. Celiac disease in Portuguese children and adults with Down syndrome. Eur J Gastroenterol Hepatol 2010 Jul;22(7):868–71.

[17] Wouters J, Weijerman ME, van Furth AM, Schreurs MW, Crusius JB, von Blomberg BM, et al. Prospective human leukocyte antigen, endomysium immunoglobulin A antibodies, and transglutaminase antibodies testing for celiac disease in children with Down syndrome. J Pediatr 2009 Feb;154(2):239–42.

[18] Pogribna M, Melnyk S, Pogribny I, Chango A, Yi P, James SJ. Homocysteine metabolism in children with Down syndrome: in vitro modulation. Am J Hum Genet 2001 Jul;69(1): 88–95.

[19] Locke AE, Dooley KJ, Tinker SW, Cheong SY, Feingold E, Allen EG, et al. Variation in folate pathway genes contributes to risk of congenital heart defects among individuals with Down syndrome. Genet Epidemiol 2010 Sep;34(6):613–23.

[20] Gumus H, Ghesquiere S, Per H, Kondolot M, Ichida K, Poyrazoglu G, et al. Maternal uniparental isodisomy is responsible for serious molybdenum cofactor deficiency. Dev Med Child Neurol 2010 Sep;52(9):868–72.

[21] Sun F, Oliver-Bonet M, Liehr T, Starke H, Ko E, Rademaker A, et al. Human male recombination maps for individual chromosomes. Am J Hum Genet 2004 Mar;74(3):521–31.

[22] Kong A, Thorleifsson G, Gudbjartsson DF, Masson G, Sigurdsson A, Jonasdottir A, et al. Fine-scale recombination rate differences between sexes, populations and individuals. Nature 2010 Oct 28;467(7319):1099–103.

[23] Slatkin MA. Bayesian method for jointly estimating allele age and selection intensity. Genet Res (Camb) 2008 Feb;90(1):129–37.

[24] Mills RE, Walter K, Stewart C, Handsaker RE, Chen K, Alkan C, et al. Mapping copy number variation by population-scale genome sequencing. Nature 2011 Feb 3;470(7332):59–65.

[25] Deuve JL, Avner P. The coupling of X-chromosome inactivation to pluripotency. Annu Rev Cell Dev Biol 2011 Nov 10;27:611–29.

[26] Hanna JH, Saha K, Jaenisch R. Pluripotency and cellular reprogramming: facts, hypotheses, unresolved issues. Cell 2010 Nov 12;143(4):508–25.

[27] Burdge GC, Lillycrop KA. Nutrition, epigenetics, and developmental plasticity: implications for understanding human disease. Annu Rev Nutr 2010 Aug 21;30:315–39.

[28] Dhawan S, Georgia S, Tschen SI, Fan G, Bhushan A. Pancreatic beta cell identity is maintained by DNA methylation-mediated repression of Arx. Dev Cell 2011 Apr 19;20(4): 419–29.

[29] Fenech MF. Dietary reference values of individual micronutrients and nutriomes for genome damage prevention: current status and a road map to the future. Am J Clin Nutr 2010 May;91(5):1438S–54S.

[30] Stidley CA, Picchi MA, Leng S, Willink R, Crowell RE, Flores KG, et al. Multivitamins, folate, and green vegetables protect against gene promoter methylation in the aerodigestive tract of smokers. Cancer Res 2010 Jan 15;70(2):568–74.

[31] Kim HS, Smithies O, Maeda N. A physical map of the human salivary proline-rich protein gene cluster covers over 700 kbp of DNA. Genomics 1990 Feb;6(2):260–7.

[32] Lyon MF. Gene action in the X-chromosome of the mouse (Mus musculus L.). Nature 1961 Apr 22;190:372–3.

[33] Tattermusch A, Brockdorff N. A scaffold for X chromosome inactivation. Hum Genet 2011 Aug;130(2):247−53.

[34] Lopez V, Kelleher SL. Zinc transporter-2 (ZnT2) variants are localized to distinct subcellular compartments and functionally transport zinc. Biochem J 2009;422(1):43−52.

[35] Roll-Hansen N. The genotype theory of Wilhelm Johannsen and its relation to plant breeding and the study of evolution. Centaurus 1979;22:201−35.

[36] Reed DR, Zhu G, Breslin PA, Duke FF, Henders AK, Campbell MJ, et al. The perception of quinine taste intensity is associated with common genetic variants in a bitter receptor cluster on chromosome 12. Hum Mol Genet 2010 Nov 1;19(21):4278−85.

[37] Hayes JE, Wallace MR, Knopik VS, Herbstman DM, Bartoshuk LM, Duffy VB. Allelic variation in TAS2R bitter receptor genes associates with variation in sensations from and ingestive behaviors toward common bitter beverages in adults. Chem Senses 2011 Mar;36(3):311−19.

[38] Palmieri VO, De Rasmo D, Signorile A, Sardanelli AM, Grattagliano I, Minerva F, et al. T16189C mitochondrial DNA variant is associated with metabolic syndrome in Caucasian subjects. Nutrition 2011 Jul-Aug;27(7−8):773−7.

[39] Liou CW, Lin TK, Huei Weng H, Lee CF, Chen TL, Wei YH, et al. A common mitochondrial DNA variant and increased body mass index as associated factors for development of type 2 diabetes: Additive effects of genetic and environmental factors. J Clin Endocrinol Metab 2007 Jan;92(1):235−9.

[40] Mueller EE, Eder W, Ebner S, Schwaiger E, Santic D, Kreindl T, et al. The mitochondrial T16189C polymorphism is associated with coronary artery disease in Middle European populations. PLoS One 2011;6(1):e16455.

[41] Wilson FH, Hariri A, Farhi A, Zhao H, Petersen KF, Toka HR, et al. A cluster of metabolic defects caused by mutation in a mitochondrial tRNA. Science 2004 Nov 12;306(5699): 1190−4.

[42] Donkena KV, Yuan H, Young CY. Vitamin Bs, one carbon metabolism and prostate cancer. Mini Rev Med Chem 2010 Dec;10(14):1385−92.

[43] Skinner MK, Manikkam M, Guerrero-Bosagna C. Epigenetic transgenerational actions of environmental factors in disease etiology. Trends Endocrinol Metab 2010 Apr;21(4):214−22.

[44] Cassidy SB, Schwartz S, Miller JL, Driscoll DJ. Prader-Willi syndrome. Genet Med 2012 Jan;14(1):10−26.

[45] Dunn GA, Bale TL. Maternal high-fat diet effects on third-generation female body size via the paternal lineage. Endocrinology 2011 Jun;152(6):2228−36.

[46] Heijmans BT, Tobi EW, Stein AD, Putter H, Blauw GJ, Susser ES, et al. Persistent epigenetic differences associated with prenatal exposure to famine in humans. Proc Natl Acad Sci U S A 2008 Nov 4;105(44):17046−9.

[47] Waterland RA, Travisano M, Tahiliani KG, Rached MT, Mirza S. Methyl donor supplementation prevents transgenerational amplification of obesity. Int J Obes (Lond) 2008 Sep;32(9): 1373−9.

[48] Tobi EW, Lumey LH, Talens RP, Kremer D, Putter H, Stein AD, et al. DNA methylation differences after exposure to prenatal famine are common and timing- and sex-specific. Hum Mol Genet 2009 Nov 1;18(21):4046−53.

[49] Kaati G, Bygren LO, Pembrey M, Sjostrom M. Transgenerational response to nutrition, early life circumstances and longevity. Eur J Hum Genet 2007 Jul;15(7):784−90.

[50] Bygren LO, Kaati G, Edvinsson S. Longevity determined by paternal ancestors' nutrition during their slow growth period. Acta Biotheor 2001 Mar;49(1):53−9.

[51] Jirtle RL, Skinner MK. Environmental epigenomics and disease susceptibility. Nat Rev Genet 2007 Apr;8(4):253−62.

[52] Moonesinghe R, Yesupriya A, Chang MH, Dowling NF, Khoury MJ, Scott AJ. A Hardy-Weinberg equilibrium test for analyzing population genetic surveys with complex sample designs. Am J Epidemiol 2010 Apr 15;171(8):932−41.

[53] Kohlmeier M, da Costa KA, Fischer LM, Zeisel SH. Genetic variation of folate-mediated one-carbon transfer pathway predicts susceptibility to choline deficiency in humans. Proc Natl Acad Sci U S A 2005 Nov 1;102(44):16025−30.

[54] 1000 Genomes Project Consortium. A map of human genome variation from population-scale sequencing. Nature 2010 Oct 28;467(7319):1061—73.

[55] Lachance J. Disease-associated alleles in genome-wide association studies are enriched for derived low frequency alleles relative to HapMap and neutral expectations. BMC Med Genomics 2010;3:57.

[56] Carroll N, Pangilinan F, Molloy AM, Troendle J, Mills JL, Kirke PN, et al. Analysis of the MTHFD1 promoter and risk of neural tube defects. Hum Genet 2009 Apr;125(3):247—56.

[57] Slatkin M. Linkage disequilibrium—understanding the evolutionary past and mapping the medical future. Nat Rev Genet 2008 Jun;9(6):477—85.

[58] Kohlmeier M, Saupe J, Schaefer K, Asmus G. Bone fracture history and prospective bone fracture risk of hemodialysis patients are related to apolipoprotein E genotype. Calcif Tissue Int 1998 Mar;62(3):278—81.

[59] Lambert JC, Brousseau T, Defosse V, Evans A, Arveiler D, Ruidavets JB, et al. Independent association of an APOE gene promoter polymorphism with increased risk of myocardial infarction and decreased APOE plasma concentrations—the ECTIM study. Hum Mol Genet 2000 Jan 1;9(1):57—61.

[60] Hanlon CS, Rubinsztein DC. Arginine residues at codons 112 and 158 in the apolipoprotein E gene correspond to the ancestral state in humans. Atherosclerosis 1995 Jan 6;112(1):85—90.

[61] Kohlmeier M, Salomon A, Saupe J, Shearer MJ. Transport of vitamin K to bone in humans. J Nutr 1996 Apr;126(Suppl. 4):1192S—6S.

[62] Corder EH, Saunders AM, Strittmatter WJ, Schmechel DE, Gaskell PC, Small GW, et al. Gene dose of apolipoprotein E type 4 allele and the risk of Alzheimer's disease in late onset families. Science 1993 Aug 13;261(5123):921—3.

[63] Ward H, Mitrou PN, Bowman R, Luben R, Wareham NJ, Khaw KT, et al. APOE genotype, lipids, and coronary heart disease risk: a prospective population study. Arch Intern Med 2009 Aug 10;169(15):1424—9.

[64] Lescai F, Chiamenti AM, Codemo A, Pirazzini C, D'Agostino G, Ruaro C, et al. An APOE haplotype associated with decreased epsilon4 expression increases the risk of late onset Alzheimer's disease. J Alzheimers Dis 2011;24(2):235—45.

[65] Excoffier L, Slatkin M. Maximum-likelihood estimation of molecular haplotype frequencies in a diploid population. Mol Biol Evol 1995 Sep;12(5):921—7.

[66] Nielsen DM, Ehm MG, Zaykin DV, Weir BS. Effect of two- and three-locus linkage disequilibrium on the power to detect marker/phenotype associations. Genetics 2004 Oct;168(2):1029—40.

[67] Martin YN, Salavaggione OE, Eckloff BW, Wieben ED, Schaid DJ, Weinshilboum RM. Human methylenetetrahydrofolate reductase pharmacogenomics: gene resequencing and functional genomics. Pharmacogenet Genomics 2006 Apr;16(4):265—77.

[68] Guenther BD, Sheppard CA, Tran P, Rozen R, Matthews RG, Ludwig ML. The structure and properties of methylenetetrahydrofolate reductase from Escherichia coli suggest how folate ameliorates human hyperhomocysteinemia. Nat Struct Biol 1999 Apr;6(4):359—65.

[69] Richter RJ, Jarvik GP, Furlong CE. Paraoxonase 1 (PON1) status and substrate hydrolysis. Toxicol Appl Pharmacol 2009 Feb 15;235(1):1—9.

[70] Chamary JV, Hurst LD. Evidence for selection on synonymous mutations affecting stability of mRNA secondary structure in mammals. Genome Biol 2005;6(9):R75.

[71] Resch AM, Carmel L, Marino-Ramirez L, Ogurtsov AY, Shabalina SA, Rogozin IB, et al. Widespread positive selection in synonymous sites of mammalian genes. Mol Biol Evol 2007 Aug;24(8):1821—31.

[72] Wang D, Sadee W. Searching for polymorphisms that affect gene expression and mRNA processing: example ABCB1 (MDR1). AAPS J 2006;8(3):E515—20.

[73] Defesche JC, Schuurman EJ, Klaaijsen LN, Khoo KL, Wiegman A, Stalenhoef AF. Silent exonic mutations in the low-density lipoprotein receptor gene that cause familial hypercholesterolemia by affecting mRNA splicing. Clin Genet 2008 Jun;73(6):573—8.

[74] Livingstone M, Atas E, Meller A, Sonenberg N. Mechanisms governing the control of mRNA translation. Phys Biol 2010;7(2):021001.

[75] Mazumder B, Seshadri V, Fox PL. Translational control by the 3'-UTR: the ends specify the means. Trends Biochem Sci 2003 Feb;28(2):91–8.

[76] Mishra PJ, Humeniuk R, Longo-Sorbello GS, Banerjee D, Bertino JR. A miR-24 microRNA binding-site polymorphism in dihydrofolate reductase gene leads to methotrexate resistance. Proc Natl Acad Sci U S A 2007 Aug 14;104(33):13513–18.

[77] Vikman S, Brena RM, Armstrong P, Hartiala J, Stephensen CB, Allayee H. Functional analysis of 5-lipoxygenase promoter repeat variants. Hum Mol Genet 2009 Dec 1;18(23):4521–9.

[78] Dwyer JH, Allayee H, Dwyer KM, Fan J, Wu H, Mar R, et al. Arachidonate 5-lipoxygenase promoter genotype, dietary arachidonic acid, and atherosclerosis. N Engl J Med 2004 Jan 1;350(1):29–37.

[79] Parle-McDermott A, Pangilinan F, O'Brien KK, Mills JL, Magee AM, Troendle J, et al. A common variant in MTHFD1L is associated with neural tube defects and mRNA splicing efficiency. Hum Mutat 2009 Dec;30(12):1650–6.

[80] Sudmant PH, Kitzman JO, Antonacci F, Alkan C, Malig M, Tsalenko A, et al. Diversity of human copy number variation and multicopy genes. Science 2010 Oct 29;330(6004):641–6.

[81] Stamoulis C, Betensky RA. A novel signal processing approach for the detection of copy-number variations in the human genome. Bioinformatics 2011 Jul 12.

[82] Varma V, Wise C, Kaput J. Carbohydrate metabolic pathway genes associated with quantitative trait loci (QTL) for obesity and type 2 diabetes: identification by data mining. Biotechnol J 2010 Sep;5(9):942–9.

[83] Sindi SS, Raphael BJ. Identification and frequency estimation of inversion polymorphisms from haplotype data. J Comput Biol 2010 Mar;17(3):517–31.

[84] Naylor JA, Nicholson P, Goodeve A, Hassock S, Peake I, Giannelli F. A novel DNA inversion causing severe hemophilia A. Blood 1996 Apr 15;87(8):3255–61.

[85] Yamaguchi-Kabata Y, Shimada MK, Hayakawa Y, Minoshima S, Chakraborty R, Gojobori T, et al. Distribution and effects of nonsense polymorphisms in human genes. PLoS One 2008;3(10):e3393.

[86] Gordon N. Ornithine transcarbamylase deficiency: a urea cycle defect. Eur J Paediatr Neurol 2003;7(3):115–21.

[87] Blau N, Scherer-Oppliger T, Baumer A, Riegel M, Matasovic A, Schinzel A, et al. Isolated central form of tetrahydrobiopterin deficiency associated with hemizygosity on chromosome 11q and a mutant allele of PTPS. Hum Mutat 2000;16(1):54–60.

[88] Roudnitzky N, Bufe B, Thalmann S, Kuhn C, Gunn HC, Xing C, et al. Genomic, genetic and functional dissection of bitter taste responses to artificial sweeteners. Hum Mol Genet 2011 Sep 1;20(17):3437–49.

[89] Hosking FJ, Papaemmanuil E, Sheridan E, Kinsey SE, Lightfoot T, Roman E, et al. Genome-wide homozygosity signatures and childhood acute lymphoblastic leukemia risk. Blood 2010 Jun 3;115(22):4472–7.

[90] Martin DI, Cropley JE, Suter CM. Epigenetics in disease: Leader or follower? Epigenetics 2011 Jul 1;6(7):843–8.

[91] Hitchins M, Owens S, Kwok CT, Godsmark G, Algar U, Ramesar R. Identification of new cases of early-onset colorectal cancer with an MLH1 epimutation in an ethnically diverse South African cohort(dagger). Clin Genet 2011 Nov;80(5):428–34.

[92] Shibui T, Higo Y, Tsutsui TW, Uchida M, Oshimura M, Barrett JC, et al. Changes in expression of imprinted genes following treatment of human cancer cell lines with non-mutagenic or mutagenic carcinogens. Int J Oncol 2008 Aug;33(2):351–60.

CHAPTER 3
Where Nutrigenetic Differences Come From

57

Terms of the Trade

- Founder effect: Population expansion from a few ancestors can explain some common variants.
- Hominins: Human species (*Homo sapiens*, *Homo ergaster*, and *Homo rudolfensis*) and their recent ancestors.
- Hominoids: Hominins plus orangutans, gorillas, and chimpanzees.
- Nutritope: An environment with a particular pattern of nutrient abundances and food toxins.
- Pleiotropic: Describing an effect that concerns diverse and unrelated body functions.
- Pseudogene: Nonfunctional genomic sequence derived by descent from a functional gene.

ABSTRACT

The nutritional requirements of humans have evolved over eons. Early primate ancestors lived mainly on fruits and leaves. Some later ancestors learned to eat starchy roots and grass seeds, and others survived on game meats and fish. Agriculture, brewing, cattle herding, and cooking opened up new sources of nutrients. All of these developments left their imprints on our genome. Our diverse abilities to digest and metabolize specific nutrients are echoes of very different nutritional environments (nutritopes). Our dependence on vitamin C from foods goes back to times when fruits and leaves where plentiful, but not everybody today needs the same amounts. Some populations have adapted to low intake levels that would be too low for people in most other parts of the world to survive on. Some of us can drink a lot of milk without discomfort; others suffer abdominal pain and bloating from even modest amounts. Many people get intoxicated when drinking alcoholic beverages and only later get a hangover. Others feel the hangover symptoms right away with little in the way of high-spirited inebriation. Where do these differences come from and how do they work? This chapter will explore the nutritional forces that have shaped our

Nutrigenetics. http://dx.doi.org/10.1016/B978-0-12-385900-6.00003-4
© 2013 Elsevier Inc. All rights reserved.

genome. You will learn that most nutrigenetic differences are not the result of random mutations and that our diverse genetic heritage carried our ancestors through feast and famine. You will also understand why there was no such thing as a specific Paleolithic (Stone Age) diet to which we are all adapted.

We now know of numerous genetic variants that reflect ancestral adaptations to particular *nutritional environments*, which we call *nutritopes*. The adaptation to milk drinking during adulthood is an easily understood and directly observable example.

3.1 WHEN OUR ANCESTORS COULD STILL MAKE THEIR OWN VITAMINS

3.1.1 Nutritional relativism

Each species has its own menu of preferred and indispensable foods and nutrients. Let's take vitamin D (cholecalciferol), which is actually a sterol, not an amine. It is also not an essential nutrient for humans because we can make a year's worth of vitamin D from a cholesterol precursor within a couple of weeks (Figure 3.1). All we have to do is spend 15—20 minutes each day around noon at a subtropical beach without covering our skin or marinating it with sunscreen lotion. Cats also like to laze in the sun, but no matter how much they try, they still need to get most of their vitamin D from food [1]. The reason seems to be that they do not have enough of the 7-dehydrocholesterol precursor in their skin, so even shaving would not help (it has been tried).

Cats also need taurine (2-aminoethane sulfonic acid). Your feline companions will go blind if you don't feed them enough taurine. Cat food manufacturers know this and often add taurine to their products. But that should not be

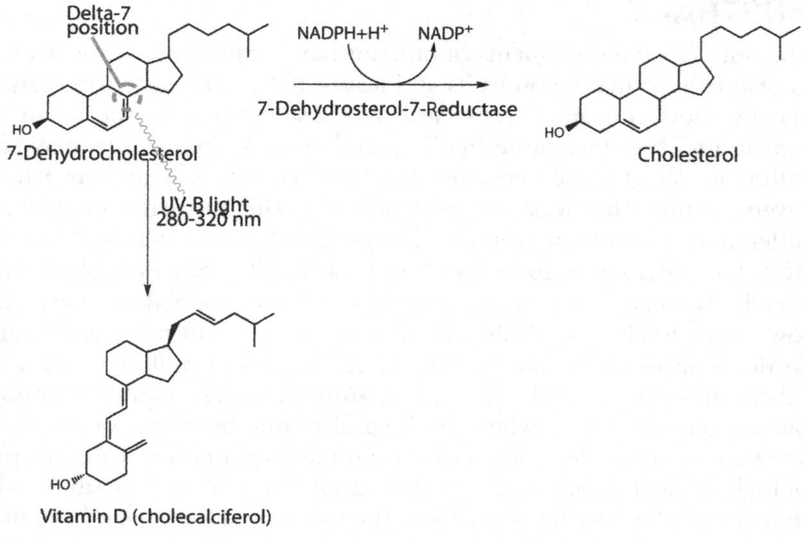

FIGURE 3.1

The ultraviolet component (UV-B; 280—320 nm) of sunlight opens up one of the rings of the cholesterol precursor 7-dehydrocholesterol (note the circle around the delta-7 double bond where the light cracks the ring) and thereby powers vitamin D synthesis in skin. This will only work if there is enough UV-B light and enough 7-dehydrocholesterol.

a reason for humans to turn to cat food, because we can make enough taurine from the amino acid cysteine. Dogs would also have a hard time adopting a vegetarian lifestyle because they are not very good at making carnitine [2] and a few other nutrients. Human vegetarians have no such difficulty, as long as they get enough of the nutrients that are essential for them [3].

This is just to illustrate that each species fits into its own nutritional niche. Carnivores get many of their essential nutrients from their prey and do not need to make them on their own. Herbivores do not need to make nutrients that are abundantly supplied in fruits and leaves. Recent carnivore converts, such as humans, may have difficulty keeping up with the new lifestyle because it takes millions of generations to fully adapt genetically. In other words, we humans still need to eat our fruits and greens, because the time since our prehominin days was not long enough to turn us from gentle herbivores and fruitarians into truly self-sufficient meat-and-potato eaters.

3.1.2 Vitamin C

We cannot make one of the simplest metabolites that most other mammals have no trouble producing, ascorbic acid (vitamin C). Our ancestors lost the ability to make their own vitamin C from glucose many millions of years ago (Figure 3.2). One can pinpoint this momentous event quite precisely. The lemurs, which branched off from the family tree of primates about 62 million years ago, still have a fully functional *L-gulono-1,4-lactone oxidase* gene and happily make their own vitamin C. The tarsiers, which branched off 45 million years ago, and all primates that branched off later share the same type of crippling changes [4,5] that disrupt production of the enzyme needed for the final step of vitamin C synthesis. This means the gene must have broken sometime between 45 and 62 million years ago.

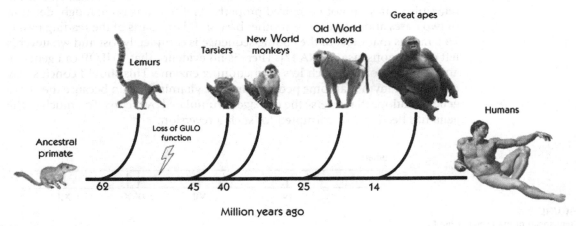

FIGURE 3.2
Human family tree going back to a hypothetical ancestral primate shows that the *GULOP* gene became corrupted in an ancestor more than 45 million years ago (lightning bolt). Loss of this gene function is the reason that humans and most current-day primates need to get their ascorbic acid (vitamin C) from food.

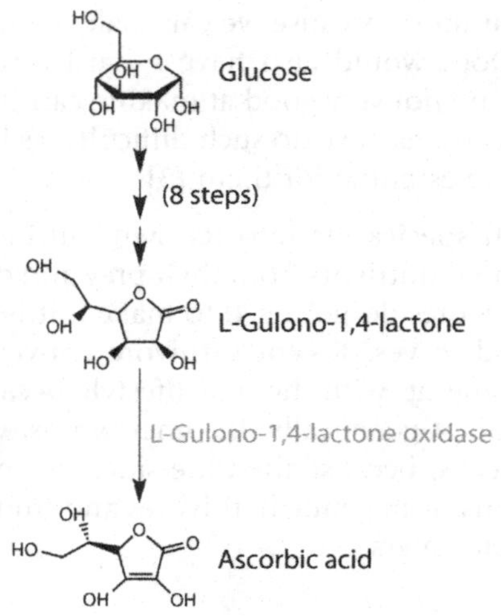

HO
OH O
OH   OH   Glucose
OH

↓ (8 steps)

OH
OH   O
O
L-Gulono-1,4-lactone
HO   OH

L-Gulono-1,4-lactone oxidase

OH
HO   O
O
Ascorbic acid
OH   OH

FIGURE 3.3
L-Gulono-1,4-lactone oxidase catalyzes the last enzymatic step of vitamin C synthesis from glucose. Humans and most primates lack this enzyme activity.

L-gulonolactone oxidase (EC 1.1.3.8) catalyzes the ninth, critical step of ascorbic acid synthesis from glucose (Figure 3.3). The precursor L-gulono-1,4-lactone is a readily available product of glucose metabolism through the pentose phosphate cycle.

GULOP is the human remnant of the gene that encodes L-gulonolactone oxidase in most other mammals. GULOP (Figure 3.4) cannot generate a functional enzyme product [6] and is therefore called a *pseudogene*.

This ruin of a gene has lost seven of its 12 original exons and at least two of the remaining ones cannot be spliced properly. As if that was not enough, deletion of two bases and insertion of another base shift large parts of the reading frame. This means that most of the coding sequence is completely lost and whatever is left creates nonsense mRNA [7]. There is no evidence that GULOP can generate any kind of protein, much less a functioning enzyme. This should conclusively debunk the myth that some people can make vitamin C again because they have gene mutations that reverse the damage accumulated over eons. Too much of the gene has been lost or corrupted for such a reversion.

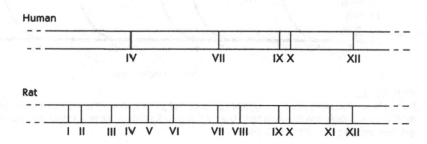

FIGURE 3.4
Comparison of the exons of the L-gulonolactone oxidase gene in rats and the crippled pseudogene in humans [6].

How to explore a gene

Many tools and data resources are available online to explore human and other genomes. The website of the National Center for Biotechnology Information (NCBI; http://www.ncbi.nlm.nih.gov/guide/) is a good starting point.

For instance, select "Gene" from the drop-down menu and type in "gulonolactone oxidase" to learn about the gene structure and sequence data of the gene. Select the rat gene and view the exonic and intronic structure shown in the figure below. The arrow representing the *Gulo* gene points from the right to the left, because the coding sequence is on the negative strand.

junctions where exons are spliced together to form the mRNA template for protein synthesis.

Exon 4 of the rat *Gulo* gene with its intronic boundaries has the sequence:

AGGTGGACAAGGAGAAGAAGCAGGTAACAGTG-
GAAGCCGGTATCCTCCTGGCTGACCTGCACCCACAG
CTGGATGAGCATGGCCTGGCCATGTCCAAGT

You can now use that information to search for the homologous human sequence. This time click on "DNA & RNA" from the choices on the left-hand side of the NCBI

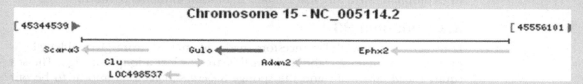

On the lower right select "Open Full View," then click "Sequence" on the lower left to open a window with the annotated coding sequence in it. The coding exons and the noncoding introns are shown in different colors. The introns are easily identifiable, because they start with a GT dinucleotide and end with AG, following the usual rules for exon/intron boundaries [8]. These four nucleotides define the

site and select "BLAST" to get the Basic Local Alignment Search Tool. On the upcoming screen select "nucleotide blast" on the center left and copy the rat exon 4 sequence into the search window. Select "Human genomic plus transcript" as the database to be searched and press "BLAST" on the bottom. The search result (partial view below) will show what is left of the original exon 4 on chromosome 8 of the human genome.

```
>□ref|NT_167187.1| D Homo sapiens chromosome 8 genomic contig, GRCh37.p2 reference
primary assembly
Length=31697033


Features flanking this part of subject sequence:
    15756 bp at 5' side: epoxide hydrolase 2
    37868 bp at 3' side: clusterin isoform 1

Score =  117 bits (63),  Expect = 9e-25
Identities = 86/97 (89%), Gaps = 2/97 (2%)
Strand=Plus/Plus

Query  1         AGGTGGACAAGGAGAAGAAGCAGGT-AACAGTGGAAGCCGGTATCCTCCTGGCTGACCTG  59
                 |||||||  ||||||||||||||||| | || |||| || || ||||||||||   |||||
Sbjct  15275941  AGGTGGATGAGGAGAAGAAGCAGGTGACCA-TGGAGGCTGGCATCCTCCTGGCCAACCTG  15275999

Query  60        CACCCACAGCTGGATGAGCATGGCCTGGCCATGTCCA  96
                 |||||||||||||||||||||||||||||||| |||||
Sbjct  15276000  CACCCACAGCTGGATGAGCATGGCCTGGCCCTGTCCA  15276036
```

3.1.3 Niacin

Most mammals, including humans, can convert tryptophan into nicotinic acid. The efficiency of the conversion varies and depends on their genes and on availability of riboflavin, vitamin B6, and iron. Species with high 3-hydroxy-anthranilic acid oxidase (3-HAOO; EC 1.13.11.6) and low amino-carboxymuconate semialdehyde decarboxylase (ACMSD; EC 4.1.1.45) activities convert tryptophan most efficiently into nicotinic acid and are least dependent on niacin intakes. Rats do not need niacin from food, because their high 3-HAOO:ACMSD activity ratio will usually divert enough tryptophan to niacin synthesis to meet needs. Cats are much less efficient and need significant amounts from food. Humans are somewhere between those extremes. That said, there is great variation in human niacin requirements and many people may need no niacin at all [9]. Tryptophan constitutes a slightly higher than average percentage of protein in milk and eggs; protein in corn (maize) is relatively low in tryptophan. Meat and fish contain significant amounts of niacin.

3.1.4 Vitamin B6

Eons ago, our unicellular ancestors made vitamin B6 from glyceraldehyde 3-phosphate or other metabolites, but this art was lost a long time ago. The sea squirt *Ciona intestinalis* and the sponge *Suberites domuncula* appear to be our closest cousins in the animal world that still have the ability to make pyridoxine [10]. Every vertebrate needs vitamin B6, but none has all the genes to make it. This means they have to get it, directly or indirectly, from someone else who does. Good food sources of vitamin B6 include potatoes, bananas, garbanzo beans, oatmeal, chicken, and pork.

3.2 RECENT EVOLUTION OF GENETIC DIFFERENCES

3.2.1 Permanent change

Adaptation to the world around us is an ongoing process. The Greek philosopher Heraclitus complained 2500 years ago that everything keeps changing (παντα ρει = everything flows), and there is no denying that humans themselves change constantly. The existence of common polymorphisms is strong evidence for the fact that the diversity of our genomes is part of what defines us. The vast majority of polymorphic variants arose in Africa and virtually all human populations worldwide now carry these variants [11]. We even share the occasional poly-morphism with other primates [12]. These millions of common differences together ensured that humans survived constant change, because there were always a few individuals with just the right genetic tools to deal with whatever nutritional adversity the environment threw at them. The human genome is fluid and responds to nutritional challenges from one generation to the next.

3.2.2 The genetics of nutritopes

Most species have a very limited geographical range. Over thousands or even millions of generations they have become well adapted to a particular environment. Other neighboring areas are less hospitable for them because of any number of reasons. Some wild flowers can only thrive in the climate typical of a certain area on a mountain range, and when the weather gets warmer, they start growing at a higher altitude with the original climate characteristics.

We are very familiar with the concept of natural habitat or *biotope* (from the Greek words *bios* = life and *topos* = place), regions with shared climate and geological features that support similar forms of life. The term was originally proposed by the German zoologist Ernst Haeckel. A somewhat related term *ecotope* was proposed much later by the Danish botanist Thorvald Sørensen 'to designate the fundamental unit of ecological plant sociology' [13].

> **KEY CONCEPT**
> Genetic adaptation of the human genome to the nutritional environment has never stopped and is still ongoing.

So is there evidence for different nutritional environments that could have shaped the human genome? The question almost answers itself. Humans are fairly unique among mammalian species, because they have settled in almost the entire world (with the exception of Antarctica) during prehistoric times. We can call the specific nutritional circumstances in a geographically defined region a nutritope in analogy to the terms for other biologically, ecologically, or climatically defined zones. Even today we can see rich evidence for diverse culinary cultures across the globe reflecting the local conditions and availability of foods. It is natural to associate tropical locations with ample representation of fruits in dishes. But does everybody realize that chocolate is used more widely in Europe than in West Africa, where a large percentage of the cocoa beans for chocolate manufacturing are grown? An important reason is that the cocoa butter in chocolate turns into a liquid at 34–36°C, which lets it melt deliciously on the tongue at body temperature. But that makes chocolate somewhat inconvenient to keep when the ambient temperatures are already that high, as they are for much of the year in tropical climates.

> **KEY CONCEPT**
> A nutritope is a local environment defined by its availability of particular foods and nutrients.

Our ancestors lived in hot and cold rain forests, savannas, frozen tundras, cold and hot deserts, at the sea shore, on barren islands, and in the high mountains. Starvation was a recurring threat for many. Because they were our ancestors, they lived at least long enough to get the next generation started. This means that their food supplies provided enough energy and essential nutrients to sustain them for many years and then to raise one or more children. They were the nutritional success stories and we inherited their genes, not the ones from the many others who did not make it or who evolved into a different species. They started digging up tubers and eating game animals when they came to regions that did not provide the fruits, lush leaves, and other easy pickings that the tropical rain forests provide for our closest living primate relatives, the bonobos (pygmy chimpanzees; *Pan paniscus*). Some of them cultivated plants whose grains, tubers, or roots are brimming with energy-rich starch. Others bred animals that could convert wild grasses, leaves, acorns, and seeds into protein-rich meat, milk, or eggs. They all found a way to survive and thrive, but it was their own way, not necessarily shared by everybody else.

3.2.3 Paleolithic diets

One cannot expect to reconstruct prehistoric diet patterns with accuracy. Qualitative trends are sometimes apparent from recovered foods, tools, and storage vessels. Another approach is the analysis of isotope distribution in teeth or bones. This kind of analysis can provide some answers even when no other evidence is available, for instance when a few bone fragments are all that is left of our prehominid ancestors. The ratio of ^{13}C to ^{12}C isotopes can tell us something about the major sources of plant foods [14]. The ratio of ^{15}N to ^{14}N isotopes indicates whether protein comes from animal sources or from plants [15]. Our distant ancestors 5–6 million years ago in the tropical forests of Central Africa might have eaten mainly fruits, leaves, herbs, tree bark, and the occasional termite, caterpillar, monkey, or other animal food source just like bonobos do today [16]. Later hominid ancestors, such as *Australopithecus afarensis* (*Lucy* was the name given to the original find), then evolved to have stronger molars, which might have helped them to chew seeds from grasses and other savanna plants [17]. *Australopithecus* species also used starchy tubers and roots [18]. Then, more than a million years ago, they learned to cook [19]. Roasting makes meats and plant foods more digestible. These early gourmets could eat more in one sitting, freeing them from the need to forage and eat all the time. Jungle-dwelling bonobos spend almost half of their waking time eating, because the leaves, bark and other foods contain so little sugar and other bioavailable nutrients. Later hominids spent probably less than a quarter as much time [20]. This may have been particularly important for people who traveled a lot. Our modern lifestyle has very ancient roots.

Their largely vegetarian menu seems to have satisfied early modern humans for a while until a few dozen restless souls, possibly driven by a *thrill-seeking* mutation in their dopamine receptor D4 [21], wanted more from life and emigrated to the meat pots of Europe and Asia about 40,000–50,000 years ago. Maybe they got word from their cousins, the Denisovans and the Neanderthals, who had left Africa hundreds of thousands years earlier for a life of eating giant aurochs, horses, deer, wooly rhinos, and the occasional mammoth or mastodon [22]. The new immigrants certainly found the meat pots, or what was left of them, and did a bit of mingling with their cousins [23,24]. The mingling part added genetic adaptations of these nonhuman hominins to the human repertoire and probably helped them survive daunting Ice Age conditions. They also ate enough berries, seeds, roots and other plants to provide vitamin C and other critical micronutrients [25].

As people became more sophisticated and learned to grow grains, vegetables, and fruits, they moved again to eating more plant-based foods. We know some details from the serendipitous discovery of the well-preserved mummy of an early European (Oetzi the Ice Man), who died 5200 years ago on a mountaintop in the Alps. Isotope analysis of his hair suggests that his usual diet was predominantly plant-based [26]. But we also know that his last two meals were entrecote of deer on a pilaf of einkorn (an early form of wheat), barley, and diverse roots, followed by a parfait of forest berries and, just a couple of hours before his death, a sandwich with chamois and ibex meat on whole-wheat bread with flax and poppy seeds [27].

3.2.4 Foods in a frozen desert

The mostly carnivorous diet of people in the Arctic a few hundred years ago may reflect conditions of much earlier arrivals to Ice Age Eurasia. They had no lush forests to forage or fields to farm; only sparse tundra. Plant consumption of the people in the furthest North appears to have been limited to the rumen content of caribou and musk oxen, and a few eagerly collected roots, tubers, leaves, and berries [28]. But they had access to sea mammals, fish, a few land animals, and birds. So, they lived on meat from seals, whales, caribou, a lot of fish, and little else. This ancient lifestyle has been preserved in native peoples living above the Arctic Circle for millennia until less than 100 years ago [29].

Hallmarks of the ancient Arctic food pattern are a high intake of protein, omega-3 fatty acids, cholesterol, and iron, and a low intake of carbohydrates, dietary fiber, and vitamin C.

3.2.5 Foods in a hot desert

The original settlers arrived in Australia more than 40,000 years ago. The *bush tucker* of these nomadic hunter-gatherers consisted of a wide range of plant and animal foods that was available. Most parts of kangaroos and other animals were eaten. The fat of bone marrow was especially prized, because there is so little fat in the wild [30]. Even better, the fat that was available in this and other foods tended to be rich in arachidonic acid and other essential omega-6 fatty acids. Insects were an important source of fat and protein. Many different plant foods were regularly on the menu, including fruits, roots, tubers, nuts, and seeds [31]. Finding these foods in the wild and then processing them into an edible dish was never easy and semistarvation was probably frequent [32]. For instance, cheeky yams (*Dioscorea bulbifera*), one of the more commonly consumed tubers, had to be dug up, peeled, sliced, soaked in water for 2 days, and then baked. The seeds from *Acacia aneura* (locally called mulga) would be roasted, shelled, and ground into a paste with water. Not exactly the food for a quick bite. On the other hand, the commonly consumed plant foods in their prepared form tend to have a low glycemic index [33], which makes them more filling. But the wild also had something for the gourmet's sweet tooth: honey ants. These insects collect sugary juices from plants and store them in their abdomen. This was a quick treat for those who knew how to find them. Many of these foods are still used by some communities and some restaurants have bush tucker on their menu for connoisseurs.

It seems likely that aborigines are genetically adapted to food scarcity. This makes them particularly susceptible to obesity and diabetes when following the prevalent Western-style diet and reversion to a more traditional diet has been found to improve their lipid and carbohydrate metabolism [34].

Shared dietary patterns of people historically living in the Australian deserts include high intakes of dietary fiber and micronutrients, and a low intake of energy and fat.

3.2.6 At the dairy farm

The use of animals for large-scale milk production was a major revolution that allowed people to live in fairly barren areas. This innovation happened in many different areas, using cows, buffalos, camels, horses, goats, and sheep, but only a few populations adopted this lifestyle. The Fulani in Mali (Africa), for instance, a cattle-tending people used to drinking a lot of milk, live side by side with the Dogons, who drink milk only occasionally [35]. It is not always clear why some groups have adopted high milk consumption and others have not, but the distinction is clear. The milk-drinking groups often have the genetically fixed ability to tolerate a high milk intake. Even if their neighbors wanted to start a dairy operation and use this protein- and micronutrient-rich food, they would struggle with bloating and other digestive discomfort, thus reinforcing their existing cultural aversions to milk.

Distinct dietary characteristics of habitual milk consumers are high intakes of saturated fat, cholesterol, and calcium.

3.2.7 Designated drinkers

Another innovation in many populations was the extensive production of alcoholic beverages. More than 4000 years ago, the Babylonians recorded recipes for making different beers. Many other cultures in Asia, Europe, Africa, and the prehistoric Americas made various alcoholic brews part of their regular diet at one time or another. Unlike other recreational stimulants, alcohol was also an important source of energy that could be stored safely without refrigeration. Yeast-containing brews also provided vitamins and other important micronutrients. And, not least, these alcoholic beverages were a relatively safe source of water in a time when streams and wells often carried disease. The millennia of alcohol consumption demonstrably left their mark on the human genome in the shape of polymorphisms altering the function of genes involved in alcohol metabolism.

3.2.8 Variation defines the human species

These are just a few of the nutritopes that different people were exposed to and which they often shaped themselves. Such diverse nutritional environments are major driving forces for the evolution of human variation. The genetic variation, in turn, ensured the resilience of human populations when they encountered a change in their environment (such as draught or collapse of food sources) or when they migrated into a different environment. Seen in this light, genetic diversity is an important strength and a defining characteristic of the human species. Proper understanding of human genetic diversity should also debunk the myth of a single type of Paleolithic diet that is best for all of us. Stone Age people adapted to the food of the land, which depended on where they lived. Different Paleolithic people had very different Paleolithic diets! Some ate almost exclusively meat and other animal products; others followed a nearly vegetarian lifestyle. Some got a lot of iron, others had trouble getting enough. And it goes on like this for just about every conceivable combination of lifestyle, food, and nutrient. Nutrition patterns shaped the people and the people shaped their nutrition patterns; it went both ways.

> **KEY QUESTION**
> Your genome carries echoes of many different Paleolithic nutrition patterns, but which ones?

3.3 MECHANISMS OF EVOLUTIONARY ADAPTATION

3.3.1 Fresh milk

As mentioned before, people can create their own nutritope by developing new foods. And they certainly did that by starting to use the milk of domesticated animals. Fresh milk contains large amounts of lactose (10–12 g/cup). The brush border in the small intestine expresses lactase (EC 3.2.1.108, encoded by LCT, OMIM 603202) during infancy to support digestion of the mother's milk. The lactase enzyme cleaves the disaccharide lactose into its constituent monosaccharides glucose and galactose (Figure 3.5). The sodium-glucose transporter (*SGLT1*; OMIM 182380) can then move both glucose and galactose into the enterocytes of the small intestine. Expression of lactase normally stops a few

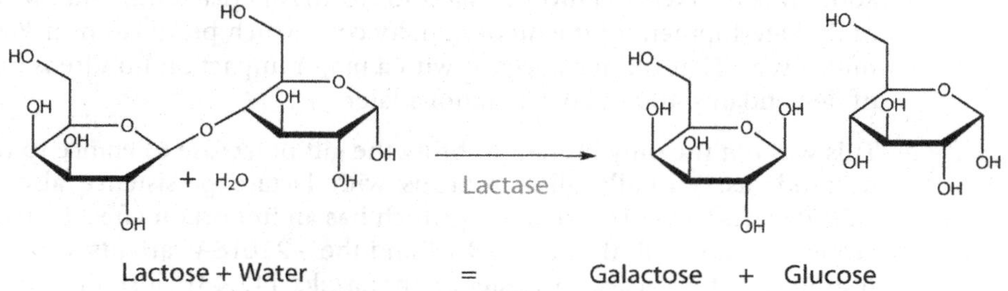

Lactose + Water   =   Galactose + Glucose

FIGURE 3.5
The digestive enzyme lactase in the brush border of enteral cells lining the small intestine splits lactose into the simple sugars glucose and galactose, which are readily absorbed.

years after weaning and lactose is not digested anymore. At this point, consumption of a cup of milk becomes unpleasant for many people because bacteria in the ileum and colon break down the undigested lactose and produce hydrogen and methane gas as by-products. People in the early Stone Ages could not absorb the lactose in milk because the genetic adaptation to this new invention had not yet occurred [36]. That happened much later, about 7000 years ago in Central Europe, and then several times again more recently in Africa and Asia [37].

The earliest physical traces of firmly established milk production in Anatolia and the Balkans go back more than 8000 years [38]. The Linear Band Ceramic Culture in Central Europe had established dairying 6500 years ago [37]. And when archeologists examined hundreds of pottery vessels from British farm sites dating to the same period, they found chemical traces of milk on more than half of them and evidence that milk was a major source of food for the people living there [39]. The discovery of prehistoric milk production raises the question how these early dairy farmers could tolerate milk without suffering cramps and diarrhea. In an incredible feat of scientific sleuthing investigators concluded that in the earliest times most of the milk was stored as cheese [38]. The fermentation of milk by bacteria during cheese production splits much of the lactose into glucose and galactose. Both of these simple sugars are easily absorbed and the problem was solved. Other fermented milk products, such as yogurt and kefir, have similarly low lactose content and are favored in Turkey, Greece, and other regions where most people are lactose intolerant.

Eventually the European dairy farmers did adapt genetically to the consumption of fresh milk. This adaptation happened in Central Europe, where milk will not go sour for a day or two, or even longer during the cold season. There, someone was born with a mutation ($-13910C>T$; rs4988235) located right in the middle of the MCM6 gene (OMIM 601806). This change kept lactase expression persistently turned on, even after infancy. This lucky mutant could drink several cups of milk without any discomfort and was probably taller, stronger and healthier than others in the community because of the extra calcium and protein [40]. But the real stunner is that this one mutant is the direct ancestor of most adult milk drinkers in Europe, Asia and North America today and the root cause of the development of the modern dairy cow, which produces nearly 20 tons of milk a year. That is quite a legacy, with a major impact on hundreds of millions of descendants 400 or so generations later.

This was not the only mutant to bring the gift of lactose tolerance to numerous descendants. Virtually all Europeans with lactase persistence also have the $-22018A$ (rs182549) variant [41] which has an important effect by itself on the lactase promoter. Both the $-13910T$ and the $-22018A$ variants also seem to be responsible for lactase persistence in Kazaks [42,43], just like in Northern Europeans. At least another 7 distinct lactase persistence-inducing variants are now known to occur in various places of the world and they have been

Table 3.1	Genetic Variants Linked to Lactase Persistence in Populations Around the World
Variant	**Population**
−13779*C	Toda and others in Kerala, India [46]
−13838*A	Tibetans [47]
−13907*G rs41525747	Somalis [48] and Sudanese [49]
−13910*T rs4988235	Northern Europeans [37], Fulani [35], Fulbe, Hausa [50], Northwest Indians [46], Kazaks, and Tajiko-Uzbeks [43]
−13913*C rs41456145	Rare in Middle Easterners [51]
−13914*A	Occasionally in Finns [52]
−13915*G rs41380347	Somalis [48], Middle Easterners, Arabs, Ethiopians, Sudanese, and Muslims in South India
−13937*A rs4988234	Xhosa [53]
−14009*G	Somalis [48]
−14010*C	Kenyans, Tanzanians, Xhosa [54]
−14011*T rs4988233	Occasionally in Somalis [48]
−14091*T	Xhosa [53]
−14176*C	Xhosa [53]
−22018*A rs182549	Northern Europeans [41], Kazaks [42], Japanese−Brazilians [55]

functionally characterized (Table 3.1). More are waiting to be discovered [44]. Genetic variants linked to lactase persistence are found across the globe in herder populations with a long history of milk drinking. Many of these have roots in the dawn of history. Beautifully descriptive rock paintings and modern isotopic analyses leave no doubt that cattle herders in the once green Sahara were milk drinkers more than 7000 years ago. We also find that about 17% of Afghans have lactase persistence [45]. They might have inherited this useful trait from Macedonian or Central Asian conquerors, or from a homegrown ancestor.

There is still the mystery of why mammals turn off lactase expression after weaning. When it is defeated in humans by a single base change, apparently nothing bad happens. Does that mean that the elaborate molecular mechanism to turn off lactase expression is a waste? Probably not. Turning off lactase expression may be a protective mechanism. Many toxins in plants are glucosides and are absorbed only if the attached glucose is cleaved off. Lactase is the only major enzyme in the intestinal tract that can cleave glucosides [56] and its activity would increase the risk of poisoning when foraging in the wild. As with most other polymorphisms, the different alleles are finely balanced between benefit and harm. If the balance was strongly tilted in one direction, the less beneficial form would disappear within a few generations. This suggests that one should always look at the benefit:harm ratios for all alleles. If we can see only a benefit, we are probably missing critical information about the adverse side of the balance.

How to locate a known polymorphism in the human genome

To get a graphical view of a polymorphism and its neigh-bors, go to the NCBI Map Viewer (http://www.ncbi.nlm.nih.gov/mapview/). Enter the species (*Homo sapiens*) and rs (reference single nucleotide polymorphism [SNP]) number (for example rs4988235 for the *lactase enhancer* variant −13910) and then select the map element (in this example, rs4988235). Scroll down to the sequence graphic and click on "Open Full View," then right-click over the rs4988235 marker and select "Zoom to Sequence at Marker." You should then see a graphic like the one below. This view is very useful for examining the sequence around the variant and finding out what other common variants surround the locus of interest.

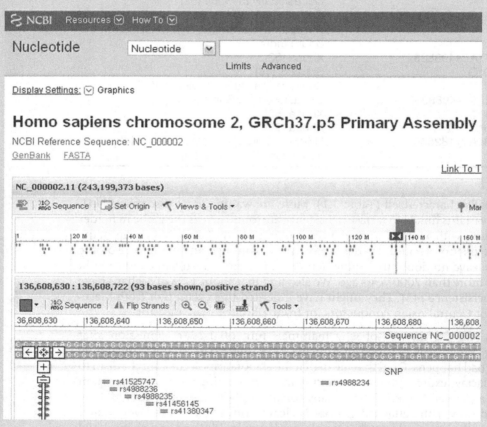

3.3.2 The taste of starch

When we chew a raw piece of potato, it first has a chalky taste, but with some chewing it starts to get a bit of a sweet flavor and tastes less abrasive. This is because amylase (EC 3.2.1.1), an enzyme in saliva, cleaves alpha-bonds between the glucose units of starch and releases maltose, a disaccharide. This is probably an evolutionary adaptation that helped our hominine ancestors to use starchy foods,

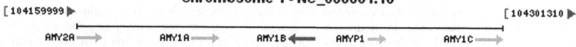

Chromosome 1 - NC_000001.10

[104159999 ▶   [104301310 ▶

AMY2A ⟶   AMY1A ⟶   AMY1B ⟵   AMYP1 ⟶   AMY1C ⟶

FIGURE 3.6
Map of the *amylase* gene region on chromosome 1 containing variable numbers of functional amylase genes.

like grass seeds and tubers, when they moved into the savannahs from the lush rain forests with their fruits and leaves. The fact is that bonobos and chimpanzees have little or no amylase in their saliva [57]. The much higher enzyme activity in humans comes from extra copies of the *amylase* gene (*AMY1*; OMIM 104700), which produce more active protein than a single gene copy does. But there is a quirk in the story, because the number of the active amylase gene copies varies considerably between people. Several copies can be chained together on each DNA strand of chromosome 1 (Figure 3.6). Some DNA strands contain as many as ten copies (Figure 3.7); some strands have none [57,58]. And the more copies people have on both strands combined, the more amylase they have in saliva. And the faster the breakdown of starch in their mouths, the more pleasant will be the taste of starchy foods such as custard or a cracker [59,60].

But why would *AMY1* copy numbers vary so widely? Comparison of the copy numbers in different native populations suggests a link to their starch consumption [57]. Populations with traditionally high starch intake tend to have larger copy numbers than groups who eat less. Finding the people with high starch intake is easy, because this includes most Europeans and Asians. The people with traditionally low starch intake, on the other hand, are few and far between because they are small minorities who get their food mostly from hunting, fishing, and raising cattle. This difference is another example of the genetic adaptation of human populations to their nutritope. Existing, presumably very old, lineages of people with large numbers of *AMY1* copies seem to have been favored in a nutritional environment with a high availability of starchy foods (such as in rice-eating Japan) and expanded to become the dominant group. People from other, equally old lineages (such as Yakut fishermen and reindeer herders) with few *AMY1* copies apparently did especially well in an environment with little starch.

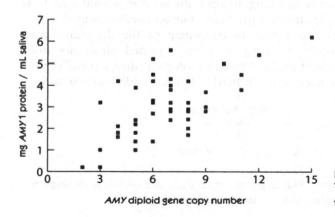

FIGURE 3.7
The number of amylase gene copies correlates well with the amount of enzyme in saliva [57].

This is all very interesting, you might say, but should we care? The answer is yes, because salivary amylase activity has a considerable impact on glucose metabolism [61]. Individuals with genetically low enzyme activity respond to starch intake with much higher glucose and insulin concentrations than people with ample enzyme activity. This means that people with few *amylase* gene copies experience a particularly unfavorable metabolic response to high-starch foods, which is likely to increase their vulnerability to obesity and diabetes in a nutritope (nutritional environment) with lots of starchy rice, pasta, pizza, buns, and cakes.

3.3.3 Home brews

Alcohol is a particularly energy-rich nutrient (7 kcal/g), providing an extra 130 calories in a pint of beer (445 mL) to those who can handle that much. Some fermented fruits may occasionally provide a natural buzz, but alcohol intake was probably negligible until people started brewing fermented beverages from fruits and grains. Alcohol dehydrogenases (ADH; EC 1.1.1.1) in the gastrointestinal tract from the stomach [62] lining to the small intestine, the liver, and other tissues metabolize ethanol (Figure 3.8). This step is rapidly followed by the action of aldehyde dehydrogenases (ALDH; EC 1.2.1.3), which convert the fairly toxic intermediate acetaldehyde into acetate. Some people metabolize ethanol much faster than others, such as those with the C/C genotype (rs1693482) of *ADH1C* (OMIM 103730) [62] and the −121C/A (rs886205) promoter variant of *ALDH2* (OMIM 100650) [63], both of which occur mainly in Caucasian populations. The unavoidable question should be, before even pondering health consequences, where these differences are coming from. The answer, once we know it, will not be simple because so many local and historical circumstances are in play. But the following research highlight may be instructive.

More than 7000 years ago, a couple of thousand years after Stone Age people in Southeast China adopted rice farming and started making fermented beverages with the newly develop grain, a mutation (48His, previously called 47His, rs1229984) in the *ADH1B* gene (OMIM 103720) increased maximal alcohol dehydrogenase activity by about 60-fold in one of them [64]. This overeager enzyme variant converts imbibed ethanol immediately into acetaldehyde, taking all the intoxicating fun out of drinking. Instead, the poor miscreant only feels the wrenching acetaldehyde effect on the brain, like an instant hangover. This effect has been compared to a built-in equivalent of the drug disulfiram (Antabuse), which is used commonly to keep reformed alcoholics from relapsing. There is little doubt that the variant protects against alcohol dependence and all its consequences [65]. It is hard to know the biological benefit of

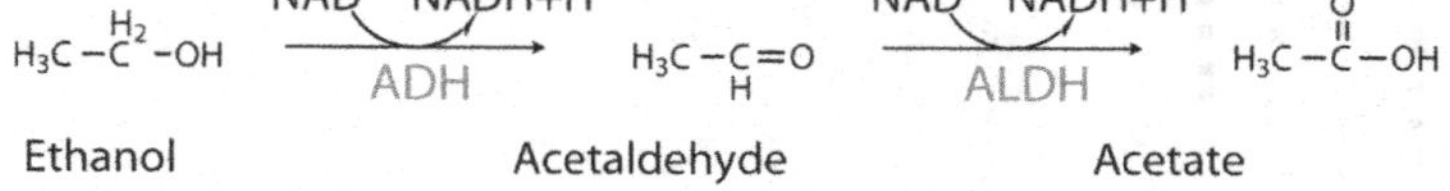

$$H_3C-\overset{H_2}{C}-OH \quad \xrightarrow[\text{ADH}]{NAD^+ \ NADH+H^+} \quad H_3C-\underset{H}{C}=O \quad \xrightarrow[\text{ALDH}]{NAD^+ \ NADH+H^+} \quad H_3C-\overset{O}{\underset{\|}{C}}-OH$$

Ethanol   Acetaldehyde   Acetate

FIGURE 3.8
Alcohol dehydrogenase (ADH) converts ethanol to acetaldehyde, which causes nausea, headache, and other hangover symptoms. Conversion of this toxic metabolite to acetic acid depends on acetaldehyde dehydrogenase (ALDH).

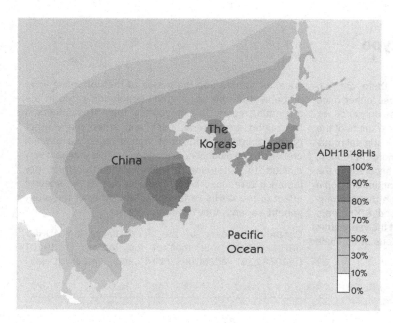

FIGURE 3.9
This contour map of the current *ADH1B* 48His gene variant prevalence in Asia (nearly everyone in the area with the darkest shade carries it) strongly suggests that a mutation causing alcohol intolerance (Oriental flush syndrome) occurred about 8000 years ago in a person living in Southeast China. This person then handed down the variant to most people in contemporary China, Japan, and the Koreas [64].

the variant on a historical scale, but the amazingly rapid propagation of the variant across Asia makes it clear that it must be considerable. Today's geographic distribution of people with the ADH1B 48His variant (Figure 3.9) suggests strongly that this teetotaler's descendants have taken over the neighborhood. Almost everybody in Zhejiang province is related to this first virtuous Stone Age man or woman and the clan now easily constitutes a majority throughout China, the Koreas, and Japan.

The abstinence-inducing effect of the *ADH1B* gene variant may be reinforced by the *ALDH2* variant 1510G>A (Lys487Glu; rs4646778). This variant encodes an enzyme form with low activity, which slows the removal of acetaldehyde generated from ethanol. Acetaldehyde is a probable carcinogen [67] and toxin for the liver and other tissues [68]. It makes the mouth dry, triggers nausea, headaches, and low blood pressure, and causes hot flushes to the face. You may have seen people get red cheeks and a headache right after drinking alcohol. This effect is often called the Oriental flushing syndrome, because it is common in Asians but uncommon in Caucasians [69]. Together, the *ADH1B* and *ALDH2* variants make sure that alcohol drinking is followed by a wave of mind-numbing acetaldehyde of fast onset and extended duration.

The contrast between the historically entrenched beer- and wine-drinking culture in Europe and the more limited alcohol consumption in many North African, Middle-Eastern and Asian cultures could not be greater. The historical alcohol consumption patterns track fairly well with genetically determined ADH:ALDH activity ratios; populations with low ratios tend to prefer alcohol consumption much more than peoples with high ratios. Indeed, people around

The extended haplotype

When we encounter common variants, there are two distinct possibilities: that the variant arose either once only or more than once. If the variant arose only once, all carriers of the variant allele are direct descendants of the one more or less recent ancestor with the original mutation. If the variant arose independently more than once, carriers do not necessarily share a recent ancestor. In the case of the ADH1B 48His (rs1229984) variant of people in Southeast Asia, we can be quite certain that the original mutation occurred in single person about 8000 years ago and that the sequence in the vicinity of the variant has not changed much over the 400 generations or so since then.

ancestor's mate and the mates of their offspring were probably from the same small population group in the originating region. This means that the 48His carriers in the region are truly relatives and share many other genetic predispositions, not just in regard to ethanol processing by alcohol dehydrogenase. This makes it very difficult to determine whether a health effect (for instance a reduced cancer risk) is due to the specific effect of the 48His variant or to any other of the many genetic variants they have in common.

Contrast the case of the Southeast Asian ADHB1 48His variant with that of the *lactase* −13910T (rs4988235) variant described earlier. Initially it was

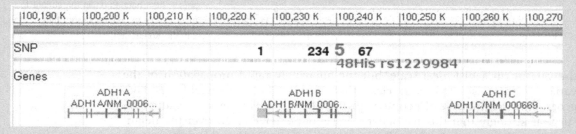

How can we know that? Today we do this by spot-checking the upstream and downstream sequence of 48His variant carriers to find out whether they all have the same haplotype. As can be seen in the gene map, they do. What you see here are the positions of seven polymorphisms (rs1042026, rs2066701, rs2075633, rs4147536, rs1229984, rs79366838, and rs28913916), labeled 1 through 7 (number 5 is the 48His variant) used to spot-check the ADH1B gene [66]. These seven polymorphisms cover a stretch of 15,845 base pairs across and beyond the entire gene. In the near future, we will probably get the full sequence for everybody in such a study, but until that is practical and affordable these spot-checks will give us almost as much information. The shared haplotype TGAGGTG in most of these Southeast Asian carriers of the 48His variant tells us that they share the entire *ADH1B* gene sequence and probably in most cases the sequences of the neighboring genes, *ADH1A* and *ADH1C*, as well. But they have even more in common, since the original

thought that this *LAC* −13910T variant was characteristic of Northern Europeans, but recent investigations have shown that many people of Fulani heritage in Mali carry exactly the same variant with the same effect of causing lactase persistence [35]. In other words, adult carriers digest the lactose in milk and avoid abdominal discomfort. In this case, we can assume that the original mutation in these Africans arose independently from the European mutation, and possibly even at multiple times in different populations. This kind of parallel development is called *convergent evolution*. That means that carriers of the LAC -13910T do not necessarily share other genetic traits.

The key lesson here is that genotype-specific associations of nutrient intake with health outcomes need extra careful scrutiny because they depend on the population context.

the world with various *ALDH2* variants encoding high ALDH activity have an increased risk of binge drinking [70,71].

3.3.4 Streamlining nutrient metabolism

Imagine an island about the size of Staten Island, except this one is not just a 25-minute ferry ride away from Manhattan. In fact, the nearest continental coast is further away from this island than North America is from Europe. And the nearest major island group is even farther away. Shortly after Easter of 1722, the first European visitors chanced upon the island unexpectedly and called it Easter Island. But they were not the first people to get to this, the most remote island (Figure 3.10) in the world.

Polynesian settlers arrived around AD 1200 on their double canoes [72]. Finding such a small island in the vast expanse of the Pacific was quite a feat [73]. But how could they escape vitamin C deficiency (scurvy), the lethal bane of long-distance sea voyagers at the time? To find Easter Island and the other outlying Pacific islands, the Polynesians must have crisscrossed the vast ocean many times, staying at sea for many months without a chance for stocking up on vitamin C-rich foods. Ferdinand Magellan lost at least 30 of his 160 sailors to scurvy during a single crossing of the Pacific Ocean [74]. And this disaster happened despite the best preparations these experienced European explorers could make.

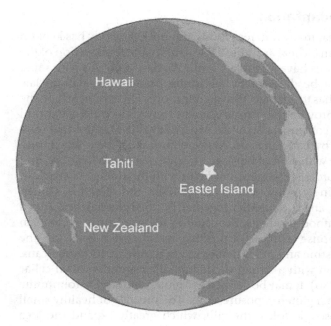

FIGURE 3.10
Easter Island is located more than 3500 km off the coast of South America and more than 4200 km from Tahiti in the Central Polynesian Islands. This extremely remote island was first settled around AD 1200 by Polynesian sea voyagers.

The likely answer is that they had on board some people with unusually low vitamin C requirements and they were the ones who made it to the island alive. But how did that work? An investigation of Easter Island natives found that about three-quarters of them carried only the Hp1 form of haptoglobin, a much higher proportion than is found among natives of the central Polynesian islands [75]. Haptoglobin is a protein that captures free hemoglobin from broken red blood cells and steers it for iron recycling to the spleen. Oxygen-carrying hemoglobin would be too dangerous to float around freely because it can oxidize cell membranes and blood proteins upon contact. Of the two common variants, haptoglobin 1 (Hp1) binds hemoglobin much more efficiently than haptoglobin 2 (Hp2) [76]. Since hemoglobin is a strong oxidant that ultimately depletes vitamin C, people with the Hp1 variant have lower vitamin C requirements than do Hp2 carriers [77]. In practical terms this means that the ones with the Hp1 variant stayed on the boat while their less fortunate companions with the Hp2 variant keeled over with scurvy and did not make it to that elusive island. It is a stark example of survival of the fittest of a few ancestors during a short historic time, and it still shows in the far more numerous modern descendants of the survivors. *Founder effect* is the apt term for the phenomenon of temporary genetic restriction to a particular genotype in a few pioneers and the transmission of this genotype to later generations of a greatly expanded population. The same selective survival of Hp1 carriers probably took place in many other nutritopes with scarce vitamin C. However, long-term survival of a modern lifestyle may be another matter, since there is evidence that people with the Hp1 variant are more prone to die from heart disease [78]!

3.3.5 Incomplete adaptation

Mankind has evolved to eat many different foods, but not everybody has kept up with all of these adaptations. There are quite a few items in the food section of the supermarket that some of us have a problem with. We have seen this with milk, where the lactose cannot be digested by all adults. There is no natural law requiring that everybody has to adapt to all available food sources or perish. Some of the offending foods provide good nourishment, but also contain harmful substances that can catch consumers unaware. Most of the novel grains that were developed from grasses by early man in Mesopotamia and have fed untold millions since then come with a lot of baggage for some people. Wheat, rye, and barley contain gluten, a protein complex that helps with the storage of starch in grains. The gluten protein complex consists of gliadins and glutelins. Gluten makes bread, pizza, cakes, and cookies stick together. Purified gluten is also added to lotions and other nonfood products. Exposure to the gliadins in gluten can trigger an immune response, with CD4-positive lymphocytes attacking the mucosa of the small intestine and other tissues in about one in 100 Americans. This multiorgan condition with a strong genetic predisposition is called celiac disease (CD; OMIM 212750). It may be defined as a gluten-induced autoimmune disease that clears up when gluten exposure stops. The mucosa of healthy small intestine contains microscopic folds, the villi, which greatly expand the area available for food absorption. It is here that lactose from milk is cleaved and sugars are absorbed. Fats are broken up and packaged into chylomicrons. Iron and other micronutrients get taken up and moved into the bloodstream. Very small amounts (as little as 10–50 mg) of gluten can cause the loss of villi and the flattening of the small intestinal lumen in people with CD. Tissue transglutaminase (EC 2.3.2.13) in the small intestinal wall modifies specific segments of the gluten protein [79]. The modified gluten peptides can then interact with tissue compatibility complex-specific CD4-positive T-cell lymphocytes and induce them to attack tissues in the small intestines, pancreas, thyroid, skin, and other organs. Attacks of gluten-sensitive lymphocytes on skin result in a characteristic blistering rash on many parts of the body (most often affecting the head, elbows, forearms, buttocks, and knees) called dermatitis herpetiformis [80].

CD is a serious condition with a significantly increased risk of death [81]. Much of the excess mortality risk comes from malignant tumors. People with CD have an approximately 16-fold increased risk of lymphoma (T-cell non-Hodgkin lymphoma). CD also increases the risk of kidney failure [82] and promotes bone mineral loss and the risk of developing osteoporosis. More than a quarter of all people with newly diagnosed CD have biochemical signs indicating some degree of liver damage [83]. Avoiding gluten intake will bring the liver back to health in most cases. There is reason to expect that gluten avoidance can also reduce several of the other health risks.

CD can be difficult to recognize because less than half of the truly affected people experience the classical symptoms of diarrhea, malabsorption, and anemia, sometimes with only minimal damage to the villi [84]. Diagnosis of CD is usually made on the basis of blood tests and follow-up biopsy of the small intestinal lining. IgA-tGT (immunoglobulin A antibodies to tissue transglutaminase) and IgA-EMA (antibodies to the endomysium, the connective tissue around bundles of smooth muscles fibers in the intestinal wall) have good predictive power. The combination of positive antibodies and a finding of flattened villi in the biopsy is conclusive. However, that is not the end of it. When CD patients have already started to avoid gluten on their own, their villi may appear normal at the time of biopsy. Even people with positive antibody test results, who did not deliberately scratch gluten from their menu, can have normal biopsy results [84, 85] and some of them will develop more definite signs of CD within a few years [86].

KEY POINT
Most people with CD are not aware of their condition.

People with type 1 diabetes mellitus (diabetes with childhood onset), autoimmune thyroiditis, unexplained infertility, and migraine headaches have an increased likelihood of having CD, even with no obvious intestinal symptoms. Take the example of type 1 diabetes. More than 10% of all cases appear to be due to CD. The current barrier to effective prevention is that we don't definitely know which newborn infants are at risk of developing CD. If there were reliably predictive tests for CD risk, infants testing positive could be given a gluten-free diet from birth. Having one or two parents with CD is a strong risk factor, but many CD patients do not have parents with CD. More than 90% of CD patients have the human leukocyte antigens (HLA) DQ2 or DQ8, but so do about 40% of people without the condition. A double dose of the variant giving rise to the DQ2 antigen is usually associated with a more severe form of CD, whereas the absence of DQ2 and DQ8 occurs more often in milder or latent forms [87]. About 40 additional genetic variants increase predictive power but still capture only about 50% of the risk for sensitization upon gluten exposure [88]. Efforts are continuing to find a combination of genetic variants that can predict most of the risk and become a practical tool for prevention. Even in the absence of primary prevention, we need to recognize that the avoidance of gluten can improve blood sugar regulation in affected patients. Children with type 1 diabetes and CD will have much better glycemic control when they start avoiding gluten [89].

USUALLY CONTAIN GLUTEN	OFTEN CONTAIN GLUTEN	USUALLY DO NOT CONTAIN GLUTEN
Barley	Candies	Amaranth
Beer	Corn chips	Buckwheat
Bread	French fries	Job's tears
Bulgur	Gravy	Legumes
Cakes	Imitation seafood	Maize (corn)
Cookies	Malt	Millet
Farina	Matzo	Potato
Kamut	Oats	Quinoa
Pasta	Potato chips	Rice
Rye	Processed meats	Sorghum
Semolina	Rice mixes	Tapioca
Spelt	Salad dressings	Teff
Triticale	Soup mixes	Wild rice
Wheat	Soy sauce	

3.4 PROTECTION AGAINST INFECTIONS

3.4.1 The problem with beans

A 6-year-old boy is brought to the emergency room by his parents because he has been throwing up and running a fever since the previous day. He is complaining of a bad headache. The family had risotto with fresh mozzarella, prosciutto, and fava beans (http://www.myrecipes.com/recipe/fava-bean-risotto-with-fresh-mozzarella-prosciutto-10000001173808/) 2 days earlier, using a highly recommended recipe from a popular cooking site. None of the other family members felt ill after eating the risotto dish.

The boy is pale and has a fast heart rate. His hemoglobin is very low with 60 g/L, total bilirubin greatly elevated with 105 µmol/L. He has hemoglobin (+++) in his urine. The physicians conclude that he has severe acute hemolysis.

How did this happen? It turns out that he has a previously undetected low-activity variant of glucose-6-phosphate 1-dehydrogenase (G6PD; EC 1.1.1.49). We now know about numerous polymorphisms which can occur in the gene encoding the G6PD enzyme; many of these polymorphisms decrease enzyme activity [90]. G6PD, the key enzyme of the pentose phosphate pathway, is critical for maintaining nicotinamide adenine dinucleotide phosphate (NADPH) and indirectly for glutathione in the reduced state [91]. Since the gene is located on this boy's single X chromosome, his low-activity variant cannot be balanced by a second copy with normal activity. His parents probably had no previous exposure to the condition themselves. The mother, from whom he inherited the variant, has a normal *G6PD* gene on her other X chromosome that compensates for the defective one. And the father probably does not have a defective gene.

About 5% of the world population has a G6PD deficiency (OMIM 305900), which makes this the most common potentially life-threatening nutrigenetic variant [92,93]. Because this is an X-linked condition, males are affected most commonly. But females can also be affected. The most obvious reason is that they carry two defective alleles, which is not rare in high-prevalence regions in Africa and Asia. But even female heterozygotes are affected in about 14% of cases [92], because their X-chromosome inactivation pattern is imbalanced and their defective X chromosome is overrepresented. There is a reason why so many people, especially in tropical and subtropical countries around the world, have low G6PD activity: infections with the malaria parasite *Plasmodium falciparum* tend to be less severe in carriers of the deficiency variants than in people with normal G6PD activity [93].

> **IMPORTANT FACT**
> Inherited G6PD deficiency puts 5% of the world population at risk of acute and chronic hemolysis.

Fava beans contain the glycosides vicine and convicine (Figure 3.11), which can damage red blood cells and often break them [94]. One reason for the high vulnerability of people with G6PD deficiency may be their reduced ability to protect red blood cells against oxidation. The oxygen-carrying hemoglobin is a corrosive oxidant that can attack phospholipids and other membrane structures. Without the

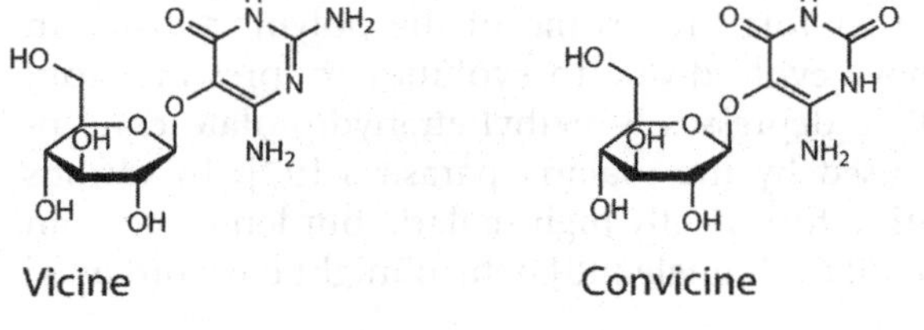

Vicine   Convicine

FIGURE 3.11
The glycosides vicine and convicine in fava beans hemolyze red blood cells of individuals with glucose-6-phosphate dehydrogenase deficiency, but not in people with normal enzyme activity.

antioxidant protection from G6PD-powered enzymes, it is only a question of time until the damaged membranes break and spill their hemoglobin. There is still debate about whether some pea varieties also cause problems in people with G6PD deficiency [95]. The consequence of fava bean ingestion is severe hemolysis, which can be life threatening due to loss of functioning red blood cells. The condition is often worsened when the released hemoglobin clogs up renal tubules and causes acute kidney failure.

> **KEY CONCEPT**
> Selective pressure from debilitating or fatal infections has shaped the way our genome interacts with nutritional factors.

The Pythagoreans would have warned the patient's parents not to eat beans, because they were opposed to the practice on philosophical grounds [95]. Their concern was not hemolysis, however. They just did not want to have their peace of mind disturbed by flatulence. At least that is what Marcus Tullius Cicero wrote about them 2000 years ago:

EX QUO ETIAM PHYTHAGOREIS INTERDICTUM PUTATUR NE FABA VESCERENTUR QUOD HABET INFLATIONEM MAGNAM IS CIBUS TRANQUILLITATI MENTIS QUAERENTI VERA CONTRARIAM.

DE DIVINATIONE, Book I, XXX

The now obsolete antimalarial drug quinine is one of the medications that can trigger hemolysis in people with G6PD deficiency. The quinine amounts in brand-name tonic waters are small, because by US law they cannot contain more than 83 mg/L. But amounts may add up with extensive use and tonics should be avoided by people with G6PD deficiency.

3.4.2 Is this folate for you or your malaria parasite?

Plasmodium malaria parasites can make their own folate, but they also use whatever they can get from the infected host's blood and tissues. Malaria has exerted massive evolutionary pressure on human populations during the millions of years since it started infecting our human ancestors and is still doing so to this day.

> **FOOD FOR THOUGHT**
> Humans and their parasites are in an eternal tug-of-war over essential nutrients. Genetic adaptations on both sides constantly change the balance.

There is some evidence from animal and human studies that folate from the host promotes parasite proliferation [96]. It is quite possible, therefore, that some of the polymorphisms in genes for folate metabolism have evolved due to evolutionary pressure from malaria. The *MTHFR* 677T allele depresses 5-methyltetrahydrofolate concentrations, the metabolite most used by the malaria parasites [97]. In various Mediterranean populations with a historically high malaria burden more than half of the adults carry the *MTHFR* 677T allele [98], which might have provided some disease protection.

3.4.3 Cystic fibrosis

The cholera epidemic of 2010 killed 5500 people in Haiti, many of them young children, and made several hundred thousand people ill among a population of 9 million. This horrific epidemic serves as a reminder that infectious diarrhea is still the second most common cause of infant death today. In the past, epidemics of cholera, typhoid fever, shigellosis, amebiasis, and all kinds of other parasitic, bacterial, and viral infectious diarrheas regularly swept human populations and killed many young children and other vulnerable groups. Unchecked loss of fluids and minerals causes rapid dehydration, electrolyte imbalances, low blood pressure, and eventually shock. When the body loses more than 15% of its water, which can happen within a few days, death becomes inevitable.

These bugs cause diarrhea through diverse mechanisms. One such mechanism is the abnormal stimulation of an adenosine triphosphate (ATP)-binding, cyclic adenosine monophosphate (cAMP)-regulated chloride channel, the cystic fibrosis transmembrane conductance regulator (*CTFR*; OMIM 602428) [99]. Epithelial cells lining the infected colon can secrete several liters of fluid/day when cholera toxin acts on them. Blocking CTFR can quickly put an end to the diarrhea [100].

It should be no surprise then to find that common genetic variants of *CFTR* have evolved that slow fluid and mineral loss. About 4% of all Caucasians carry a variant with a three-nucleotide deletion, the delta F508 allele. Growing evidence indicates that carriers of this and similarly acting variants have a reduced risk of typhoid fever [101] and probably some other enteric infections [102].

The downside of this adaptation is that people inheriting the delta F508 allele from both parents suffer from a serious disease, cystic fibrosis (CF). As far as we know, for Caucasians CF is the most frequent genetic disease with life-threatening consequences early in life [103]. The most easily recognized change in CF is the high (> 60 mmol/L) chloride content of sweat, which increases electrical conductance [104]. Sweat electrolyte or electric conductance analysis is a very simple procedure that helps in recognizing the condition in newborns and others suspected of having the disease. The real problem for CF patients is, however, that secretions in the lungs and pancreas are too thick, don't flow well, and clog up glands. The low output of digestive enzymes from the pancreas greatly limits digestion of foods. Without enzyme substitution, CF patients produce fatty, foul-smelling stools due to the lack of amylase, proteases, and lipase that depend on normal pancreas function. Poor fat absorption makes it particularly difficult to get enough essential fatty acids, vitamin K, and other fat-soluble nutrients from foods [105]. Careful dietary treatment can offset many of the severe nutritional symptoms and allow a better focus on prevention and treatment of severe infections, lung disease, and other organ dysfunctions.

3.5 THE GENETICS OF TASTE AND SMELL

3.5.1 Subjective taste

Humans have molecular sensors that detect the four classical taste qualities, sweet, sour, salty, and bitter, and also the umami (savory) and fatty [106,107] taste qualities. While we eat and drink, additional sensors constantly gather information about fat content, pungent properties, and probably also the content of various nutrients. Many of these sensors vary significantly between individuals. To begin with, people differ greatly in taste bud density. Individuals with a lot of taste buds are called super tasters because they tend to have a greater ability to detect bitter, sweet and salty tastes. Genetically encoded differences in the various receptor genes influence taste perception even more. In the end, this means that we can never really know what a food tastes like to somebody else. With our increasingly detailed understanding of the numerous components of our taste organ, we can determine whether an individual can perceive a particular taste quality, such as the bitterness in grapefruit. But, just like with color blindness, it is hard to understand what it is like to be blind to the taste of a compound. Blockers of some taste qualities can give some limited, but fascinating, insights into individual differences. Adenosine-5'monophosphate is one of these miracle compounds [108]. It blocks the ability to taste many bitter compounds. The Indian herb gurmar (*Gymnema sylvestre*) is called the *sugar destroyer* in the ayurvedic tradition for the simple reason that it stops sweet tastes cold [109]. The peptide gurmarin in the herb selectively inhibits the T1R3 sweet taste receptor [110]. But these chemical inhibitors cannot truly recreate the subtle differences of selective insensitivity that some genetic variants come with. We should therefore respect the judgment of others who do not like broccoli, because we might hate it even more than they do if we had their taste receptors. Or maybe we have taste receptors just like theirs and at long last realize why we have never liked broccoli.

Finally, it is important to recognize that many of these receptors are also expressed in the small and large intestines, where they activate specific cell functions. Sweet taste receptors in enterocytes and enteroendocrine (hormone-secreting) cells of the small intestine, for instance, respond to sugars, amino acids, proteins, and other macronutrient molecules. Many of these food compounds trigger secretion of hormones and promote nutrient uptake [111]. Artificial sweeteners act on sweet taste receptors in the intestines just as they do in the mouth. Bitter taste receptors, on the other hand, can activate transporters, anion exchangers, and water channels. The increased water and electrolyte secretion helps to flush out potential toxins. Transporters such as *ABCB1* (OMIM 171050) can pump poisons back into the intestinal lumen for quick removal [112].

Genetic variation in taste receptors is likely to modify such nutrient- and toxin-related signaling. However, as important as differences in individual responsiveness to particular foods might be for conditions such as irritable bowel syndrome and cancer, such nutrigenetic interactions have not yet received much attention.

3.5.2 Bitter taste is in the mouth of the beholder

The very complex and fairly comprehensive bitter taste system is a vital protector against eating harmful substances. Bitter taste receptors and other sensors tell us which foods to avoid. Bitter-tasting compounds include peptides in spoiled foods, antithyroid compounds, and substances that can cause bleeding, as well as leaves, seeds, berries, and roots with deadly poisons. Young children quickly spit out such bitter items without ever being told about their danger. So why do we make them eat spinach and other foods that taste bitter to them? Why do we regularly berate people for their dislike of broccoli or Brussels sprouts? Can we not trust our built-in food warning system? The Roman philosopher Lucretius knew about the dilemma of a dinner host when his guests have such different tastes. He wrote more than 2000 years ago: 'this may taste like food to some, but more like bitter poison to others' (ut quod ali cibus est fuat acre venenum; Titus Lucretius Carus, ca. 99 BC—ca. 55 BC, De rerum natura, Book IV, line 637). The roots of clashing food preferences are firmly rooted in our nutritional ancestry. When population groups occupy a particular nutritope for many generations, the sensors for common poisons in the food supply are sharpened. It goes back to the undeniable fact that none of our ancestors died during childhood from eating harmful foods, not a single one. The children, who did not have the optimal gene variants and died fom eating harmful foods never had the chance to produce offspring and become the ancestor of someone alive today. The genes for unimportant sensors, on the other hand, will slowly decay because crippling mutations can be transmitted to the next generation without great disadvantage. Different sensors will be strengthened or weakened in different nutritopes. Our genome is necessarily the product of the many different nutritopes of our long-forgotten ancestors. It is very unlikely that we share their nutritopes from hundreds and thousands of years ago because our nutritional circumstances today are almost certainly different.

> **KEY CONCEPT**
> None of your ancestors died during childhood from eating harmful foods, not a single one.

To understand how genetic diversity applies to food dislikes, we need to consider the functional elements of the bitter taste system. There are at least 25 functional members in the TAS2R class of G protein-mediated bitter taste receptors. These receptors require the proper function of the gustatory G proteins, alpha-gustducin and alpha-transducin, sarcoplasmic and endoplasmic reticulum Ca-ATPases [113], and other genes. Each of these genes is probably subject to genetic variation in bitter taste acuity, but so far only a few of the *TAS2R* genes have been explored in any detail. A limited survey of individuals from across the world has revealed much higher genetic variation in bitter taste receptor genes than in other genes, which is consistent with adaptation to numerous local nutritopes [114].

Table 3.2 Genes and Ligands Related to Bitter Taste

Taste receptor	Food/Ingredient	Polymorphism	More > Less Sensitive
TAS2R1	Spoilt or fermented foods (bitter peptides in protein hydrolysates), thiamine	–	–
TAS2R3	Coffee	rs765007 5′ UTR	R3-
TAS2R4	Colchicine, coffee, denatonium benzoate, PROP, and spoiled or fermented foods (bitter peptides in protein hydrolysates), steviol glycosides	Val96Leu rs2234001	R5haploblock CC-CC-GG-TT > TT-GG-AA-GG
TAS2R5	Coffee and cycloheximide	rs2234012 5′ UTR rs2227264 S26I	
TAS2R7	Chloroquine, and *Hoodia gordonii*, papaverine, and strychnine	Met304Iso rs619381	–
TAS2R8	Saccharin	–	–
TAS2R9	Ofloxacin, pirenzapine, and procainamide	Ala187Val rs3741845	–
TAS2R10	Strychnine		
TAS2R13	–	–	
TAS2R14	Absinthe ([-]-α-thujone), aristolochic acid, fishberries (picrotoxinin), *Hoodia gordonii*, and spoiled or fermented foods (bitter peptides in protein hydrolysates), steviol glycosides	rs11610105 rs3741843	–
TAS2R16	Bearberry (arbutin), ethanol, gentiobiose, spoilt or fermented foods (bitter peptides in protein hydrolysates), and willow bark and meadowsweet (salicin)	G>C rs1308724 C>A rs846672 Lys172Asn rs846664	G > CC C > AA N > KK
TAS2R19	Grapefruit and tonic water (quinine)	Cys299Arg rs1077242	CC > R
TAS2R20	Diphenidol, cromolyn	–	–
TAS2R30	Absinthin, andrographolide, and amarogentin	–	–
TAS2R31	Aristolochic acid, parthenolide, quinine, acesulfame K, saccharin	–	–
TAS2R38	Broccoli, Brussels sprouts, cabbages, and watercress (goitrin and other glucosinolates), chard, ethanol, and PROP	Ala49Pro rs713598 Val262Ala rs1726866 Iso296Val rs10246939	PAV > AVI
TAS2R39	Green tea (catechins), quinine	–	–
TAS2R40	Dapsone, diphenidol, diphenhydramine	–	–
TAS2R41	–	–	–

(Continued)

Table 3.2	Genes and Ligands Related to Bitter Taste—cont'd		
Taste receptor	**Food/Ingredient**	**Polymorphism**	**More > Less Sensitive**
TAS2R42	–	–	–
TAS2R43	Acesulfame K, aloe (aloin), aristolochic acid, caffeine, and saccharin	Copy number variation	–
TAS2R44	Acesulfame K, aloe (aloin), aristolochic acid, and saccharin	–	–
TAS2R45	Acesulfame K, saccharin	Copy number variation	–
TAS2R46	Caffeine, clerodane diterpenoids, colchicine, and strychnine	–	–
TAS2R47	Denatonium	–	–
TAS2R50	Amarogentin and andrographolide	–	–
TAS2R55	–	–	–
TAS2R60	Grapefruit?	C>T rs4595035	C > T
alpha-gustducin	–	–	–
alpha-transducin	–	–	–

PROP, propylthiouracil; UTR, untranslated region.

Most bitter taste receptors respond to chemically diverse compounds and many of these compounds stimulate more than one receptor (Table 3.2). The signals from individual taste buds, each carrying one particular receptor, are bundled and integrated as they travel to the brain, where they generate the perception of a complex taste experience. This means that variation in a receptor will not necessarily turn off perception of a taste entirely but will modify the overall impression. Consider coffee, for example. At least three different receptor genes, *TAS2R3* (OMIM 604868), *TAS2R4* (OMIM 604869) and *TAS2R5* (OMIM 605062), encode proteins that sense the burnt taste of roasted coffee and another two bitter taste receptor genes, *TAS2R43* (OMIM 612668) and *TAS2R46* (OMIM 612774), sense caffeine. Together with still more taste sensors they create a complex pattern that corresponds to our favorite brew of coffee. But the detected pattern will taste different for each individual consumer depending on his or her bitter taste receptor haplotypes. Some will love a particular brand of coffee and others hate it, simply because everybody tastes a different flavor pattern. No wonder that there are so many varieties of coffee beans, roasting processes, and brewing techniques, each with their own dedicated following.

On the other hand, there are compounds that interact mostly with a single receptor. If someone has only low-function variants of this receptor, they may not be able to sense the bitter taste of the compound at all. For instance, propylthiouracil (PROP, a commonly used test substance; Figure 3.12) acts on the extensively investigated TAS2R38 with a narrow specificity. It is not surprising,

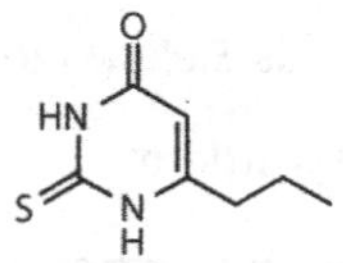

FIGURE 3.12
Propylthiouracil.

therefore, that a combination of three common *TAS2R38* (OMIM 607751) polymorphisms strongly determines whether people perceive the bitter taste of PROP or not [115,116]. People without at least one copy of a high-functioning variant are virtually blind to the galling bitterness of PROP and they are also less sensitive to the bitter taste of broccoli and other cruciferous vegetables [117,118]. Genetic variation in the *TASR38* gene is very old, since humans share these polymorphisms with Neanderthals [119]. This means that the variants probably emerged well before the Neanderthals diverged from modern humans hundreds of thousands of years ago. Goitrin (5-vinyloxazolidine-2-thione; Figure 3.13) is a natural chemical in cruciferous vegetables (broccoli, Brussels sprout, cabbages, kale, collards, arugula, cress, radishes and more), which promotes the development of goiter (enlargement of the thyroid gland), particularly in people with low iodine status. This compound might also increase the risk that mothers with iodine deficiency give birth to a child with severe mental retardation due to cretinism. It would make a lot of sense if populations at risk developed a warning system that steered them away from foods that could cause such risks. The high-function variants of TAS2R38 and probably additional bitter taste receptors do precisely that [120]. People with the most sensitive TAS2R38 haplotype, PAV, can detect the unpleasant bitter taste of goitrin, but people with the least sensitive haplotype, AVI, cannot. But many cruciferous vegetables are also rich sources of folate and other helpful nutrients. Maternal folate deficiency greatly increases the risk of having a child with birth defects. Whether avoidance of cruciferous vegetables is beneficial for PAV carriers would depend therefore, among other factors, on their iodine status and on their access to alternative food sources of folate. Cabbage dislike may have worked well in iodine-deficient environments, such as remote mountain valleys, as long as the adapted population stayed there. But with extensive migrations to and from such areas and with the reliance on store-bought foods, it is unlikely that such innate preferences can still guide our personal nutritional choices. After all, since iodine deficiency has become uncommon in industrialized countries, most people should now ignore the message from their TAS2R38 receptors and go for the

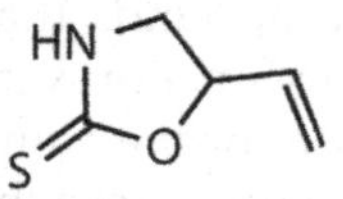

FIGURE 3.13
Goitrin.

folate-rich broccoli. However, on second thoughts, maybe pervasive consumption of folate-supplemented foods may shift that balance again.

Copy number variation is another common mechanism for varying bitter taste sensitivity. Both copies of the *TAS2R43* (OMIM 612668) gene are missing in a quarter of all European adults and only one-quarter has two functioning copies [121]. Individuals without functioning TAS2R43 are much less likely to detect the bitter aftertaste of the artificial sweeteners saccharin and acesulfame K. The number of functioning copies of *TAS2R45* (OMIM 613967) also varies due to deletions.

3.5.3 A wholesome taste

The notion of sweetness evokes powerful emotional associations that go way beyond simple food tasting. It is an overwhelmingly positive feeling that prompts people to talk about sweet girls, the sweet spot on a racket, or home sweet home. Sweet is a flavor that comes with goodness written all over it. How could anybody not like lots of it? But the reality is that not everybody likes it very sweet all the time. Common genetic variants partially determine who does and who does not. By the way, this affinity to sweets is a very human feature that is not shared by our feline companions. They just would not know what we are talking about when we are calling them sweet, because they do not have functioning sweet taste receptors [122]. And while we are at it, the artificial sweetener aspartame also works only for us and our closest primate cousins [123], not for any of our pets.

The molecules that make foods taste sweet are mainly sugars, a few amino acids, and peptides (including the artificial sweetener aspartame, *N*-(L-α-aspartyl)-L-phenylalanine, 1-methyl ester, and an assortment of other natural and synthetic compounds). The exact three-dimensional structure of the food molecules makes all the difference. For example, isomaltose, which consists of two glucose molecules connected by an α-$(1 \rightarrow 6)$ linkage, has a pleasantly sweet taste. But if you bend the isomaltose molecule a bit and connect the two glucose moieties through β-$(1 \rightarrow 6)$ linkage (Figure 3.14), you get gentiobiose, which has a distinctly bitter taste detected by *TAS2R16* (OMIM 604867) [124]. Gentiobiose is one of the products that form when glucose is caramelized by heating, thus tempering the sweetness of the sugar mass with its distinctive bitter note.

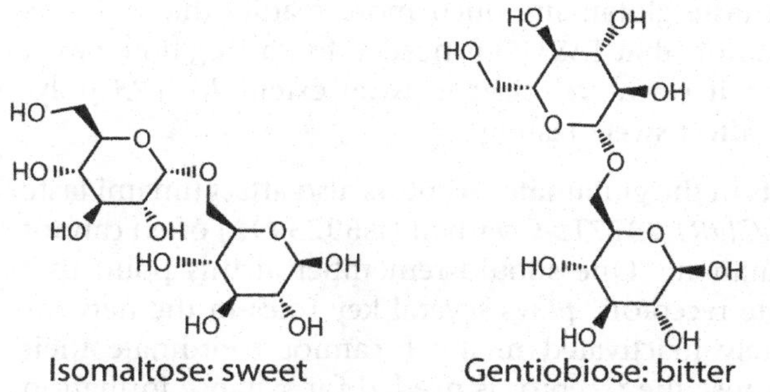

Isomaltose: sweet   Gentiobiose: bitter

FIGURE 3.14
Isomaltose and gentiobiose both consist of two glucose molecules, but one tastes sweet and the other one bitter.

Two human sweet taste receptors, *TAS1R2* (OMIM 606226) and *TAS1R3* (OMIM 605865), functioning as heterodimers in taste buds, are the predominant molecules for sensing sweet substances in food [125]. The G protein alpha-gustducin (*GNAT3*; OMIM 139395) mediates the response signal downstream. Common variants occur in the receptor genes as well in GNAT3.

A common variant, -1572C>T (rs307355), in the promoter region of the *TAS1R3* gene appears to reduce sucrose sensitivity [126]. A blinded taste test found that people with the homozygous C/C genotype were twice as good at detecting sucrose in water than those with other genotypes. The T allele occurs almost exclusively in people of African origin. Fushan et al. [126] suggested that the high-sensitivity allele T helped early populations migrating out of Africa to overcome the difficulty that cold-climate plants had much lower sugar content than the plants in their original African environment.

The *TAS1R2* polymorphism, Ile191Val (rs35874116), has also been found to affect habitual sugar consumption [127]. Overweight and obese Canadians with the 191Val allele consume less sugar than otherwise matched individuals without this variant.

GNAT3 polymorphisms explain 13% of the variation in sucrose perception [128].

3.5.4 A meat eater's taste buds

Some taste buds are tuned to detect meaty flavors called *umami* (a Japanese word for savory). They do this using heterodimers consisting of the taste receptors *TAS1R1* (OMIM 606225) and *TAS1R3* (OMIM 605865), several metabotropic glutamate receptors, including mGluR1 (OMIM 604473) and mGluR4 (OMIM 604100), and several ionotropic glutamate receptors. Glutamic acid (or sodium monoglutamate; MSG) and 5'-ribonucleotides are particularly effective stimulators of the umami sensors.

The sensitivity of umami detection varies 10—20-fold between people and much of that variation is related to several common sequence variants in both *TAS1R1* and *TAS1R3* genes [129,130]. For example, people with the *TAS1R1* 372T variant (rs34160967) need only half as much glutamate to taste the umami flavor compared to people with the 372A allele. Individuals with the *TAS1R3* variant 757Arg (rs307377) also can taste glutamate much more readily than 757Cys carriers [129,131]. We should note that TAS1R3 is needed for tasting both sweet and umami flavors. However, it is not yet clear to what extent *TAS1R3* polymorphisms, such as 757Arg, affect sweet tasting.

As might be expected, variants in the glutamate receptors also affect umami taste sensitivity. People with the *mGluR1* 2977T>C variant (rs6923492) often cannot detect small amounts of glutamate. One should remember at this point that mGluR1, like other glutamate receptors, plays several key roles in the nervous system. Mice with deliberately inactivated mGluR1 cannot coordinate their movements well (ataxia) because the receptor is needed for synapse formation in the cerebellum [132]. This glutamate receptor is also involved in pain perception [133] and cognition [134]. The evolutionary fate of any change in the *mGluR1* sequence will therefore depend on the balance of harm and benefit not only to umami taste sensitivity but also to motor skills, cognition, and pain threshold. Is it worth getting a new preference for meat when it comes with increased clumsiness in handling your spear, slower speed when moving away from wounded prey, and not realizing that taking on the mastodon all by yourself was a bad idea to begin with? The 2977C variant is so finely tuned that it does not seem to affect the nervous system, but only the umami taste threshold [135]. Multiple effects of a polymorphism on unrelated functions is called *pleiotropy*. Because of the high integration of many nutrition functions into the regulation of other body functions, pleiotropic effects of variants are common.

> **KEY CONCEPT**
> Nutrigenetic variants often have pleiotropic actions, affecting diverse body functions.

Variation in umami taste sensitivity is very interesting because without it meat would taste more or less like cardboard. When, a few millions years ago, the *TAS1R1* gene of the giant pandas (*Ailuropoda melanoleuca*), a bear species, lost its function, they took meat off their menu and switched to bamboo shoots [136]. There may be a lesson there, somewhere. It remains to be seen, though, whether people with low umami sensitivity actually eat less meat.

3.5.5 They don't know what they eat

We now understand that taste buds in the circumvallate papillae of the tongue measure how much fat is in a meal. The glycoprotein CD36 (OMIM 173510) works like a receptor for long-chain fatty acids. The tongue also has a special lipase (lingual lipase; EC 3.1.1.3), which needs neither bile acids nor lipase cofactor to work, that makes sure that fatty acids are released from the fats in food and can be tasted. After eating a good serving of a fatty food, expression of the sensor drops off and we get tired of having another serving [137]. At least this is how we hope it would work. But some people apparently do not get the message. They are not good at telling the difference between high-fat and low-fat foods. A common genetic variant (rs1761667) in the *CD36* fatty acid sensor can predict who is most likely to have a blind spot for fat in their food [138]. People with two copies of the relatively common rs1761667A allele were reported to need about 4% fat in a solution to taste the fat in it (Figure 3.15). That is more than the amount of fat in whole fresh milk. Those with two copies of the G allele were able to taste the fat in a solution with 0.6% fat. That is a more than sixfold difference in sensitivity [139].

It should not surprise us, therefore, that people without at least one copy of the gene that produces the more sensitive sensor tend to add more fat to their food than people without the variant. They also appear to consume slightly more calories and have a higher risk of becoming obese. And it can get even worse for carriers of this fat-insensitive CD36 variant: if they use the lipase inhibitor orlistat

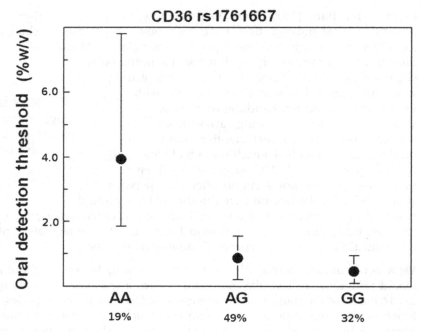

FIGURE 3.15

People with two copies of the *CD36* rs1761667A allele (group on the left) appear to have a particularly high fat-sensing threshold [139], which means that they can only experience the creamy taste of fat if the food contains a lot of it.

(trade name Xenical) to lose weight, they become even more oblivious to the fat in their food. This is because the medication blocks lingual lipase action and prevents the release of the fatty acids that are the basis of fat sensing by CD36 [139].

3.5.6 Sour and salt sensing

It is sometimes said that twins have twice the fun. Scientists at the Twins Days Festival in Twinsburg, Ohio, had a challenge for twins that was not only fun to do but also gave some real insight into the genetics of sour taste [140]. What was going on there? Visitors at the twin convention had a competition for the most sensitive palate. It was all about who was best in recognizing the taste of spiked water. Alternative choices were sweet, sour, bitter, salty, and plain water. Samples were offered, starting from the lowest concentration and increasing stepwise until the participant recognized the taste. The challenge for each competitor was to identify the correct taste. And to make sure they really got it right they then had to sort six samples by flavor intensity. Three of each set were plain water and the other ones contained the lowest flavor concentration initially recognized by the competitor. What came out of this battle of the taste buds was a clear demonstration that shared genetic factors were more important than the environment in determining sour taste sensitivity. Heritability of the sour taste threshold was over 50% in this experiment. A channel complex consisting of polycystic kidney disease-2-like 1 (*PKD2L1*; OMIM 604532) and polycystic kidney disease-1-like 3 (*PKD1L3*; OMIM 607895) proteins appears to be the main sensor. Both genes encoding these proteins contain common variants, but their impact on sour tasting threshold is still untested.

It seems that the ability to detect salt is shaped by short-term salt exposure [141] and experiences during pregnancy and infancy [142], and also significantly by genetic factors. Variants in the *TRPV1* (encoding a pain receptor in many cells) and the *SCNN1B* (encoding part of a sodium channel) genes appear to modify the ability to taste salt [157].

3.5.7 A bouquet of culinary fragrances

Sensing odors is even more important for our food choices than detecting tastes and is many times more complex. Unpleasant odors deter us from eating spoiled foods and pleasant ones invite us to a feast. The genetic complexity of these odorant receptor genes is astounding. Humans have at least 855 *odorant receptor* (OR) genes, but no individual has more than about four hundred of them in working order. The others are pseudogenes and not up to snuff. The functioning genes alone constitute more than 2% of our entire gene repertoire and all of them just for finding out what's for dinner without having to get up from the sofa.

As might be expected, variation between individuals is huge. To give a nutritionally related example, some people (about 7%) are not at all bothered by the fishy odor of trimethylamine (a natural metabolite of dietary choline) because they are a thousand times less sensitive to it than everybody else [143]. This would make them fitting companions for people with the genetic condition of fish-odor syndrome (trimethylaminuria; OMIM 602079). Individuals with that condition cannot metabolize the choline-metabolite trimethylamine and suffer from social isolation because of their offensive body odor. They can reduce odor production by avoiding choline-rich foods [144].

Much of the diversity between individuals comes from the fact that everybody has a different set of functioning OR genes [145]. Obviously, only expressed receptors can detect an odorant.

Additional diversity is due to extensive copy number variation [146], all of which has arisen in the last 3–4 million years. This is evolution in action. Gene duplications happen fairly often during recombination. Nothing constrains the extra gene to maintain its initial functional state because the original gene can still take care of business. Successive mutations can then reshape the extra gene to recognize another odorant. Comparisons of primate and human OR genes demonstrate that this kind of rebuilding to make new genes can happen very rapidly, within a few tens of thousands of generations.

A third source of OR gene variation are base changes. A case in point is one of missing asparagus odor. Most people produce urine with a characteristically pungent odor shortly after eating asparagus. But this does not happen with everybody. For many years the debate raged whether some people just do not produce the smelly metabolites from asparagus or whether a true anosmia (smell blindness) is the reason. Now we know that both explanations are correct. Some individuals do not produce enough of the smelly sulfurous metabolites of asparagusic acid to be noticed [147]. A few others cannot perceive the odor when it is clearly present, and sometimes both explanations hold [147]. The inability to perceive asparagus odor depends on the single nucleotide polymorphism rs4481887 [148,149] near the *odorant receptor 2M7* gene (*OR2M7*). People with two copies of the G allele are constitutionally oblivious to the aftereffects of this delicious vegetable. They will never know what they are missing (and probably would not care).

3.6 GENDER DIFFERENCES

3.6.1 Why women cannot be like men

Finally, we must not forget to ask how females respond differently to foods and nutritional status than men. After all, the genetic differences between XX-carrying women and XY-sporting men are much greater than any other DNA sequence differences we will find in healthy people. We all know about the war of the sexes at the dinner table, but how much of that is genetically pro-grammed? There are hints of important nutrigenetic differences between genders, but much remains to be found out.

The most obvious nutritional difference between men and women concerns the amount and distribution of stored fat. Young healthy women carry about 40% more body fat than men of comparable age and circumstances [150]. Women tend to store more of their body fat around the hips and buttocks, while any extra fat in men goes usually to the abdomen and chest. High estrogen activity in women is a major reason for the gender difference in body composition and its distribution. A significant amount of body fat is critical for conception and the successful completion of pregnancy.

Current Dietary Reference Intakes (DRI, the official recommendations for healthy nutrient intakes) differ for some nutrients on a per weight basis. The biggest gender-specific difference in intake targets concerns iron (to balance losses with menstrual blood). It is due to their much lower requirements that men are much more vulnerable to the harmful effects of iron accumulation when they carry gene variants predisposing to hemochromatosis (discussed later in more detail in the Section 4.8.1 on iron nutrition). Energy requirement per weight is also slightly lower for women than for men. When women exercise, they tend to get a slightly greater percentage of their energy from stored fat than men [151].

3.6.2 Polymorphisms in estrogen-responsive genes

One might expect that polymorphisms in estrogen-related genes would result in differential responsiveness to intake and overfeeding in females but not in males, and studies seem to indicate just that. For instance, polymorphisms in the first intron of the *estrogen receptor alpha* (*ESR1*; OMIM 133430) were associated with body composition in Dutch adolescent girls, but not in their male counterparts [152]. The lesson we need to draw from such findings is that we should never assume that the interaction between genetic polymorphisms and nutritional factors is the same for both genders. Because the activities of estrogen and other reproductive hormones change dramatically from one life stage to the next, some nutrigenetic effects may manifest only with the onset of sexual maturation or during pregnancy, and diminish with menopause.

A good example of this principle is the gender-specific impact of genetic variation on choline requirement. Choline has been recognized as an essential nutrient because many people cannot make enough on their own and have to get additional amounts with food. Endogenous synthesis depends on phosphatidylethanolamine-N-methyltransferase (EC 2.1.1.17), encoded by the estrogen-induced gene, *PEMT* (OMIM 602391). This makes a lot of sense, because during pregnancy and lactation women must supply large amounts of choline to their children. A very common variant (rs12325817) puts a brake on PEMT activity, but the effect is seen mainly in women [153]. The reason is that this locus is right in the middle of an estrogen-responsive element in the *PEMT* promoter region. Replacement of a specific guanine with a cytosine in the promoter sequence diminishes the ability of the *PEMT* gene to respond to estrogen [154]. Most men need to get choline from food, regardless of their genotype, because they do not make enough estrogen to increase PEMT activity. Young women with two copies of the G allele, on the other hand, usually do not depend on choline from food because their estrogen plays the *PEMT* gene like a well-tuned fiddle. But women with one or two C alleles will develop signs of deficiency without adequate choline intake. It should be noted that the C allele appears to be a fairly recent acquisition, which is less common in Africans and not found in any primates. This variant could well go back to early human migration to the meat pots of Eurasia requiring an adaptation to the large amounts of choline that go with high meat intake.

3.6.3 Male-specific polymorphisms

The most immediately apparent genetic characteristic of men is their Y chromosome. With about 23 million bases and 78 genes it is the smallest of all human chromosomes. Surprisingly, this is also the chromosome that has changed the most during recent evolution. It has been noted that there are about as many differences between human and chimpanzee Y chromosomes as there are between the entire genomes of humans and chickens [155]. Many of the genes on the Y chromosome contribute to sex determination and sperm function. So far, genotype-specific interactions of Y-chromosomal polymorphisms with nutritional factors have received little attention, but this could change soon.

Another group of male-specific characteristics relates to the male sexual organs. We are going to address just the prostate gland because, despite its small size (similar to a walnut), it causes a lot of problems. Most men will eventually develop prostate cancer, though only few will ever experience an aggressive and potentially lethal disease. Nutritional factors play a significant role in progression toward symptomatic disease and death. Genistein (Figure 3.16) and related isoflavones in soy have estrogenic activity that counteracts men's own tumor-promoting testosterone. It has been suggested, therefore, that increasing soy

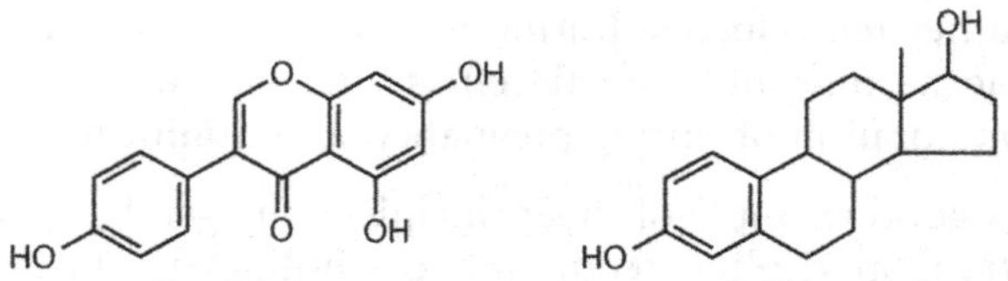

FIGURE 3.16
The soy isoflavone genistein is structurally similar to estrogens and mimics some of their actions, but with much less potency.

Genistein   17β-Estradiol

consumption may slow prostate cancer progression, but findings have been very inconsistent. These variable results may be explained by the fact that about two in five Americans respond differently to isoflavones because they carry a cytosine (rs2987983) instead of the more common thymine at position −13950 of the *estrogen receptor beta* gene (*ESR2*; OMIM 601663). High isoflavone intake appears to decrease the risk of aggressive, lethal disease in men with the −13950C allele but to increase risk in men without it [156]. Compared to men consuming almost no isoflavones, those eating more than about 1 mg/day (this is the amount in about 3 g of tofu) had a 37% lower prostate cancer risk when they carried one or two copies of the −13950C allele. But the majority, who did not have a −13950C allele, had a 47% higher cancer risk with a high isoflavone intake. So the decision to eat soy worked out really well for the C allele carriers. Men without this allele, on the other hand, might have second thoughts and want to wait for more research before taking the risk. What would you tell your father to do?

Practice questions

How can we know at what time in history a gene has become a pseudogene?

What makes broad (fava) beans more dangerous for men than for women?

To what kind of diet were Paleolithic humans adapted?

How many generations does it take a human population to adapt to a new nutritope?

What genetic change would it take for humans to adapt to a diet without vitamin C?

Why are there so many different popular varieties of chocolate?

Why is it that men tend to eat much more meat than women?

SUMMARY AND SEGUE TO THE NEXT CHAPTER

We have traced how some of the human nutritional idiosyncrasies, such as the dependence on foods for vitamin C, go back millions of years and reflect an adjustment to very early nutritional environments. Many later nutritional environments, which we call nutritopes, have continued to shape the human genome. Our genes evolved in many different directions because there were so many different nutritional circumstances. The many genetic variants we see today in human populations are echoes of our diverse past. Some of us are adapted to milk drinking; others are not. Some have no problem with the occasional alcohol indulgence, while others only get something like an instant hangover. Further examples of divergent responses to particular foods and nutrients are numerous. Our own set of genetic variants makes us all very unique, giving us a very personal pattern of nutritional dos and don'ts.

Now, these examples may be interesting for the telling, but are they more than just a few anecdotes? The next section will review more systematically how individual genetic predisposition influences the metabolism of many nutrients.

References

[1] Morris JG. Idiosyncratic nutrient requirements of cats appear to be diet-induced evolutionary adaptations. Nutr Res Rev 2002;15(1):153—68.
[2] Sanderson SL, Gross KL, Ogburn PN, Calvert C, Jacobs G, Lowry SR, et al. Effects of dietary fat and L-carnitine on plasma and whole blood taurine concentrations and cardiac function in healthy dogs fed protein-restricted diets. Am J Vet Res 2001;62(10):1616—23.
[3] Rebouche CJ, Lombard KA, Chenard CA. Renal adaptation to dietary carnitine in humans. Am J Clin Nutr 1993;58(5):660—5.
[4] Pollock JI, Mullin RJ. Vitamin C biosynthesis in prosimians: evidence for the anthropoid affinity of Tarsius. Am J Phys Anthropol 1987;73(1):65—70.
[5] Cui J, Pan YH, Zhang Y, Jones G, Zhang S. Progressive pseudogenization: vitamin C synthesis and its loss in bats. Mol Biol Evol 2011;28(2):1025—31.
[6] Nishikimi M, Yagi K. Molecular basis for the deficiency in humans of gulonolactone oxidase, a key enzyme for ascorbic acid biosynthesis. Am J Clin Nutr 1991;54(Suppl. 6):1203S—8S.
[7] Inai Y, Ohta Y, Nishikimi M. The whole structure of the human nonfunctional L-gulono-gamma-lactone oxidase gene—the gene responsible for scurvy—and the evolution of repetitive sequences thereon. J Nutr Sci Vitaminol (Tokyo) 2003;49(5):315—19.
[8] Burge CB, Karlin S. Finding the genes in genomic DNA. Curr Opin Struct Biol 1998;8(3):346—54.
[9] Horwitt MK, Harper AE, Henderson LM. Niacin-tryptophan relationships for evaluating niacin equivalents. Am J Clin Nutr 1981;34(3):423—7.
[10] Tanaka T, Tateno Y, Gojobori T. Evolution of vitamin B6 (pyridoxine) metabolism by gain and loss of genes. Mol Biol Evol 2005;22(2):243—50.
[11] Hrdlickova B, Westra HJ, Franke L, Wijmenga C. Celiac disease: moving from genetic associations to causal variants. Clin Genet 2011 Sep;80(3):203—313.
[12] Mountain JL, Cavalli-Sforza LL. Inference of human evolution through cladistic analysis of nuclear DNA restriction polymorphisms. Proc Natl Acad Sci U S A 1994;91(14):6515—19.
[13] Sørensen Thorvald. Some ecosystemical characteristics determined by Raunkiær's circling method. Helsinki: Skandinaviska naturforskaremöten; 1936. pp. 474—475.
[14] Sponheimer M, Codron D, Passey BH, de Ruiter DJ, Cerling TE, Lee-Thorp JA. Using carbon isotopes to track dietary change in modern, historical, and ancient primates. Am J Phys Anthropol 2009;140(4):661—70.
[15] Fuller BT, Marquez-Grant N, Richards MP. Investigation of diachronic dietary patterns on the islands of Ibiza and Formentera, Spain: Evidence from carbon and nitrogen stable isotope ratio analysis. Am J Phys Anthropol 2010;143(4):512—22.
[16] Oelze VM, Fuller BT, Richards MP, Fruth B, Surbeck M, Hublin JJ, et al. Exploring the contribution and significance of animal protein in the diet of bonobos by stable isotope ratio analysis of hair. Proc Natl Acad Sci U S A 2011;108(24):9792—7.

[17] Teaford MF, Ungar PS. Diet and the evolution of the earliest human ancestors. Proc Natl Acad Sci U S A 2000;97(25):13506–11.

[18] Laden G, Wrangham R. The rise of the hominids as an adaptive shift in fallback foods: plant underground storage organs (USOs) and australopith origins. J Hum Evol 2005;49(4): 482–98.

[19] Luca F, Perry GH, Di Rienzo A. Evolutionary adaptations to dietary changes. Annu Rev Nutr 2010;30:291–314.

[20] Organ C, Nunn CL, Machanda Z, Wrangham RW. Phylogenetic rate shifts in feeding time during the evolution of Homo. Proc Natl Acad Sci U S A 2011;108(35):14555–9.

[21] Matthews LJ, Butler PM. Novelty-seeking DRD4 polymorphisms are associated with human migration distance out-of-Africa after controlling for neutral population gene structure. Am J Phys Anthropol 2011;145(3):382–9.

[22] Balter V, Simon L. Diet and behavior of the Saint-Cesaire Neanderthal inferred from biogeochemical data inversion. J Hum Evol 2006;51(4):329–38.

[23] Yotova V, Lefebvre JF, Moreau C, Gbeha E, Hovhannesyan K, Bourgeois S, et al. An X-linked haplotype of neandertal origin is present among all non-African populations. Mol Biol Evol 2011;28(7):1957–62.

[24] Reich D, Green RE, Kircher M, Krause J, Patterson N, Durand EY, et al. Genetic history of an archaic hominin group from Denisova Cave in Siberia. Nature 2010;468(7327): 1053–60.

[25] Le Bras-Goude G, Binder D, Zemour A, Richards MP. New radiocarbon dates and isotope analysis of Neolithic human and animal bone from the Fontbregoua Cave (Salernes, Var, France). J Anthropol Sci 2010;88:167–78.

[26] Macko SA, Engel MH, Andrusevich V, Lubec G, O'Connell TC, Hedges RE. Documenting the diet in ancient human populations through stable isotope analysis of hair. Philos Trans R Soc Lond B Biol Sci 1999;354(1379):65–75.

[27] Holden TG. The Food Remains from the Colon of the Tyrolean Ice Man. Oxford, England: Oxbow Books; 2002.

[28] Porsild A. Edible plants of the Arctic. Arctic 1953;6:15–34.

[29] Milman N, Laursen J, Mulvad G, Pedersen HS, Pedersen AN, Saaby H. 13Carbon and 15nitrogen isotopes in autopsy liver tissue samples from Greenlandic Inuit and Danes: consumption of marine versus terrestrial food. Eur J Clin Nutr 2010;64(7):739–44.

[30] Naughton JM, O'Dea K, Sinclair AJ. Animal foods in traditional Australian aboriginal diets: polyunsaturated and low in fat. Lipids 1986;21(11):684–90.

[31] O'Dea K. Traditional diet and food preferences of Australian aboriginal hunter-gatherers. Philos Trans R Soc Lond B Biol Sci 1991;334(1270):233–40. discussion, 40–1.

[32] O'Dea K, White NG, Sinclair AJ. An investigation of nutrition-related risk factors in an iso-lated Aboriginal community in northern Australia: advantages of a traditionally-orientated life-style. Med J Aust 1988;148(4):177–80.

[33] Thorburn AW, Brand JC, Truswell AS. Slowly digested and absorbed carbohydrate in traditional bushfoods: a protective factor against diabetes? Am J Clin Nutr 1987;45(1): 98–106.

[34] O'Dea K. Marked improvement in carbohydrate and lipid metabolism in diabetic Australian aborigines after temporary reversion to traditional lifestyle. Diabetes 1984;33(6):596–603.

[35] Lokki AI, Jarvela I, Israelsson E, Maiga B, Troye-Blomberg M, Dolo A, et al. Lactase persis-tence genotypes and malaria susceptibility in Fulani of Mali. Malar J 2011;10:9.

[36] Burger J, Kirchner M, Bramanti B, Haak W, Thomas MG. Absence of the lactase-persis-tence-associated allele in early Neolithic Europeans. Proc Natl Acad Sci U S A 2007;104(10): 3736–41.

[37] Gerbault P, Liebert A, Itan Y, Powell A, Currat M, Burger J, et al. Evolution of lactase persistence: an example of human niche construction. Philos Trans R Soc Lond B Biol Sci 2011;366(1566):863–77.

[38] Evershed RP, Payne S, Sherratt AG, Copley MS, Coolidge J, Urem-Kotsu D, et al. Earliest date for milk use in the Near East and southeastern Europe linked to cattle herding. Nature 2008;455(7212):528–31.

[39] Copley MS, Berstan R, Dudd SN, Docherty G, Mukherjee AJ, Straker V, et al. Direct chemical evidence for widespread dairying in prehistoric Britain. Proc Natl Acad Sci U S A 2003; 100(4):1524–9.

[40] Campbell CD, Ogburn EL, Lunetta KL, Lyon HN, Freedman ML, Groop LC, et al. Demonstrating stratification in a European American population. Nat Genet 2005;37(8): 868–72.

[41] Waud JP, Matthews SB, Campbell AK. Measurement of breath hydrogen and methane, together with lactase genotype, defines the current best practice for investigation of lactose sensitivity. Ann Clin Biochem 2008;45(Pt 1):50–8.

[42] Xu L, Sun H, Zhang X, Wang J, Sun D, Chen F, et al. The -22018A allele matches the lactase persistence phenotype in northern Chinese populations. Scand J Gastroenterol 2010;45(2): 168–74.

[43] Heyer E, Brazier L, Segurel L, Hegay T, Austerlitz F, Quintana-Murci L, et al. Lactase persistence in central Asia: phenotype, genotype, and evolution. Human Biology 2011;83(3): 379–92.

[44] Itan Y, Jones BL, Ingram CJ, Swallow DM, Thomas MG. A worldwide correlation of lactase persistence phenotype and genotypes. BMC Evol Biol 2010;10:36.

[45] Rahimi AG, Delbruck H, Haeckel R, Goedde HW, Flatz G. Persistence of high intestinal lactase activity (lactose tolerance) in Afghanistan. Hum Genet 1976;34(1):57–62.

[46] Gallego Romero I, Basu Mallick C, Liebert A, Crivellaro F, Chaubey G, Itan Y, et al. Herders of Indian and European cattle share their predominant allele for lactase persistence. Mol Biol Evol 2012 Jan;29(1):249–60.

[47] Peng MS, He JD, Zhu CL, Wu SF, Jin JQ, Zhang YP. Lactase persistence may have an independent origin in Tibetan populations from Tibet, China. J Human Genet 2012 Jun;57(6): 394–7.

[48] Ingram CJ, Raga TO, Tarekegn A, Browning SL, Elamin MF, Bekele E, et al. Multiple rare variants as a cause of a common phenotype: several different lactase persistence associated alleles in a single ethnic group. J Mol Evol 2009;69(6):579–88.

[49] Tishkoff SA, Reed FA, Ranciaro A, Voight BF, Babbitt CC, Silverman JS, et al. Convergent adaptation of human lactase persistence in Africa and Europe. Nat Genet 2007;39(1): 31–40.

[50] Mulcare CA, Weale ME, Jones AL, Connell B, Zeitlyn D, Tarekegn A, et al. The T allele of a single-nucleotide polymorphism 13.9 kb upstream of the lactase gene (LCT) (C-13.9kbT) does not predict or cause the lactase-persistence phenotype in Africans. Am J Hum Genet 2004;74(6):1102–10.

[51] Ingram CJ, Elamin MF, Mulcare CA, Weale ME, Tarekegn A, Raga TO, et al. A novel polymorphism associated with lactose tolerance in Africa: multiple causes for lactase persistence? Hum Genet 2007;120(6):779–88.

[52] Khabarova Y, Torniainen S, Savilahti E, Isokoski M, Mattila K, Jarvela I. The -13914G>A variant upstream of the lactase gene (LCT) is associated with lactase persistence/non-persistence. Scand J Clin Lab Invest 2010;70(5):354–7.

[53] Torniainen S, Parker MI, Holmberg V, Lahtela E, Dandara C, Jarvela I. Screening of variants for lactase persistence/non-persistence in populations from South Africa and Ghana. BMC Genet 2009;10:31.

[54] Jensen TG, Liebert A, Lewinsky R, Swallow DM, Olsen J, Troelsen JT. The -14010*C variant associated with lactase persistence is located between an Oct-1 and HNF1alpha binding site and increases lactase promoter activity. Hum Genet 2011 Oct;130(4):483–93.

[55] Mattar R, Monteiro Mdo S, Silva JM, Carrilho FJ. LCT-22018G>A single nucleotide polymorphism is a better predictor of adult-type hypolactasia/lactase persistence in Japanese-Brazilians than LCT-13910C<T. Clinics (Sao Paulo) 2010;65(12):1399.

[56] Nemeth K, Plumb GW, Berrin JG, Juge N, Jacob R, Naim HY, et al. Deglycosylation by small intestinal epithelial cell beta-glucosidases is a critical step in the absorption and metabolism of dietary flavonoid glycosides in humans. Eur J Nutr 2003;42(1):29–42.

[57] Perry GH, Dominy NJ, Claw KG, Lee AS, Fiegler H, Redon R, et al. Diet and the evolution of human amylase gene copy number variation. Nat Genet 2007;39(10):1256–60.

[58] Mandel AL, Peyrot des Gachons C, Plank KL, Alarcon S, Breslin PA. Individual differences in AMY1 gene copy number, salivary alpha-amylase levels, and the perception of oral starch. PLoS One 2010;5(10):e13352.

[59] Ferry AL, Mitchell JR, Hort J, Hill SE, Taylor AJ, Lagarrigue S, et al. In-mouth amylase activity can reduce perception of saltiness in starch-thickened foods. J Agric Food Chem 2006;54(23): 8869—73.

[60] de Wijk RA, Prinz JF, Engelen L, Weenen H. The role of alpha-amylase in the perception of oral texture and flavour in custards. Physiol Behav 2004;83(1):81—91.

[61] Mandel AL, Breslin PA. High endogenous salivary amylase activity is associated with improved glycemic homeostasis following starch ingestion in adults. J Nutr 2012;142(5): 853—8.

[62] Husemoen LL, Jorgensen T, Borch-Johnsen K, Hansen T, Pedersen O, Linneberg A. The association of alcohol and alcohol metabolizing gene variants with diabetes and coronary heart disease risk factors in a white population. PLoS One 2010;5(8):e11735.

[63] Ginsberg G, Smolenski S, Neafsey P, Hattis D, Walker K, Guyton KZ, et al. The influence of genetic polymorphisms on population variability in six xenobiotic-metabolizing enzymes. J Toxicol Environ Health B Crit Rev 2009;12(5—6):307—33.

[64] Peng Y, Shi H, Qi XB, Xiao CJ, Zhong H, Ma RL, et al. The ADH1B Arg47His polymorphism in east Asian populations and expansion of rice domestication in history. BMC Evol Biol 2010;10:15.

[65] Li D, Zhao H, Gelernter J. Strong association of the alcohol dehydrogenase 1B gene (ADH1B) with alcohol dependence and alcohol-induced medical diseases. Biol Psychiatry 2011;70(6):504—12.

[66] Li H, Gu S, Cai X, Speed WC, Pakstis AJ, Golub EI, et al. Ethnic related selection for an ADH Class I variant within East Asia. PLoS One 2008;3(4):e1881.

[67] Homann N, Stickel F, Konig IR, Jacobs A, Junghanns K, Benesova M, et al. Alcohol dehydrogenase 1C*1 allele is a genetic marker for alcohol-associated cancer in heavy drinkers. Int J Cancer 2006;118(8):1998—2002.

[68] Eriksson CJ. The role of acetaldehyde in the actions of alcohol (update 2000). Alcohol Clin Exp Res 2001;25(Suppl. 5 ISBRA):15S—32S.

[69] Shibuya A, Yoshida A. Frequency of the atypical aldehyde dehydrogenase-2 gene (ALDH2(2)) in Japanese and Caucasians. Am J Hum Genet 1988;43(5):741—3.

[70] Takeshita T, Morimoto K. Self-reported alcohol-associated symptoms and drinking behavior in three ALDH2 genotypes among Japanese university students. Alcohol Clin Exp Res 1999;23(6):1065—9.

[71] Fischer M, Wetherill LF, Carr LG, You M, Crabb DW. Association of the aldehyde dehydrogenase 2 promoter polymorphism with alcohol consumption and reactions in an American Jewish population. Alcohol Clin Exp Res 2007;31(10):1654—9.

[72] Wilmshurst JM, Hunt TL, Lipo CP, Anderson AJ. High-precision radiocarbon dating shows recent and rapid initial human colonization of East Polynesia. Proc Natl Acad Sci U S A 2011;108(5):1815—20.

[73] Finney BR. Voyaging canoes and the settlement of Polynesia. Science 1977;196(4296): 1277—85.

[74] Fitzpatrick S, Callaghan R. Magellan's crossing of the Pacific. The Journal of Pacific History 2008;43(2):145—65.

[75] Nagel R, Etcheverry R, Guzman C. Haptoglobin Types in Inhabitants of Easter Island. Nature 1964;201:216—7.

[76] Melamed-Frank M, Lache O, Enav BI, Szafranek T, Levy NS, Ricklis RM, et al. Structure-function analysis of the antioxidant properties of haptoglobin. Blood 2001;98(13): 3693—8.

[77] Langlois MR, Delanghe JR, De Buyzere ML, Bernard DR, Ouyang J. Effect of haptoglobin on the metabolism of vitamin C. Am J Clin Nutr 1997;66(3):606—10.

[78] De Bacquer D, De Backer G, Langlois M, Delanghe J, Kesteloot H, Kornitzer M. Haptoglobin polymorphism as a risk factor for coronary heart disease mortality. Atherosclerosis 2001;157(1):161—6.

[79] Tye-Din JA, Stewart JA, Dromey JA, Beissbarth T, van Heel DA, Tatham A, et al. Comprehensive, quantitative mapping of T cell epitopes in gluten in celiac disease. Sci Transl Med 2010;2(41):41ra51.

[80] Caja S, Maki M, Kaukinen K, Lindfors K. Antibodies in celiac disease: implications beyond diagnostics. Cell Mol Immunol 2011;8(2):103—9.

[81] Tio M, Cox MR, Eslick GD. Meta-analysis: coeliac disease and the risk of all-cause mortality, any malignancy and lymphoid malignancy. Aliment Pharmacol Ther 2012 Mar;35(5): 540—51.

[82] Welander A, Prutz KG, Fored M, Ludvigsson JF. Increased risk of end-stage renal disease in individuals with coeliac disease. Gut 2012;61(1):64—8.

[83] Sainsbury A, Sanders DS, Ford AC. Meta-analysis: coeliac disease and hypertransaminasaemia. Aliment Pharmacol Ther 2011 Jul;34(1):33—40.

[84] Walker MM, Murray JA, Ronkainen J, Aro P, Storskrubb T, D'Amato M, et al. Detection of celiac disease and lymphocytic enteropathy by parallel serology and histopathology in a population-based study. Gastroenterology 2010;139(1):112—9.

[85] Mones RL, Yankah A, Duelfer D, Bustami R, Mercer G. Disaccharidase deficiency in pediatric patients with celiac disease and intact villi. Scand J Gastroenterol 2011;46(12): 1429—34.

[86] Sperandeo MP, Tosco A, Izzo V, Tucci F, Troncone R, Auricchio R, et al. Potential celiac patients: a model of celiac disease pathogenesis. PLoS One 2011;6(7):e21281.

[87] Biagi F, Bianchi PI, Vattiato C, Marchese A, Trotta L, Badulli C, et al. Influence of HLA-DQ2 and DQ8 on severity in celiac disease. J Clin Gastroenterol 2012;46(1):46—50.

[88] Trynka G, Hunt KA, Bockett NA, Romanos J, Mistry V, Szperl A, et al. Dense genotyping identifies and localizes multiple common and rare variant association signals in celiac disease. Nat Genet 2011;43(12):1193—201.

[89] Pascolo P, Faleschini E, Tonini G, Ventura A. Type 1 diabetes mellitus and celiac disease: usefulness of gluten-free diet. Acta Diabetol 2011 [Epub ahead of print].

[90] Cappellini MD, Fiorelli G. Glucose-6-phosphate dehydrogenase deficiency. Lancet 2008;371(9606):64—74.

[91] Mehta A, Mason PJ, Vulliamy TJ. Glucose-6-phosphate dehydrogenase deficiency. Baillieres Best Pract Res Clin Haematol 2000;13(1):21—38.

[92] Nkhoma ET, Poole C, Vannappagari V, Hall SA, Beutler E. The global prevalence of glucose-6-phosphate dehydrogenase deficiency: a systematic review and meta-analysis. Blood Cells Mol Dis 2009;42(3):267—78.

[93] Hedrick PW. Population genetics of malaria resistance in humans. Heredity 2011 Oct;107(4): 283—304.

[94] McMillan DC, Schey KL, Meier GP, Jollow DJ. Chemical analysis and hemolytic activity of the fava bean aglycon divicine. Chem Res Toxicol 1993;6(4):439—44.

[95] Brandt O, Rieger A, Geusau A, Stingl G. Peas, beans, and the Pythagorean theorem—the relevance of glucose-6-phosphate dehydrogenase deficiency in dermatology. J Dtsch Dermatol Ges 2008;6(7):534—9.

[96] Chango A, Abdennebi-Najar L. Folate metabolism pathway and *Plasmodium falciparum* malaria infection in pregnancy. Nutr Rev 2011;69(1):34—40.

[97] Asawamahasakda W, Yuthavong Y. The methionine synthesis cycle and salvage of methyltetrahydrofolate from host red cells in the malaria parasite (*Plasmodium falciparum*). Parasitology 1993;107(Pt 1):1—10.

[98] Wilcken B, Bamforth F, Li Z, Zhu H, Ritvanen A, Renlund M, et al. Geographical and ethnic variation of the 677C>T allele of 5,10 methylenetetrahydrofolate reductase (MTHFR): findings from over 7000 newborns from 16 areas world wide. J Med Genet 2003;40(8): 619—25.

[99] Thiagarajah JR, Verkman AS. New drug targets for cholera therapy. Trends Pharmacol Sci 2005;26(4):172—5.

[100] Sonawane ND, Hu J, Muanprasat C, Verkman AS. Luminally active, nonabsorbable CFTR inhibitors as potential therapy to reduce intestinal fluid loss in cholera. FASEB J 2006;20(1): 130—2.

[101] van de Vosse E, Ali S, de Visser AW, Surjadi C, Widjaja S, Vollaard AM, et al. Susceptibility to typhoid fever is associated with a polymorphism in the cystic fibrosis transmembrane conductance regulator (CFTR). Hum Genet 2005;118(1):138–40.

[102] Dean M, Carrington M, O'Brien SJ. Balanced polymorphism selected by genetic versus infectious human disease. Annu Rev Genomics Hum Genet 2002;3:263–92.

[103] Cohen-Cymberknoh M, Shoseyov D, Kerem E. Managing cystic fibrosis: strategies that increase life expectancy and improve quality of life. Am J Respir Crit Care Med 2011;183(11): 1463–71.

[104] Bombieri C, Claustres M, De Boeck K, Derichs N, Dodge J, Girodon E, et al. Recommendations for the classification of diseases as CFTR-related disorders. J Cyst Fibros 2011;10(Suppl. 2):S86–102.

[105] Sathe MN, Patel AS. Update in pediatrics: focus on fat-soluble vitamins. Nutr Clin Pract 2010;25(4):340–6.

[106] Liu P, Shah BP, Croasdell S, Gilbertson TA. Transient receptor potential channel type M5 is essential for fat taste. J Neurosci 2011;31(23):8634–42.

[107] Galindo MM, Voigt N, Stein J, van Lengerich J, Raguse JD, Hofmann T, et al. G protein-coupled receptors in human fat taste perception. Chem Senses 2012 Feb;37(2): 123–39.

[108] Ming D, Ninomiya Y, Margolskee RF. Blocking taste receptor activation of gustducin inhibits gustatory responses to bitter compounds. Proc Natl Acad Sci U S A 1999;96(17): 9903–8.

[109] Kamei K, Takano R, Miyasaka A, Imoto T, Hara S. Amino acid sequence of sweet-taste-suppressing peptide (gurmarin) from the leaves of Gymnema sylvestre. J Biochem 1992;111(1):109–12.

[110] Yasumatsu K, Ohkuri T, Sanematsu K, Shigemura N, Katsukawa H, Sako N, et al. Genetically-increased taste cell population with G(alpha)-gustducin-coupled sweet receptors is associated with increase of gurmarin-sensitive taste nerve fibers in mice. BMC Neurosci 2009;10:152.

[111] Kojima I, Nakagawa Y. The role of the sweet taste receptor in enteroendocrine cells and pancreatic beta-cells. Diabetes Metab J 2011;35(5):451–7.

[112] Finger TE, Kinnamon SC. Taste isn't just for taste buds anymore. F1000 Biol Rep 2011;3:20.

[113] Iguchi N, Ohkuri T, Slack JP, Zhong P, Huang L. Sarco/endoplasmic reticulum Ca-ATPases (SERCA) contribute to GPCR-mediated taste perception. PLoS One 2011;6(8):e23165.

[114] Kim U, Wooding S, Ricci D, Jorde LB, Drayna D. Worldwide haplotype diversity and coding sequence variation at human bitter taste receptor loci. Hum Mutat 2005;26(3):199–204.

[115] Reed DR, Zhu G, Breslin PA, Duke FF, Henders AK, Campbell MJ, et al. The perception of quinine taste intensity is associated with common genetic variants in a bitter receptor cluster on chromosome 12. Hum Mol Genet 2010;19(21):4278–85.

[116] Feeney E, O'Brien S, Scannell A, Markey A, Gibney ER. Genetic variation in taste perception: does it have a role in healthy eating? Proc Nutr Soc 2011;70(1):135–43.

[117] Sacerdote C, Guarrera S, Smith GD, Grioni S, Krogh V, Masala G, et al. Lactase persistence and bitter taste response: instrumental variables and Mendelian randomization in epidemiologic studies of dietary factors and cancer risk. Am J Epidemiol 2007;166(5):576–81.

[118] Duffy VB, Hayes JE, Davidson AC, Kidd JR, Kidd KK, Bartoshuk LM. Vegetable intake in college-aged adults is explained by oral sensory phenotypes and TAS2R38 genotype. Chemosens Percept 2010;3(3–4):137–48.

[119] Lalueza-Fox C, Gigli E, de la Rasilla M, Fortea J, Rosas A. Bitter taste perception in Neanderthals through the analysis of the TAS2R38 gene. Biol Lett 2009;5(6):809–11.

[120] Wooding S, Gunn H, Ramos P, Thalmann S, Xing C, Meyerhof W. Genetics and bitter taste responses to goitrin, a plant toxin found in vegetables. Chem Senses 2010;35(8):685–92.

[121] Roudnitzky N, Bufe B, Thalmann S, Kuhn C, Gunn HC, Xing C, et al. Genomic, genetic and functional dissection of bitter taste responses to artificial sweeteners. Hum Mol Genet 2011;20(17):3437–49.

[122] Li X, Li W, Wang H, Bayley DL, Cao J, Reed DR, et al. Cats lack a sweet taste receptor. J Nutr 2006;136(Suppl. 7):1932S–4S.

[123] Li X, Glaser D, Li W, Johnson WE, O'Brien SJ, Beauchamp GK, et al. Analyses of sweet receptor gene (Tas1r2) and preference for sweet stimuli in species of Carnivora. J Hered 2009;100(Suppl. 1):S90–100.

[124] Sakurai T, Misaka T, Ueno Y, Ishiguro M, Matsuo S, Ishimaru Y, et al. The human bitter taste receptor, hTAS2R16, discriminates slight differences in the configuration of disaccharides. Biochem Biophys Res Commun 2010;402(4):595–601.

[125] Bachmanov AA, Bosak NP, Floriano WB, Inoue M, Li X, Lin C, et al. Genetics of sweet taste preferences. Flavour Fragr J 2011;26(4):286–94.

[126] Fushan AA, Simons CT, Slack JP, Manichaikul A, Drayna D. Allelic polymorphism within the TAS1R3 promoter is associated with human taste sensitivity to sucrose. Curr Biol 2009;19(15):1288–93.

[127] Eny KM, Wolever TM, Corey PN, El-Sohemy A. Genetic variation in TAS1R2 (Ile191Val) is associated with consumption of sugars in overweight and obese individuals in 2 distinct populations. Am J Clin Nutr 2010;92(6):1501–10.

[128] Fushan AA, Simons CT, Slack JP, Drayna D. Association between common variation in genes encoding sweet taste signaling components and human sucrose perception. Chem Senses 2010;35(7):579–92.

[129] Shigemura N, Shirosaki S, Sanematsu K, Yoshida R, Ninomiya Y. Genetic and molecular basis of individual differences in human umami taste perception. PLoS One 2009;4(8):e6717.

[130] Raliou M, Grauso M, Hoffmann B, Schlegel-Le-Poupon C, Nespoulous C, Debat H, et al. Human genetic polymorphisms in T1R1 and T1R3 taste receptor subunits affect their function. Chem Senses 2011;36(6):527–37.

[131] Chen QY, Alarcon S, Tharp A, Ahmed OM, Estrella NL, Greene TA, et al. Perceptual variation in umami taste and polymorphisms in TAS1R taste receptor genes. Am J Clin Nutr 2009;90(3):770S–9S.

[132] Ichise T, Kano M, Hashimoto K, Yanagihara D, Nakao K, Shigemoto R, et al. mGluR1 in cerebellar Purkinje cells essential for long-term depression, synapse elimination, and motor coordination. Science 2000;288(5472):1832–5.

[133] Simon L, Toth J, Molnar L, Agoston DV. MRI analysis of mGluR5 and mGluR1 antagonists, MTEP and R214127 in the cerebral forebrain of awake, conscious rats. Neurosci Lett 2011;505(2):155–9.

[134] Sun H, Neugebauer V. mGluR1, but not mGluR5, activates feed-forward inhibition in the medial prefrontal cortex to impair decision making. J Neurophysiol 2011;106(2):960–73.

[135] Downey PM, Petro R, Simon JS, Devlin D, Lozza G, Veltri A, et al. Identification of single nucleotide polymorphisms of the human metabotropic glutamate receptor 1 gene and pharmacological characterization of a P993S variant. Biochem Pharmacol 2009;77(7):1246–53.

[136] Zhao H, Yang JR, Xu H, Zhang J. Pseudogenization of the umami taste receptor gene Tas1r1 in the giant panda coincided with its dietary switch to bamboo. Mol Biol Evol 2010;27(12):2669–73.

[137] Martin C, Passilly-Degrace P, Gaillard D, Merlin JF, Chevrot M, Besnard P. The lipid-sensor candidates CD36 and GPR120 are differentially regulated by dietary lipids in mouse taste buds: impact on spontaneous fat preference. PLoS One 2011;6(8):e24014.

[138] Keller KL, Liang LC, Sakimura J, May D, van Belle C, Breen C, et al. Common variants in the CD36 gene are associated with oral fat perception, fat preferences, and obesity in african americans. Obesity (Silver Spring) 2012 May;20(5):1066–73.

[139] Pepino MY, Love-Gregory L, Klein S, Abumrad NA. The fatty acid translocase gene, CD36, and lingual lipase influence oral sensitivity to fat in obese subjects. J Lipid Res 2012 Mar;53(3):561–6.

[140] Wise PM, Hansen JL, Reed DR, Breslin PA. Twin study of the heritability of recognition thresholds for sour and salty taste. Chem Senses 2007;32(8):749–54.

[141] Ayya N, Beauchamp GK. Short-term effects of diet on salt taste preference. Appetite 1992;18(1):77–82.

[142] Stein LJ, Cowart BJ, Epstein AN, Pilot LJ, Laskin CR, Beauchamp GK. Increased liking for salty foods in adolescents exposed during infancy to a chloride-deficient feeding formula. Appetite 1996;27(1):65–77.

[143] Amoore J, Forrester L. Anosmia to trimethylamine: The primary fishy odor. J Chem Ecol 1976;2:49–56.

[144] Rehman HU. Fish odor syndrome. Postgrad Med J 1999;75(886):451–2.

[145] Menashe I, Man O, Lancet D, Gilad Y. Different noses for different people. Nat Genet 2003;34(2):143–4.

[146] Hasin Y, Olender T, Khen M, Gonzaga-Jauregui C, Kim PM, Urban AE, et al. High-resolution copy-number variation map reflects human olfactory receptor diversity and evolution. PLoS Genet 2008;4(11):e1000249.

[147] Waring RH, Mitchell SC, Fenwick GR. The chemical nature of the urinary odour produced by man after asparagus ingestion. Xenobiotica 1987;17(11):1363–71.

[148] Pelchat ML, Bykowski C, Duke FF, Reed DR. Excretion and perception of a characteristic odor in urine after asparagus ingestion: a psychophysical and genetic study. Chem Senses 2011;36(1):9–17.

[149] Eriksson N, Macpherson JM, Tung JY, Hon LS, Naughton B, Saxonov S, et al. Web-based, participant-driven studies yield novel genetic associations for common traits. PLoS Genet 2010;6(6):e1000993.

[150] Borrud LG, Flegal KM, Looker AC, Everhart JE, Harris TB, Shepherd JA. Body composition data for individuals 8 years of age and older: US population, 1999–2004. Vital Health Stat 2010;11(250):1–87.

[151] Dasilva SG, Guidetti L, Buzzachera CF, Elsangedy HM, Krinski K, De Campos W, et al. Gender-based differences in substrate use during exercise at a self-selected pace. J Strength Cond Res 2011;25(9):2544–51.

[152] Voorhoeve PG, van Mechelen W, Uitterlinden AG, Delemarre-van de Waal HA, Lamberts SW. Estrogen receptor-alpha gene polymorphisms and body composition in children and adolescents. Horm Res Paediatr 2011;76(2):86–92.

[153] da Costa KA, Kozyreva OG, Song J, Galanko JA, Fischer LM, Zeisel SH. Common genetic polymorphisms affect the human requirement for the nutrient choline. FASEB J 2006;20(9):1336–44.

[154] Resseguie ME, da Costa KA, Galanko JA, Patel M, Davis IJ, Zeisel SH. Aberrant estrogen regulation of PEMT results in choline deficiency-associated liver dysfunction. J Biol Chem 2011;286(2):1649–58.

[155] Hughes JF, Skaletsky H, Pyntikova T, Graves TA, van Daalen SK, Minx PJ, et al. Chimpanzee and human Y chromosomes are remarkably divergent in structure and gene content. Nature 2010;463(7280):536–9.

[156] Hedelin M, Balter KA, Chang ET, Bellocco R, Klint A, Johansson JE, et al. Dietary intake of phytoestrogens, estrogen receptor-beta polymorphisms and the risk of prostate cancer. Prostate 2006;66(14):1512–20.

[157] Dias AG, Rousseau D, Duizer L, Cockburn M, Chiu W, Nielsen D, El-Sohemy A. Genetic variation in putative salt taste receptors and salt taste perception in humans. Chem Senses 2013 [Epub ahead of print] doi:10.1093/chemse/bjs090.

CHAPTER 4

How Nutrients are Affected by Genetics

Terms of the Trade

- Allozyme: Enzyme with altered amino acid sequence corresponding to a variant allele.
- Expressivity: Range of symptoms and outcomes associated with a particular genotype.
- Haploinsufficiency: Loss of one of the two gene copies normally available.
- Homolog: Gene that is functionally and structurally similar to a gene in a different species.
- Inborn error of metabolism: Referring to the grossly defective processing of a metabolite.
- Penetrance: Percentage of people with a particular genotype that show an expected outcome.

ABSTRACT

For all practical purposes, nutrigenetics is the science of inborn responsiveness to specific nutritional factors. This chapter will highlight examples of genetic variants that modify the effects of nutrient and other food components. It is not the intention, nor would it be possible within this book's limited space and the reader's available time, to provide a comprehensive list of all reported interactions. There are just too many of them and the list grows daily. Instead, there will be a review of selected pathways and mechanisms that explain how people with one genetic variant respond in a certain way to nutrients and food ingredients, while those without the variant respond differently. You might want to remember a simple rule of thumb: if a variant is rare, something probably went wrong and the variant did not spread more widely. If a variant is common, something went right and an advantage of some kind helped to spread the variant. It is important to recognize that some of the genotype-specific responses to nutrients mentioned have been observed in only one study and need further investigation.

Nutrigenetics. http://dx.doi.org/10.1016/B978-0-12-385900-6.00004-6

4.1 IF SOMETHING CAN GO WRONG IT WILL

4.1.1 Inborn errors of metabolism versus common genetic variants

Genetic susceptibility and nutrition exposure together cause many diseases and health problems. The frequency of genetic predispositions with nutritional implications ranges from extremely rare to extremely common. There is a simple reason why some genetic variants are rare and others are common. The fidelity of our DNA duplication mechanisms is amazingly high, with more than 99.999998% correctly copied bases with each cell division. However, this still means that in the hours after conception more than 50 mutations will occur in each of us. These random mutations can affect any of the more than 6.6 billion bases in our genome, and the resulting sequence changes in genes and other important parts of our genome will be transmitted to the next generation.

A REMINDER

A mutation that occurs after the first few cell divisions will be present in only a small minority of cells of the adult body.

Each of the genes we need for the processing, use, and elimination of nutrients and their waste products will be found to be broken in some rare cases. In fact, many of these genes were originally identified because they were defective in somebody with a rare condition. Defects in some of these nutritionally relevant genes have such catastrophic consequences that the pregnancy comes to a premature end or causes ultimately fatal illness during childhood. Since these mutations are not transmitted to the next generation, they can never become common. Mutations in the X chromosomes of males tend to fall into this category of most harmful mutations because there is no second backup gene copy. Most other mutations have little or no direct effect on the health of the first-generation carrier. A detrimental effect occurs only when someone in later generations happens to inherit defective copies from both mother and father. This is the reason why children of parents with closely shared ancestry are at an increased risk of inherited disease. As explained earlier, the child of first cousins, for instance, has a 12.5% chance of inheriting from both parents the very same mutation that arose originally in the parents' shared grandparent. This is the unfortunate time when the consequences of a mutation may come to bear. The pioneers of human genetic research observed that the majority of people with inherited disease were the offspring of first cousins [1]. It is easy to see that disadvantageous mutations tend to die out after a number of generations if they limit the carrier's number of children and grandchildren. It follows that polymorphisms, which by definition occur in at least 1% of a sizeable population, must have provided an advantage to ancestral generations (or be closely linked to such an advantageous polymorphism). The benefit usually accrues to heterozygous carriers because they are the most common and therefore most relevant at a population level. Homozygous carriers may also have an advantage but serious, sometimes even fatal, disadvantages are not uncommon.

Archibald Garrod laid the foundation of modern nutrigenetics when he spoke before the Royal College of Physicians in 1908. In his first Croonian Lecture he called it *chemical individuality* and said, "Even those idiosyncrasies with regard to drugs and articles of food which are summed up in the proverbial saying that what is one man's meat is another man's poison presumably have a chemical basis." [2] One of the genetic abnormalities presented in his lecture was alkaptonuria (OMIM 203500). The fact that skin and ear cartilage of affected individuals have a bluish tint and that their urine conspicuously darkens when left standing drew attention to this rare medical condition.

Among the more curious features is that patients tend to have black earwax. Garrod thought that 'alkaptonuria is not the manifestation of disease, but is rather of the nature of an alternative course of metabolism, harmless

> Even those idiosyncrasies with regard to drugs and articles of food which are summed up in the proverbial saying that what is one man's meat is another man's poison presumably have a chemical basis.

Garrod's prescient sentence ushered in the era of nutrigenetic science.

and lifelong.' [1]. Thus, the concept of inborn error of metabolism was born. These are conditions with an evident change in the normal processing of important chemicals in the human body. We now know that many of these metabolic alterations come from changes in genes needed for the processing and use of nutrients and their breakdown products. Garrod was unfortunately wrong about the harmlessness of the condition, which causes painful and debilitating cartilage damage in the spine, shoulders, and hips [3]. To this day, the nutritional options for minimizing the harmful effects of alkaptonuria are limited. A generous intake of vitamin C (as much as 1000 mg/day) can prevent formation of the toxic metabolite benzoquinone acetate [4]. Limiting the consumption of tyrosine reduces the production of homogentisic acid [5], but is difficult to implement in practice.

The scientific significance is that alkaptonuria was the first human disease to which Mendel's laws of inheritance were explicitly applied. A recessive mode of inheritance modeled the genetic transmission closely since the alkaptonuria was observed neither in the parents nor in children of individuals with the disease. Alkaptonuria is caused by the lack of homogentisate 1,2-dioxygenase (EC 1.13.11.5) activity (encoded by the gene *HGD*; OMIM 203500). People with two defective *HGD* copies [6] cannot properly metabolize the amino acid tyrosine (Figure 4.1). The tyrosine metabolite, homogentisic acid (called alkapton in Garrod's time), and its melanin-like oxidation products (ochronotic pigment) are excreted in significant amounts in urine and accumulate in connective tissues.

Key concept

The main question concerning us here is whether either rare genetic defects or common genetic variants respond to specific nutrition interventions. In its own way, each nutrition-response abnormality illustrates critical steps for the processing of the body's fuel and building blocks.

In regard to the rare recessive defects, we need to ask whether the heterozygous carriers have a mildly altered response to nutrition. Often we simply don't yet know this and should be interested in further exploration of that possibility.

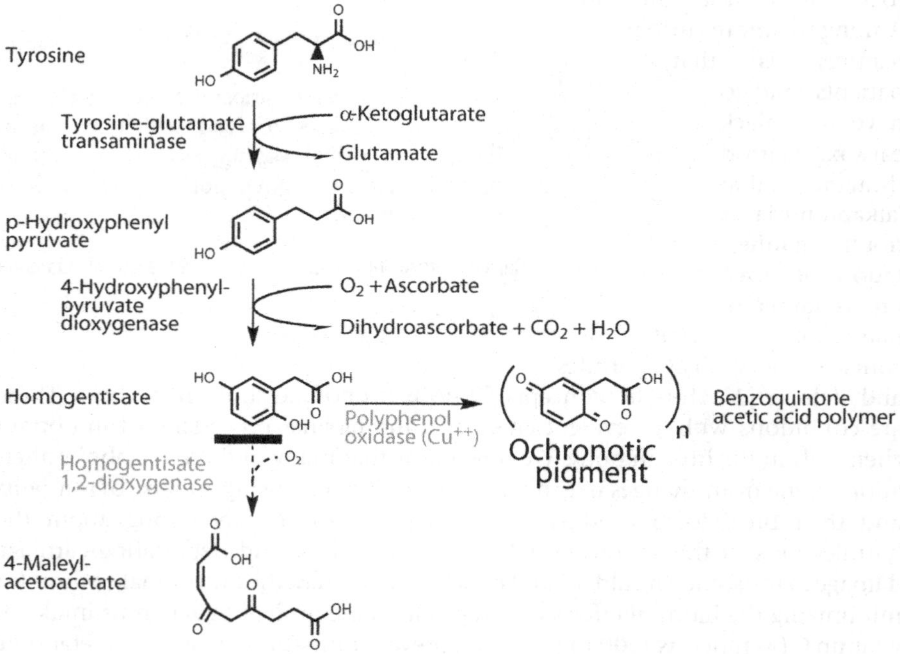

FIGURE 4.1
Alkaptonuria was recognized early on because the urine of carriers turns dark brown after a while. The urine color is due to ochronotic pigment that develops when people with defective homogentisate 1,2-dioxygenase excrete large amounts of the tyrosine metabolite homogentisate.

4.1.2 Newborn metabolic testing

Currently, all states (with a very small number of exceptions) in the USA mandate newborn screening for all forms of biotinidase deficiency, cystic fibrosis (CF), galactosemia (GALT), homocystinuria, maple syrup urine disease (*MSUD*; OMIM 248600), medium-chain acyl-CoA dehydrogenase (MCAD), phenylketonuria (PKU) and hyperphenylalaninemia, three more amino acid disorders, another four fatty acid disorders, and nine organic acid disorders (http://genes-r-us. uthscsa.edu/). The screening programs are usually organized by the state health departments. Many other countries around the world have similarly extensive newborn screening programs. Most of the inborn errors of metabolism that may be detected are very rare but respond to some degree to nutrition therapy, prevent lifelong suffering, and save money. More than four million newborn infants per year are screened in the USA alone [7].

Newborn screening

The goal is to test each child immediately after birth and have the test results available within a few days to prevent irreversible harm from exposure to feeding. How is it done?

Typically, the heel of the newborn infant is pricked with a retractable, single-use lancet, a few drops of blood are spotted on a test card, and the blood is then left to dry. In the laboratory, little circles are then punched from a dried blood spot on the card and used for analysis.

mass spectrometry (MS/MS, to indicate that it involves two mass spectrometric separation stages). This method measures the concentration of phenylalanine as well as the ratio of phenylalanine to tyrosine and generates virtually no false-positive results.

The same MS/MS analysis can also detect about 40 other inborn errors of metabolism. This is notable because several of them are responsive to nutritional intervention, including most cases of MSUD, short-chain

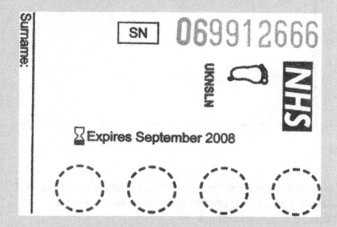

Test cards are usually made of absorbent material and have designated areas for blood drops and the infant's information.

The Guthrie test, a semiquantitative bacterial inhibition assay, was the original method to determine whether a child has PKU. This time-honored test has limited sensitivity and specificity. The more recently introduced fluorometric test does better on both accounts. The most recent standard, which is now routinely used for newborn screening in the USA and many other countries, is tandem

acyl-CoA dehydrogenase deficiency (SCADD), and medium-chain acyl-CoA dehydrogenase deficiency (MCADD). One of the major advances with this form of screening is that it can detect many milder forms of these metabolic disorders, which remained undetected and untreated with previous methods.

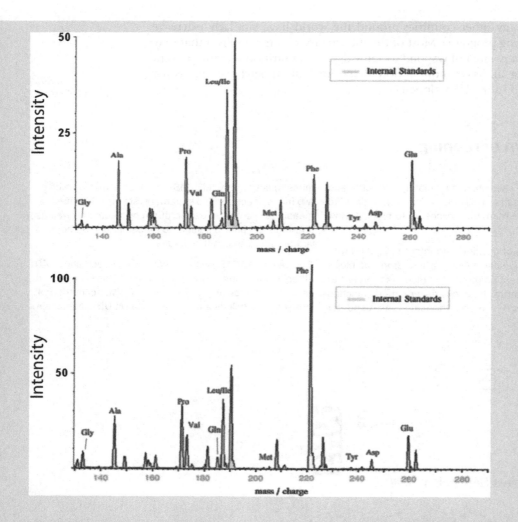

Tandem mass spectrometry separates compounds in blood by their molecular mass. This permits the quantitative measurement of more than a hundred important metabolites in blood. In this example from a child with PKU (spectrum at the bottom), the phenylalanine concentration is much higher than in an unaffected child (spectrum on top).
Mass spectra courtesy of Dr. David Millington, Duke University, North Carolina.

4.2 PROTEINS AND AMINO ACIDS
4.2.1 Protein preference

It has been known for a long time (Figure 4.2) that some people just don't like to eat a lot of protein. Aversion to protein-rich foods due to low ornithine transcarbamoylase activity (described in detail below) is a possible, but rare, explanation. A much more likely reason may be a common variant (102 T>C; rs6313) of the *serotonergic receptor 2A* (*HTR2A*; OMIM 182135), which is known to play a key role in appetite regulation. The approximately 20% of individuals with the *HTR2A* T/T genotype appear to consume more protein and more essential amino acids than others [8].

Jack Sprat could eat no fat,
His wife could eat no lean,
And so between them both,
They lick'd the platter clean.

FIGURE 4.2
Did people centuries ago really know something about inborn nutritional differences or did they just refer to different habits? Did Mrs. Sprat have a strong aversion to protein-rich foods because of a urea cycle defect which could have killed her, or was it maybe an *HTR2A* variant? *Unknown author. The life of Jack Sprat, his wife & cat. Batchelar, Printer, London, 1810.*

But the importance of *HTR2A* extends far beyond simple intake regulation and defines our personality to some extent. This gene influences many emotional traits, including mood, obsessive-compulsive predisposition, risk behavior, and sexual preferences [9]. The *HTR2A* T/T genotype also predisposes to smoking tobacco [10]. This is mentioned just to emphasize how entrenched the mechanisms are that give rise to food preferences and that they go beyond just a simple choice between meat and potatoes.

4.2.2 Urea cycle defects

Ornithine transcarbamoylase (OTC; EC 2.1.3.3) deficiency (OMIM 311250) is the most common genetic cause of defective ammonia removal through the urea cycle. The enzyme combines carbamoyl phosphate with ornithine to generate citrulline. Because the gene is located on the X chromosome, the metabolic effects of a gene defect are much more serious in males and cause their death in early childhood. If females have a defective *OTC* gene, they can still fall back on the unaffected copy on their second X chromosome. But because X chromosome inactivation is often unbalanced, they may have more liver cells with the defective variant than with the normally functioning *OTC* copy. As many as one in 14,000 otherwise healthy women appear to have genetically low OTC activity [11]. In most of them the condition goes unrecognized and untreated until a metabolic decompensation with excessive ammonia accumulation occurs. Such metabolic decompensation tends to be very dramatic and is often fatal. Episodes of ammonia accumulation (hyperammonemia) may be triggered by unaccustomed high protein intake, such as with the Atkins diet [12]. Many people with reduced OTC activity have an aversion to meat and other protein-rich foods, often preferring a vegetarian lifestyle [13]. The abnormal sensitivity to a high protein load can be assessed by a standardized challenge with 35 g protein per m^2 body surface and measurement of orotic acid excretion in urine [14]. Orotic acid synthesis starts from carbamoyl phosphate and when OTC works inefficiently, excess orotic acid spills over into urine.

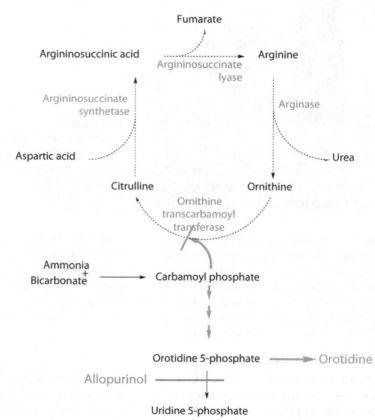

FIGURE 4.3
The allopurinol test uses the fact that this antigout medication inhibits the conversion of orotidine 5-phosphate to uridine 5-phosphate, which makes the alternative metabolite orotidine overflow in people with a urea cycle defect and becoming detectable in urine.

Detection of low OTC activity status is difficult without full sequencing of the *OTC* gene and neighboring regions. Measurements of metabolites in blood or urine have limited sensitivity on their own. An alternative metabolic test requires that patients with suspected OTC deficiency take the antigout medication allopurinol to enhance orotidine excretion and thereby get a stronger metabolite signal [15]. Low citrulline concentration in blood has been considered for newborn screening, but this method will miss most late-onset urea cycle disorders [11]. The idea behind this metabolomic screen is that individuals with defective OTC or other urea cycle disorders will produce little citrulline, the product of the OTC enzyme (Figure 4.3).

Events that accelerate the breakdown of tissue protein, such as fasting [16], malnutrition, pregnancy, and surgery for weight loss, can similarly cause a sudden rise in ammonia concentrations that will lead to brain damage and death without determined and rapid intervention. Emergency treatment with intravenous benzoic acid (reduces ammonia production by combining with glycine to form directly excretable hippurate) and arginine is usually effective. Curbing protein intake and using benzoate, phenyl acetate, or phenyl butyrate [17], if necessary, can maintain the health of individuals with moderate OTC deficiency.

4.2.3 Phenylketonuria

Our daily need for the essential amino acid L-phenylalanine for protein synthesis is about 10 mg/kg body weight [18]. Phenylalanine is also the precursor of the amino acid L-tyrosine, which itself is needed for the synthesis of proteins, compounds acting as neurotransmitters and hormones (adrenaline, noradrenaline, and dopamine), and melanin. The conversion of phenylalanine to tyrosine removes excess phenylalanine and supplements tyrosine supplies. The enzyme phenylalanine hydroxylase (*PHA*; OMIM 612349), which is responsible for this conversion, uses tetrahydropterin (BH4) as a cofactor (Figure 4.4).

The condition caused by defective conversion of phenylalanine to tyrosine is PKU (OMIM 261600) because the urine of affected patients contains large amounts of the phenylketone phenylpyruvate and other abnormal Phe metabolites. About one in 15,000 infants in the USA and Canada is born with defective phenylalanine hydroxylase (PAH) [19]. Infants of Turkish and Irish ancestry are affected most often, and infants with African, Japanese, and Ashkenazi Jewish heritage are affected least often. Some PAH variants encode allozymes with some residual activity and cause less severe conditions. Classical (severe) PKU is defined by phenylalanine blood concentrations above 1200 μmol/L. Milder forms of

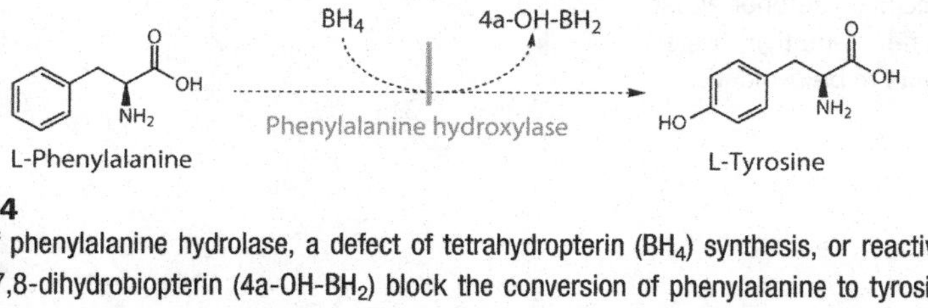

FIGURE 4.4
Disruption of phenylalanine hydrolase, a defect of tetrahydropterin (BH$_4$) synthesis, or reactivation of 4a-hydroxy-7,8-dihydrobiopterin (4a-OH-BH$_2$) block the conversion of phenylalanine to tyrosine. The resulting disorder phenylketonuria rapidly causes damage to the brain after birth.

PKU treatment

Phenylalanine Restriction

Children and adolescents with complete or partial PHA deficiency need to keep their phenylalanine intake low enough to keep their phenylalanine blood concentration under 360 µmol/L. There is no question that guidelines have to be followed strictly at least into adulthood. Current opinion is moving toward a recommendation to maintain this low-phenylalanine diet throughout life. It is particularly critical that women keep their phenylalanine blood concentration low before and during pregnancies, because phenylalanine concentrations above normal increase the risk of birth defects and stunt normal brain development [235]. Untreated PKU patients are about four times as likely as treated patients to have children with low birth weight, cognitive impairment, a small head (and brain) size (microcephaly), and a deformed face (facial dysmorphies).

Another important medical consideration is the altered response to medications. Some antipsychotic medications are less effective in PKU patients with high phenylalanine:tyrosine ratios, because they target tyrosine-derived neurotransmitters.

Low phenylalanine intake is achieved with a combination of phenylalanine free formulas (Milupa PKU 2, Periflex Infant PKU, Phenex-1, PKU Anamix, Phenyl-Free, PhenylAde, and others) and semisynthetic proteins (Camino PRO bar, Milupa lp-bar, PhenylAde Amino Acid bar, Phlexy-10 bar, and PKU tangles snack puffs) with tightly controlled amounts of low-phenylalanine foods (fruits and vegetables, and grain products). There are specialty products, including pasta and cereals, that are designed to contained very little phenylalanine. Breast milk and regular formula, dairy products, eggs, fish, meats, peas, soy, and other high-protein foods contain too much phenylalanine to be compatible with a low-phenylalanine diet. Chocolate and beer also fall into that incompatible category. The artificial sweetener aspartame (N-(L-α-Aspartyl)-L-phenylalanine 1-methyl ester) contains 56% phenylalanine and needs to be avoided.

Limiting phenylalanine intake for years on end is particularly difficult for affected children and adults. The cost of the necessary products and low-protein foods runs into several tens of thousands of dollars per year [21] and meals take extra time to prepare. It is very important to ensure extensive psychosocial support, just like for all children (and adults) with severe dietary constraints.

Amino Acid Supplementation

Adequate tyrosine intake is important, because the majority of this amino acid normally comes from conversion of phenylalanine. High phenylalanine:tyrosine ratios may reduce dopamine synthesis. This theoretical consideration has prompted some to advocate for a generous tyrosine allowance with additional supplements, but the very limited clinical evidence so far does not support adding tyrosine [26]. Some specialists measure tyrosine blood concentrations at frequent intervals (some monthly) and adjust tyrosine supplements accordingly. Additional tryptophan is also suggested by some [27].

Children need to get adequate amounts of total amino acids as well as of each individual essential amino acid. This is particularly important, as well as difficult, during diseases and growth spurts. Growth impairment is a definite risk with such restrictive diets.

Biopterin Supplementation

For patients with defects in genes related to biopterin synthesis and metabolism, supplemental biopterin may restore normal PHA activity. In many cases, patients with mild PKU and benign hyperalaninemia also respond well, and some individuals with classical PKU even show benefit. The currently approved biopterin derivative sapropterin (Kuvan) can be used at daily doses of up to 20 mg/kg. The cost of the treatment can be up to US$200,000/year [21].

PKU exist when the concentration is between 600 and 1200 µmol/L. If the phenylalanine blood concentration is below 600 µmol/L but still above the normal concentration of 200 µmol/L, we are dealing with non-PKU mild hyperphenylalaninemia [20]. Disease-indicating toxic metabolites are excreted in urine sometimes only if the phenylalanine blood concentration exceeds 600 µmol/L and only several weeks after birth [21]. The milder conditions still require careful treatment because even a moderate elevation of phenylalanine concentration is likely to cause cognitive damage as described below.

A third group of gene variants affect genes needed for the synthesis and metabolism of the PAH cofactor biopterin. It is important to distinguish this group of gene variants from PAH defects because treatment with supplemental biopterin is often particularly effective [22,23].

The abnormal phenylalanine metabolites cause irreversible damage to the central nervous system within days after birth. Unchecked high phenylalanine concentrations in the absence of dietary treatment are the most harmful, of course. However, some cognitive limitations may also develop despite rigorous treatment [24]. It has been suggested that each 100 µmol/L phenylalanine concentration excess prior to puberty cause a two-point intelligence quotient (IQ) deficit. It is important to remember that potential harm comes not only from excess phenylalanine and its toxic metabolites but at the same time from a lack of the essential phenylalanine metabolite tyrosine.

4.2.4 Cystinuria

Garrod included cystinuria among his original four inborn errors of metabolism. With a prevalence of about one in 7000 newborns, this condition is among the most common monogenic metabolic disorders. Cystinuria is characterized by abnormal loss of cystine into urine, which promotes the formation of kidney stones with a high cystine content.

The causes of the condition are variants in the genes *SLC3A1* (OMIM 104614) and *SLC7A9* (OMIM 604144), which encode the two halves of the b0,+ amino acid transport system for neutral and dibasic amino acids (the spelling of the b0,+ designation is a bit cumbersome, but we'll have to live with it).

Nutritional treatment focuses on producing dilute and slightly alkaline urine by drinking more plain water. This reduces the concentration of cystine in urine and makes crystal formation less likely. Medications for treatment of cystinuria include chelating agents (bucillamine, alpha-mercaptopropionylglycine, and D-penicillamine) and captopril.

4.2.5 Hartnup disease

About one in 30,000 infants are born with Hartnup disease (OMIM 234500). This condition arises from the insufficient function of the B(0)AT1 (*SLC6A19*; OMIM 608893) transport system for neutral amino acids in the brain, kidneys, and small intestine. This means that neutral amino acids are not absorbed well

Cystinuria treatment

Fluid Intake

The solubility of cystine in water is less than 300 mg/L at pH 7. Since patients with cystinuria excrete significantly more than 600 mg/day, the goal is to produce urine volume well in excess of 2.5 L/day. This target volume usually requires consumption of more than 4 L/day, particularly in warm weather. Consumption of low-sodium carbonated mineral water and citrus juices can be helpful. The specific gravity of urine should be kept under 1.010 kg/L. Specialty paper strips can measure both specific gravity and pH (for instance, StoneGuard II and UriDynamics).

Alkalinization of Urine

Urine pH should be maintained at 7.0–7.5 using potassium citrate supplements (60–80 mmol/day) and checking pH daily with pH indicator paper. The potassium citrate tablets should be taken with breakfast, lunch, and dinner, each time whole with a large glass of water. Potassium-based salt substitutes should not be used by people using potassium citrate. Excessive potassium concentration in blood is a potential side effect and blood potassium concentrations need to be monitored in people with reduced kidney function or with some medications such as potassium-sparing diuretics and ACE inhibitors.

L-Tryptophan

L-Phenylalanine

L-Tyrosine

from food in the small intestine and not recovered well from ultrafiltrate in the kidneys [27]. The abnormal loss makes it difficult for patients to get enough of several essential amino acids, particularly the branched-chain amino acids leucine, isoleucine, and valine, and the large neutral amino acids phenylalanine, tyrosine, and tryptophan.

Typical symptoms include uncoordinated movements (cerebellar ataxia) and photosensitive skin rash with dark (hyperpigmentation) and pale (hypopigmentation) patches. Consider the skin manifestations for a moment. They are basically the same as seen in pellagra due to niacin deficiency. This similarity cannot surprise us because tryptophan, of which patients with Hartnup disease do not have enough, is normally the starting point for niacin synthesis. Niacin supplements will usually clear up the dermatitis, which again indicates that a lack of this vitamin, or rather its tryptophan precursor, is the cause of this symptom.

Lifelong treatment aims to ensure high protein intake, use of a niacin supplement (50–300 mg/day), and protection against ultraviolet (UV) light with sunscreen, hat, and clothing.

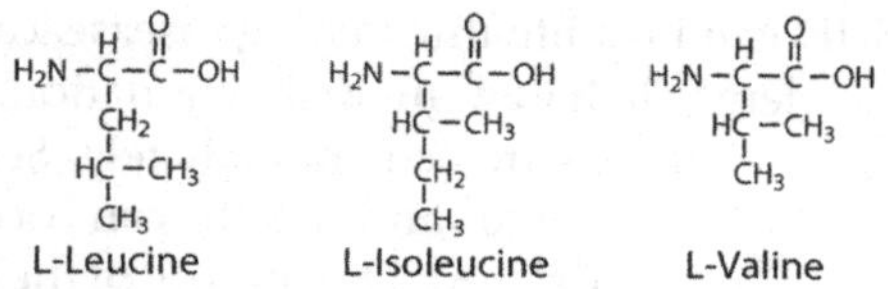

FIGURE 4.5
Leucine, isoleucine, and valine cannot replace each other, despite their chemical similarities. We have to consume adequate amounts of each individual branched-chain amino acid to avoid deficiency and malnutrition.

4.2.6 Maple syrup urine disease

Leucine, isoleucine, and valine share the structural feature of a branched carbon chain (Figure 4.5). This similarity does not change the fact that each of these branches is an essential amino acid in its own right, which means that one cannot be converted into another or replace it.

However, all three share the branched-chain alpha-keto acid dehydrogenase complex (BCKD) for one step of their breakdown (Figure 4.6). Four distinct genes, *BCKDHA* (OMIM 608348), *BCKDHB* (OMIM 248611), *DBT* (OMIM 248610), and *DLD* (OMIM 238331), encode different components of this complex. A variant in any of these genes can abolish the activity of the entire complex cause MSUD. The name of this inborn error of metabolism comes from the fact that the metabolites excreted with urine have an odor reminiscent of maple syrup. The condition occurs in about one out of 185,000 newborn children worldwide, but is much more common in Pennsylvania Mennonite and Ashkenazi Jewish communities due to founder effects [28].

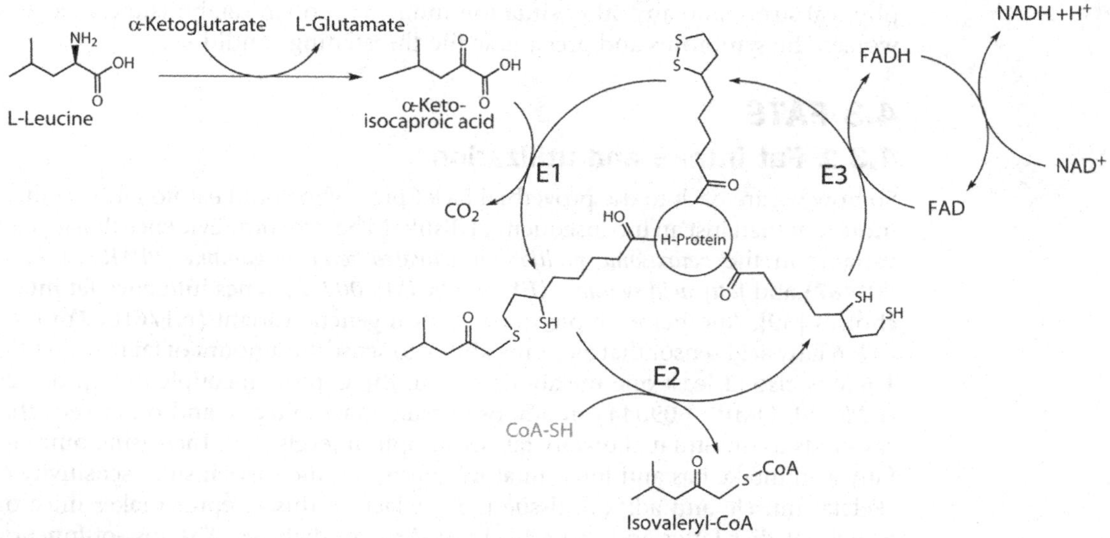

FIGURE 4.6
Maple syrup urine disease is caused by a lack of branched-chain alpha-keto acid dehydrogenase activity. Variation in any of the various protein components of the enzyme complex (E1, E2, E3, and H-protein) can disrupt its function. CO_2, carbon dioxide; FAD, flavin adenine dinucleotide; NAD, nicotinamide adenine dinucleotide.

MSUD can be a life-threatening disease, causing feeding problems, lethargy, mental retardation, and poor growth, unless treated immediately. Seizures and eventual coma and death can occur if treatment is delayed. The severity of the symptoms and their consequences depends on the specific causative variants, ranging from very severe (classic MSUD) to very mild (benign MSUD). Another important distinction is whether the condition responds to thiamine or lipoic acid supplements. In some patients (mostly among those with a variant in the E1 or E2 subunit), all metabolic and health abnormalities normalize completely with high thiamine intake.

> **FOUNDER EFFECT**
>
> An otherwise rare variant can be very common in a population because it was introduced by an early member before a phase of extensive internal expansion. Its high frequency does not necessarily mean that it provided a survival or proliferation benefit.

In a few others (some of those with a defect in the lipoic acid-binding E3 subunit), extra lipoic acid can be very effective. Because the E3 subunit also serves as a component of the pyruvate dehydrogenase and alpha-ketoglutarate dehydrogenase complexes, a genetic E3 subunit defect will diminish all three enzyme activities and cause an accumulation of lactic acid in an affected individual.

Dietary treatment has to be lifelong because neurological damage can occur at any life stage. Newborn screening for MSUD is now universal in the USA and most other Western countries, but can miss a significant minority of the milder cases [29]. Treatment needs to be tailored to the individual gene defect. Patients with the most severe forms use special amino acid mixtures without the offending branched-chain amino acids. Thiamine or lipoic acid supplements should be used to the extent they are found to be effective in the individual patient. Illness, physical stress, and any other situation inducing protein catabolism can acutely worsen the symptoms and precipitate life-threatening conditions.

4.3 FATS

4.3.1 Fat intake and utilization

So now we are back to the proverbial Jack Sprat, who could eat no fat. Was there more to it than just an inconsequential dislike? There is some evidence that genetic variants in the *peroxisome proliferator-activated receptor gamma (PPARγ, OMIM 601487)* and *fatty acid synthase (FASN, OMIM 600212)* genes influence fat intake choices [30]. Another example is a common genetic variant (rs1761667) in the *CD36* fatty acid sensor that alters the ability to sense the amount of fat in food [31]. There is also at least one metabolic sensor, the G protein-coupled receptor 120 (*GPR120*; OMIM 609044) in adipose tissue, macrophages, and other cells that responds to unsaturated dietary fat consumption levels [32]. Increasing omega-3 fatty acid intake has anti-inflammatory effects and increases insulin sensitivity of skeletal muscle and adipose tissue [33]. A lack of this receptor makes mice on a high-fat diet fatter and more likely to develop diabetes. The loss-of-function GPR120 variant Arg270His mirrors this effect in humans. The carriers of just one 270Cys allele (3% allele frequency in European populations) are more likely to be obese than individuals without this variant. This example illustrates another mechanism that makes specific individuals more vulnerable to a diet where saturated fats crowd out unsaturated fats, both in terms of quality and quantity.

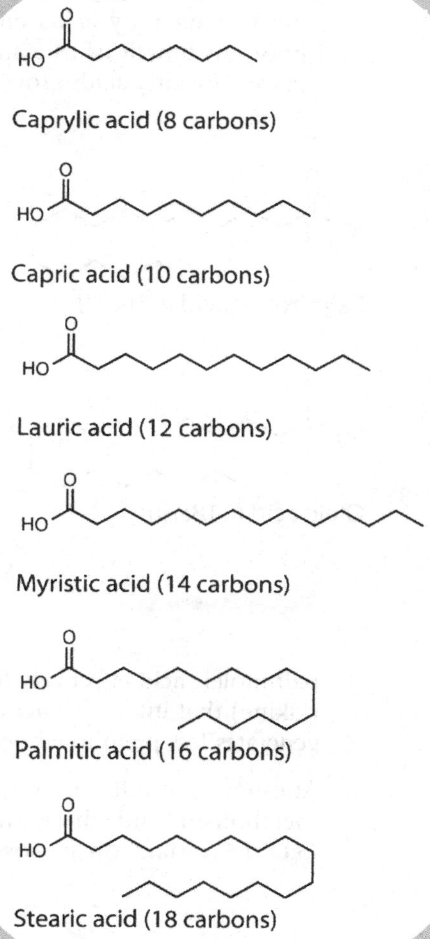

Let us consider for a moment how polymorphisms may alter the thermogenic effect of fat [34]. There comes an energy cost with the digestion, absorption, and metabolic processing of nutrients from foods. The amount of energy it takes to get macronutrients to their destination in the body is somewhere around 5% of a meal's calories, but the actual percentages vary considerably depending on size and composition of the meal, as well as other variables. Analysis of the correlation between the thermic effect of fat and variants in a number of candidate genes [34] pointed to several promising links. Consider it a special class assignment to work through them. One of them concerns the *GAD2* gene (OMIM 138275), which encodes glutamate decarboxylase 65 (GAD65; EC 4.1.1.15). Glutamate is a neurotransmitter and glutamate carboxylases end the excitation phase by inactivating glutamate. *GAD2* is also closely related to the release of insulin in the beta cells of the islets in the pancreas, where the gene is expressed at particularly high levels. Its expression is particularly high in the islet beta cells of the pancreas and antibodies against the enzyme are a recognized cause of autoimmune diabetes [35]. It is no surprise, therefore, that the *GAD2* gene has been linked to obesity and diabetes in several studies. The *GAD2*−243G (rs2236418) variant that is associated with an increased thermogenic effect of fat sits in the promoter region and may in turn ultimately affect weight balance.

Caprylic acid (8 carbons)

Capric acid (10 carbons)

Lauric acid (12 carbons)

Myristic acid (14 carbons)

Palmitic acid (16 carbons)

Stearic acid (18 carbons)

4.3.2 Saturated fats

4.3.2.1 CHAIN MODIFICATION

The saturated fats we consume differ in flavor and feel depending on their chain length. The typical chain length of fatty acids in fat from cow milk is 14−18 carbons.

The fats in goat cheese, on the other hand, contain shorter and more volatile fatty acids with 8–14 carbons, which give it a smoother feel and sharper flavor and odor. After some metabolic juggling, a large percentage of both shorter and longer chain fatty acids end up as oleic acid (C18:1ω9), which has many important functions as a structural constituent, protein modifier, and energy source. The fatty acids grow longer because of the activity of elongase for long-chain fatty acids family 6 (Elov6; EC 6.2.1.3), encoded by *ELOVL6* (OMIM 611546).

The stearoyl-CoA desaturase (EC 1.14.99.5) encoded by the *SCD1* gene (OMIM 604031) gives it the final touch by adding a double bond in the middle (Figure 4.7).

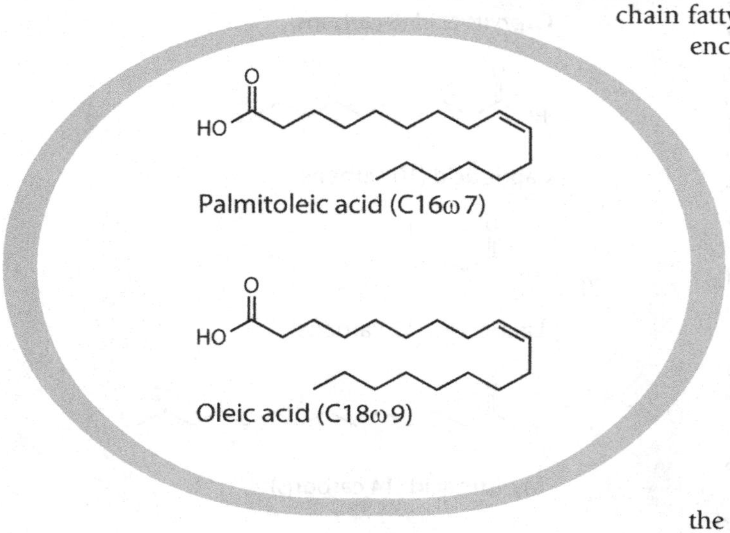

Palmitoleic acid (C16ω7)

Oleic acid (C18ω9)

Extensive polymorphic variation exists with this system for the modification of saturated fatty acids. The common variant rs603424 in the *SCD1* gene, for instance, strongly influences the proportion of the saturated 16-carbon fatty acid (palmitic acid) to the monounsaturated derivative palmitoleic acid [36]. This is of great interest because palmitoleic acid effectively functions as a kind of adipose tissue messenger (adipokine) that informs other organs about the state of fat stores. Since the variant generates less palmitoleic acid, there is less effective adipokine signaling.

Another example of a genetic interaction between saturated fatty acid metabolism and these modifying enzymes is the effect of a common *ELOVL6* variant on glucose metabolism and insulin resistance. It has been

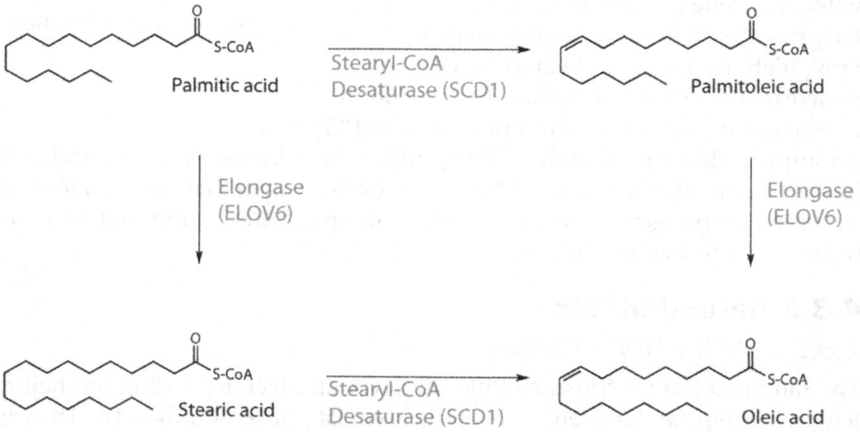

FIGURE 4.7
The elongase ELOVL6 adds a pair of carbons to medium-length fatty acids. The stearyl-CoA desaturase helps with the conversion of palmitic acid to palmitoleic acid and of stearic acid to oleic acid. Both enzymes act on coenzyme A derivatives of the fatty acids.

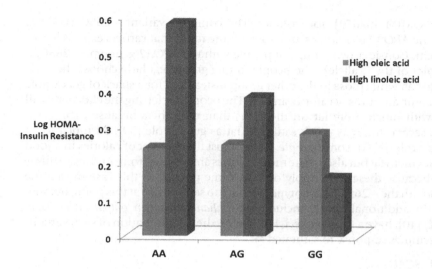

FIGURE 4.8
People with the ELOV6 rs682447 A/A genotype were found to have greater insulin resistance than people with the G/G genotype. This constitutional resistance was only seen in users of sunflower oil (high in linoleic acid), not in lovers of olive oil with its high content of oleic acid [38]. HOMA-IR, Homeostasis Model Assessment insulin resistance index.

observed in a number of studies that greater activity of this elongase makes muscle and fat cells more insulin resistant. When the gene is inactivated (knocked out) in mice, they do not become insulin resistant with over-feeding like mice with the intact gene do [37]. The G allele of the rs682447 variant is like a natural experiment that diminishes the activity of the *ELOVL6* gene. It sits almost a thousand bases upstream of the gene start site, possibly abolishing a transcription enhancer sequence. People with the rs682447 G allele do not easily develop insulin resistance (Figure 4.8). The higher production of long-chain saturated fatty acids from the shorter food-derived precursors in carriers of the A allele seems to blunt normal insulin-induced signaling. The risk of becoming insulin resistant is about twice as high in people with two A alleles compared to people with two G alleles [38]. However, this effect was only seen in people with higher than average intake of linoleic acid (C18:2 ω6). The omega (ω) notation indicates that the double bond is at the sixth carbon, counting from the acyl end of the fatty acid. Olive oil seems to be able to restore normal responsiveness to insulin. Basically, for the approximately 25% of people with two A alleles the conclusion seems clear: olive oil is the better fit for them, probably helping many of them to avoid abnormal insulin resistance.

4.3.2.2 APOA2

The smooth temptation of our favorite source of saturated fat is hard to resist, whether it is bacon, butter, or chocolate. We may know that we really don't need that dark chocolate bar with its extra 7 g of saturated fat, because it raises blood cholesterol and carries extra calories. But maybe just for once we have the right genes that make us invulnerable to the bad fat, right?

Unfortunately, none of us get a pass. When it comes to gaining weight, the satu-rated fat just seems worse for some than for others. There is *apolipoprotein A-II*

(*APOA2*; OMIM 107670), for instance. The common variant −265C (rs5082) changes the *APOA2* promoter sequence with the result that carriers express less of the protein. Consider two groups of people with the *APOA2* genotype −265C/C (two copies of the C allele). The people in one group regularly choose the extra chocolate bar, while those in the other group instead opt for a snack of guacamole with the same amounts of calories and fat. The people preferring the chocolate will end up with much more fat on their ribs than the others because their snack contains several times as much saturated fat as guacamole [39]. So much for 'a calorie is a calorie.' For some people, it is not just the number of calories in a food that makes them fat but also where those calories are coming from. And again, life is not fair because these rules apply only to some people, in this case mostly the 10−20% with the −265C/C genotype. We have to say 'mostly' at this point, because variants in additional genes, including *circadian locomotion output cycles kaput* (*CLOCK*) [40], have a similar effect. It is likely to be a combination of such variants that determines response to saturated fat.

4.3.2.3 SCAD

Short-chain saturated fatty acids differ considerably from their long-chain cousins. First, there is the short chain length, of course. More to the point, they are produced mostly by bacteria in the large intestine from dietary fiber. Acetate and propionate are important fuels for the cells of the intestinal wall. Some of the acetate and propionate and most of the butyrate are metabolized in the liver. This is where the short-chain acyl-CoA dehydrogenase (SCAD; EC 1.3.99.2), encoded by the *ACADS* gene (OMIM 606885), comes into play. A few infants are born with a defective gene. This deficiency (OMIM 201470) has come to clinical attention with the introduction of MS/MS newborn screening. The frequency of severe deficiency is about 1:40,000−1:100,000, but much higher in people of Ashkenazi-Jewish ancestry, with a heterozygous carrier frequency of about one in 15. Symptoms [41] are highly variable, ranging from severe and potentially fatal metabolic decompensation (hypoglycemia and acidosis), seizures, low muscle strength (hypotonia), and failure to thrive to slowly evolving muscle disease (myopathy) and sometimes heart disease (cardiomyopathy).

Nutritional treatment needs to focus on avoiding fasting, taking frequent meals and riboflavin supplements, and using menus with high-carbohydrate and low-fat content.

While the cases with severe symptoms are caused by variants associated with more or less complete deficiency, there are relatively common variants associated with slightly reduced activity. Two variants with 70−90% activity, 625G>A (Gly185Ser) and 511C>T (Arg147Trp), are common in people of European descent. About 6% of the Dutch population has two copies of these or other *ACADS* variants. The clinical consequences and their individual nutritional requirements need to be better understood. High doses of riboflavin can increase SCAD activity and reduce butyryl carnitine concentration in some cases, but there is no demonstrated benefit for asymptomatic individuals [42]. This might be an example of cosmetic correction of a biochemical oddity without improving health.

4.3.3 Unsaturated fatty acids

4.3.3.1 POLYUNSATURATED FATTY ACIDS

Our body can oxidize a wide range of fatty acids and use them as an energy fuel without too much distinction or preference. However, this is very different when it comes to the specific functions of fatty acids as structural building blocks and precursors of hormones and other bioactive compounds. There the requirements are very specific and the properties depend entirely on chain length, the absence of chain branching, and the position and number of double bonds. Of particular importance are two groups of unbranched, long-chain fatty acids with several double bonds: in the omega-3 group they start at the third carbon, counting from the carbon furthest away from the oxygen; and in the omega-6 group they start at the sixth carbon. The distinction is critical, because the human metabolism cannot add double bonds to convert a saturated or monounsaturated fatty acid into an omega-6 polyunsaturated fatty acid, or convert an omega-6 polyunsaturated fatty acid into an omega-3 polyunsaturated fatty acid. We need to get sufficient amounts of both from dietary sources. Fatty acids can be converted into other members of the same group by changing the length and number of double bonds. Nonetheless, there are distinct functional and metabolic differences even between fatty acids within the same group because the conversion reactions tend to be directional.

Particularly relevant in our context is the fact that several of these conversion activities differ considerably between individuals. Always remember that these desaturases and elongases act on multiple fatty acids from both the omega-3 and omega-6 groups and that the different fatty acids compete for the shared enzymes. The delta-5-desaturase (D5D; EC 1.14.19.-), encoded by *fatty acid desaturase 1* (*FADS1*; OMIM 606148), and the delta-6-desaturase (D6D; EC 1.14.19.-), encoded by *FADS2* (OMIM 606149), are probably most relevant because they limit the conversion rates [43]. An important aspect of *FADS1* and *FADS2* genetics is that the genes sit right next to each other along with a third member of this group, *FADS3*. Their proximity means that variants are more likely to be inherited together.

D6D (FADS2) converts linoleic acid into gamma-linolenic acid (GLA, C18:3 ω6), alpha-linolenic acid (ALA, C18:3 ω6) into stearidonic acid (SDA, C18:4 ω3), and tetracosapentaenoic acid (C24:5 ω3) into tetracosahexaenoic acid (C24:6 ω3; Figure 4.9). After an elongation step, D5D (FADS1) then completes conversion to arachidonic acid and eicosapentaenoic acid (EPA, 20:5 ω3), respectively. Conversion of ALA to EPA is very inefficient.

If all this variation in *FADS1* and *FADS2* seems just too much, it turns out to be quite streamlined and simple. The vast majority of people have only one of two common variant combinations or haplotypes [44]. Haplotype D is associated with higher *FADS1* gene expression and enzyme activity than the haplotype A. Individuals with haplotype D have an increased percentage of the longer, highly unsaturated fatty acids, both of the omega-3 series (EPA and docosahexaenoic

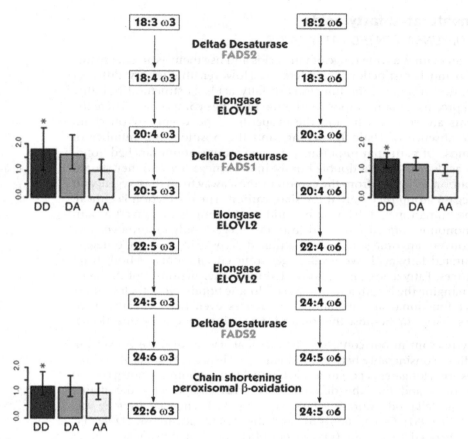

FIGURE 4.9

Some of the same desaturating and elongating enzymes act on omega-6 and omega-3 fatty acids. The haplotype D across both *FADS1* and *FADS2* is associated with higher activity than haplotype A. Haplotype D increases the proportion of both the omega-6 metabolite arachidonic acid (bar diagram on the right) and eicosapentaenoic acid (on the upper left). Haplotype D was good for prehistoric people because it boosts production of very long-chain omega-3 fatty acids for brain growth. Now this haplotype generates a lot of the proinflammatory metabolite arachidonic acid from abundant linoleic acid and may increase cardiovascular risk [45].

[DHA]) and of the omega-6 series (arachidonic acid). This overeager haplotype D arose about 200,000 years ago, well before modern humans emigrated from Africa. Haplotype D occurs neither in Denisovans, which split from the modern human line more than a million years ago, nor in Neanderthals, which started to pursue their own evolutionary destiny half a million years ago. The new haplotype seems to be an adaptation of fatty acid metabolism to the high dependency of the oversized modern human brain on DHA, particularly during infancy when brain grows at an astounding daily rate of 2–3 g, much of it in the form of DHA-containing lipids. The D haplotype may be advantageous in a nutritope where most DHA has to be made from ALA and other precursors.

Modern humans have a much higher intake of linoleic acid (omega-6 fatty acid). Under such circumstances, the excessive conversion of linoleic acid into proinflammatory arachidonic acid by the genetically overactive haplotype D FADS complex is the unfortunate consequence of a once lifesaving adaptation [44]. Individuals with (modern) African heritage carry mostly haplotype D and may be at particularly high risk for the proinflammatory effects of mayonnaise (usually high in linoleic acid).

The two fatty acid elongases ELOVL2 (EC 6.2.1.3; OMIM 611814) and ELOVL5 (EC 6.2.1.3; OMIM 611805) act in coordination with the desaturases. ELOVL5 adds two carbons to the polyunsaturated fatty acids with 18 (ALA and linoleic, gamma-linolenic, and stearidonic acids) and 20 (eicosapentaenoic acid) carbons [45]. ELOVL2 extends the fatty acids with 20 and 22 carbons.

The conversion of EPA to DHA is a bit more convoluted, which partially explains the difficulty of making enough DHA from ALA. The precursor EPA has to be elongated in two steps to the 24-carbon intermediate, imported into peroxisomes, and then shortened by beta-oxidation to the final 22-carbon fatty acid DHA. The very long-chain acyl-CoA reductase is needed for shortening the 24-carbon intermediate, which needs to be moved from the cytosol, where it is produced, into peroxisomes by the adrenoleukodystrophy protein (*ABCD1*; OMIM 300371).

It is not just differences in fatty acid processing that determine the response to polyunsaturated fatty acid (PUFA) intake. Variants in the genes encoding PUFA targets can do that, too. An important example is the variant −75G/A (rs670) in the *APOA1* (OMIM 107680) gene, which encodes the high-density lipoprotein (HDL) constituent apolipoprotein A-I. HDL cholesterol concentration rises with increasing PUFA intake in people with *APOA1* −75A, but not in people without this allele (Figure 4.10)[46]. Since both cholesterol and the apolipoprotein A-I moiety increase in this genotype-specific response, high PUFA intake may

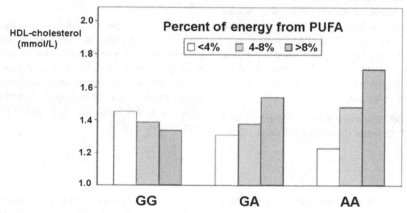

FIGURE 4.10

High polyunsaturated fatty acid (PUFA) intake increases high-density lipoprotein cholesterol concentration in carriers of the common APOA1 allele −75A and suggests an opportunity to boost risk-reducing HDL in some susceptible individuals [46]. HDL, high-density lipoprotein.

actually reduce cardiovascular risk. However, generating more definite proof with a long-term clinical trial would take many years and enormous funds.

A much-discussed disease

A defect in the *ALDP* gene causes X-linked adrenoleuko-dystrophy (*ALD*; OMIM 300100) because very long-chain fatty acids (VLCFA) are not metabolized. The VLCFA cerotenic acid (C26:0), lignoceric acid (C24:0), and behenic acid (C22:0) accumulate in many tissues and interfere there with normal cell function. These fatty acids are

a mixture of triglycerides with three oleic acid moieties and triglyceride with three erucic acid moieties) were not as successful as had been hoped [48]. The treatment quickly decreases C26:0 levels in blood, but not in the brain where the decrease is most needed.

Erucic acid (C18ω9)

Oleic acid (C18ω9)

consumed with food and are also generated in the body [47].

Symptoms can differ greatly between patients depending on the sequence change in the *ALDP* gene. Most common are attention deficit, learning difficulties, and seizures. Adrenal gland insufficiency, cognitive decline, peripheral neuropathy, progressive decline of vision, or spastic paraplegia occurs in some families. Clinical outcomes are determined by the progressive demyelination of neurons and can eventually cause death in childhood or adolescence. Heterozygous carriers (all women, of course, because it is an X-linked condition) often develop adult-onset adrenomyeloneuropathy with differing degrees of adrenal insufficiency, weakness and numbness of the limbs, and brain degenerative disease (focal central nervous system demyelination).

Efforts to treat ALDP deficiency with a mixture of long-chain unsaturated fatty acids (*Lorenzo's Oil*, consisting of

Current treatment guidelines aim primarily to keep the intake of saturated VLCFA to a minimum. Since very long-chain aldehydes, very long-chain alcohols, and VLCFA are interconvertible in the body [49], all of these sources have to be considered. This is further complicated by the fact that the body produces significant amounts of these VLCFA without help from the outside. Nonetheless, minimizing intake appears to be moderately beneficial in asymptomatic patients with some residual ALDP activity.

Increasing the intake of DHA may also be beneficial [50]. We would have expected this just based on our knowledge of the metabolic pathway, because the lack of active ALDP in these patients blocks conversion of EPA to DHA. It is important to recognize that such predictions are not always confirmed in clinical practice, and in this case the evidence is still fairly thin. Conjugated linoleic acid may also be useful [51].

Numerous variants in the two desaturases influence polyunsaturated fatty acid conversions. A common variant in intron 9 of *FADS1* (rs174547) has a particularly strong impact on fatty acid composition in blood [52]. The arachidonic acid:linoleic acid and EPA:ALA ratios are about twice as high in people with the

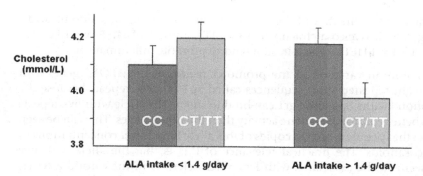

FIGURE 4.11
The T allele of the rs174546 FADS1 appears to make carriers more responsive to the cholesterol-lowering effect of dietary alpha-linolenic acid (ALA) [54].

T/T genotype than with the C/C genotype [53]. Another commonly investigated *FADS1* variant (rs174546) resides in the 3′ untranslated region (3′ UTR) just 953 bases downstream of rs174547. The question with all these variants is which of them directly modify enzyme activity and which of them just ride on the coattails of the primary actors, pulled along because they are inherited together in one haplotype package.

The impact of one of the *FADS2* variants (rs174576) appears to be almost as large, which is not surprising because it is highly correlated (in linkage disequilibrium) with the *FDAS1* variant rs174547.

These interactions are only meant to illustrate how strongly the totality of these variants, together with our fat intakes, shapes our individual fatty acid patterns. What it means in the end is that compared to someone with the rs174547T/T genotype, a C/C carrier needs to use about twice the amount of canola oil (containing about 10% ALA) with their scrambled eggs to raise the percentage of EPA to the same desired level.

These genotype-specific efficiencies of fatty acid conversion have significant health consequences. For instance, people with the T allele of the rs174546 *FADS1* variant tended to have about 5% lower serum cholesterol concentration in one study (Figure 4.11), but only if their ALA intake was greater than 1.4 g/day [54]. You can see how the cholesterol-lowering response depends both on the genotype and on the amount of ALA precursor available for conversion into more effective EPA and DHA. This kind of interaction is easily missed unless an effort is made to tease this needle out from the haystack of variables.

Various *FADS* variants have also been associated with Crohn disease, glucose metabolism (fasting glucose concentration and insulin resistance), and resting heart rate.

4.3.3.2 EICOSANOIDS

Important differences also exist in some of the activities that convert particular fatty acids into their active metabolites. We shall look here at just one of them, arachidonate 5-oxidase (ALOX5; EC 1.13.11.35), encoded by the *ALOX5* gene (OMIM 152390). The ALOX5 enzyme generates a number of hormone-like eicosanoids (the name refers to the 20 carbons of the parent compound, arachidonic acid), including 5-hydroxyeicosatetraenoate (5-HETE), 6-trans-LTB4, 5-oxo-ETE, 15-HETE, and 5,15-diHETE, which in sum tend to promote inflammation.

There is common variation in the promoter region of the *ALOX5* gene, with three to eight repeated short sequences called Sp1 tandem repeats because the transcription factors Sp1 and Egr1 can bind to them. The allele with five repeats provides better control of enzyme activity than the other alleles. This can be seen by the fact that people with two copies of this allele have lower concentrations of active eicosanoids. The practical relevance of this is that the allele with five repeats seems to be associated with less atherosclerosis in the carotid arteries, where narrowing could cause stroke or vascular dementia [55]. People with two variant alleles (with other than five repeats) were found to have a thicker inner layer (intima) of the arterial wall if they consumed more than 120 mg/day arachidonic acid, the precursor of the inflammatory eicosanoids, and also if they get very little EPA and DHA from cold-water ocean fish or fish oil capsules. How can carriers of the risk alleles avoid getting too much arachidonic acid? Basically this is done by cutting back on meats (which contain around 50–80 mg/100 g [3.5 oz.] serving), egg yolks (70 mg in one large egg), and seafood (80–400 mg/100 g [3.5 oz.] serving). This is another plausible example showing that genetic variants often define typical disease risk but are not immutable destiny.

4.3.4 Branched-chain fatty acids

Fatty acids that do not have their carbons arranged in a neat, linear chain cannot be metabolized in mitochondria by the usual beta-oxidation sequence. The peroxisomes handle these less common fatty acids with alpha-oxidation, which means starting with the first carbon. Phytanic acid is possibly the most important example of a fairly common branched-chain fatty acid. It is generated when the phytol side chain is released from chlorophyll and becomes oxidized during

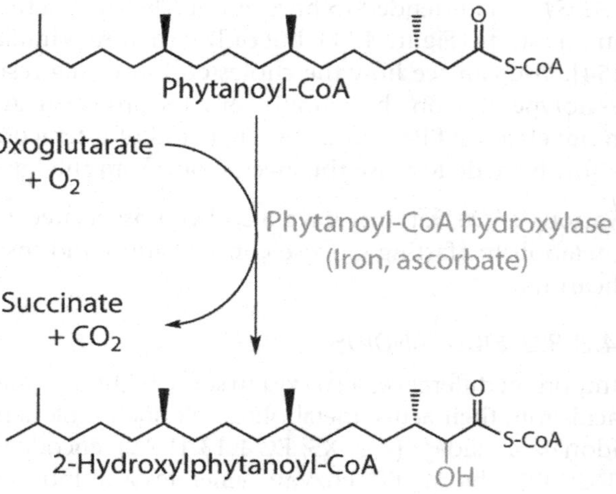

FIGURE 4.12
Individuals with Refsum disease cannot metabolize phytanic acid due to lack of phytanoyl-CoA hydroxylase activity.

digestion by ruminants (Figure 4.12). A typical mixed diet contains about 50 mg phytanic acid/day, mostly from beef and milk fat. Small amounts of phytol may also be released from the side chain of chlorophyll and oxidized to phytanic acid.

Phytanic acid

A very small number of people have defective peroxisomal alpha-oxidation, usually due to variants in the *phytanoyl-CoA hydroxylase* gene (*PHYH*; OMIM 602026) leading to lack of phytanoyl-CoA hydroxylase (EC 1.14.11.18) activity or to variants in the *peroxin 7* gene (*PEX*; OMIM 601757), which encodes a receptor needed for the import of the hydroxylase into the peroxisome.

The accumulation of phytanic acid due to this inborn error of metabolism, which is called Refsum disease (OMIM 266500), causes slowly progressive and irreversible damage to the retina and nervous system. The patients become symptomatic very slowly, often starting during later childhood and even well into adulthood. The main symptoms are dry scaling of the skin (ichthyosis), progressive loss of sight (retinitis pigmentosa), loss of the sense of smell (anosmia), loss of hearing, cerebellar ataxia, and peripheral polyneuropathy. Diminished ability to see in dim light (night blindness) and narrowing of the field of vision are early symptoms.

For the longest time it was thought that the fate of affected patients was inevitable. Recent anecdotal experience with rigorous early intervention demonstrates that prevention is effective [56]. The avoidance of foods with high phytanic acid content, such as beef, dairy foods, lamb, tuna, and soy, is most important. Stored phytanic acid can also be removed very slowly by lipid apheresis, which has to be done several times a week over many months to make a difference. The reason for this tedious procedure is that phytanic acid in the triglycerides of adipose tissue has to be mobilized before it can be eliminated by removing triglyceride-containing lipoproteins from blood.

4.4 STEROLS

4.4.1 Sterol absorption

More than a third of the cholesterol eaten with an egg is absorbed, packaged into chylomicrons, and transferred into the bloodstream. Several proteins work together to facilitate cholesterol uptake from the intestinal lumen into the mucosal enterocyte. Particularly important is an intestinal sterol transporter, called Niemann-Pick C1 like 1 protein (encoded by *NPC1L1*; OMIM 608010). The microsomal triglyceride transfer protein (*MTP*; OMIM 157147) assists with the packaging of the sterol into chylomicrons (its absence causes abetalipoproteinemia). An adenosine triphosphate (ATP)-driven transporter actively returns cholesterol back into the intestinal lumen. This transporter is a sandwich made up of ABCG5 (OMIM 605459) and ABCG8 (OMIM 605460).

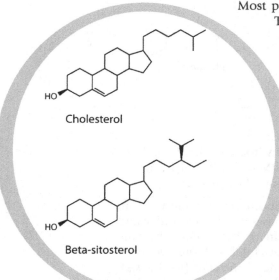

Cholesterol

Beta-sitosterol

Most plants do not make cholesterol like mammals do. They contain instead various structurally similar compounds (usually called plant sterols or phytosterols), including beta-sitosterol, campesterol, and stigmasterol. Beta-sitosterol is basically a cholesterol molecule with two extra carbons in the side chain. The extra ethyl group means that we normally absorb beta-sitosterol a hundred times less effectively than cholesterol. The human body is very particular about which sterols it likes and which ones it would rather get rid of. Cholesterol is absorbed quite well from the small intestines, but most people do not take up plant sterols very well. This is a good idea, because plant sterols disrupt the regulation of cholesterol and steroid hormone metabolism [57].

4.4.2 Phytosterolemia

The sterol uptake complex already favors cholesterol over plant sterols [58]. In addition, the ABCG5/G8 transporter complex selectively flushes plant sterols out of the mucosal cell back into the intestinal lumen. This transporter complex consists of two components, the ATP-binding cassette subfamily G member 5 (*ABCG5*; OMIM 605459) and member 8 (*ABCG8*; OMIM 605460). We know from the inherited condition, phytosterolemia (OMIM 210250), what happens when the ABCG5/G8 transporter does not work. The most obvious physical sign of this rare condition is deposits (xanthomas) of cholesterol and plant sterols in the skin and soft tissue, usually over the Achilles tendons, elbows (Figure 4.13), and knees, and less often at other sites. The plasma concentration of plant sterols, which usually constitute less

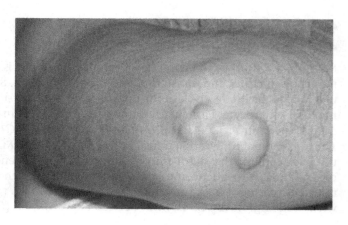

FIGURE 4.13
Large xanthomas over the elbow of a patient with phytosterolemia.

than 0.5% of all sterols in blood, is greatly elevated. The main health concern is the increased atherosclerosis risk. Loss-of-function mutations in the *ABCG5/G8* transporter genes fully explain the condition.

It is important to remember that all plant sterols in blood and tissues come from foods because the body cannot produce these compounds. A newborn infant with a ABCG5 or ABCG8 gene defect who gets only minimal amounts of plant sterols from the mother does not have high plant sterol concentration in blood. That is precisely what was observed in an 11-month-old girl with nonsense mutations in both copies of her *ABCG5* gene [59]. Her plant sterol concentration was not particularly high (0.23% of total sterols), but she had cholesterol deposits over her Achilles tendon. Her main problem was extremely high blood cholesterol concentration (1023 mg/dl). This tells us conclusively that it is the blockage of cholesterol efflux from the mucosal cell that causes the problem. The elevated plant sterol concentrations appear to be largely a side issue, very unusual but of much less consequence than the defective regulation of cholesterol metabolism.

Treatment of these patients needs to focus primarily on minimizing the amount of cholesterol available for absorption, not so much on the amount of plant sterols. The effect of dietary cholesterol on blood cholesterol concentration is normally limited because more intestinal cholesterol comes from bile than from foods. But at least some patients with defective ABCG5 or G8 transporter components have an unusually strong response to cholesterol intake and will experience a good cholesterol-lowering effect. Individuals who are heterozygous for loss-of-function variants or carry variants with moderately reduced activity are likely to have an exaggerated response to high cholesterol intake, as explained below.

4.4.3 Cholesterol hyperresponders

You have probably followed more than enough discussions about the health consequences of high cholesterol intake. The conversation quickly tends to come around to whether to eat eggs or not. The inconsistent results of large population studies have caused considerable uncertainty about the best choice. Asking about the right consumption level is a reasonable question, since eggs are a particularly cholesterol-rich food, weighing in at a hefty 190 mg for a large one. There is little doubt that cholesterol rises in some people when they start eating eggs, but not in many others [60,61]. Some of the genetic differences between people with and without a strong increase in blood cholesterol concentration after increasing their cholesterol intake (hyperresponders) are becoming clearer. As might be expected, variants in several of the genes related to cholesterol absorption explain at least some of the difference.

The common polymorphism −493G/T (rs1800591) in *MTP* alters the efficiency of sterol absorption from the small intestine [62]. As explained above, MTP helps with assembling the lipids into neat little chylomicron packages. Approximately 13% of people have the *MTP* −493T/T genotype and appear to absorb more cholesterol from a given dose than people with the G allele.

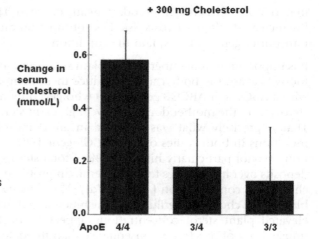

+ 300 mg Cholesterol

Change in serum cholesterol (mmol/L)

FIGURE 4.14
Added cholesterol raises serum cholesterol concentration much more in people with the *APOE* 4/4 genotype than in those with the *APOE* 3/4 or 3/3 genotypes. This graph shows what happens when healthy individuals get an additional 300 mg cholesterol on top of a standard, healthy National Cholesterol Education Program (NCEP)-style diet [64].

Apolipoprotein E (*APOE*; OMIM 107741) also contributes to the variable response. It is a small protein constituent of chylomicron remnants with newly absorbed cholesterol. Because Apo-E binds to low-density lipoprotein (LDL) receptors, it steers the chylomicron remnants to the liver cells, which are already waiting for the cargo from the small intestine. The relatively common E4 allele (rs429358 and rs7412) elicits a stronger rise of blood cholesterol (Figure 4.14) than the other alleles [63,64] because the encoded Apo-E isoform is a more effective LDL receptor ligand.

The third response modulator worth mentioning is the 1950C>G (Gln640Glu) polymorphism (rs6720173) in the *ABCG5* gene. People with two copies of the C allele can expect an increase of blood cholesterol concentration of about 0.026 mmol/L, compared to virtually no change in people with a G allele [65].

A final consideration concerns the fat-soluble compounds absorbed, transported, and delivered to the liver along with cholesterol. Let us take, for example, the carotenoids from fruits and vegetables. Cholesterol hyperresponders were found to absorb more beta-carotene and lutein from the same test meals than people with no detectable response to cholesterol [66]. At least in regard to the *ABCG5* 1950C>G polymorphism, responses to cholesterol and lutein followed a similar pattern; C/C carriers on average had a much stronger response to lutein intake than individuals with the G allele [65].

4.5 CARBOHYDRATES

4.5.1 Carbohydrate digestion

Prehistoric people relied heavily on carbohydrates as their major fuel source. Most of the fruits, leaves, and roots they ate contain both simple and complex

Table 4.1	Prehistoric Sources of Carbohydrates and the Digestive Enzymes that Act on Them		
Carbohydrates	**Carbohydrate Sources**	**Digestive Enzymes**	**Enzyme Sources**
Glucose	Fruits	—	—
Maltose	Germinating seeds	Maltase/Glucoamylase	Brush border of small intestines
Trehalose	Mushrooms, yeast	Trehalase	Brush border of small intestines
Isomaltose	Honey	Sucrase/Isomaltase	Brush border of small intestines
Starches	Fruits, leaves, roots, seeds	Salivary amylase Pancreatic amylase Maltase/Glucoamylase	Brush border of small intestines, mouth, pancreas secretion
Soluble fiber	Fruits, oats, vegetables	Bacterial enzymes	Terminal ileum, colon
Fructose	Fruits	—	—
Sucrose	Fruits	Sucrase/Isomaltase	Brush border of small intestines
Galactose	Peas	—	—
Lactose	Milk	Lactase	Brush border of small intestines

carbohydrates (sugars and starches) but are low in fat and protein. The bulk of the carbohydrate in these foods consists of glucose in a wide range of combinations and arrangements (Table 4.1). Smaller amounts come from fructose and galactose and their polymers. Glucose, fructose, and galactose are absorbed through transporters in the small intestine. The polymers become available only after they have been cleaved into their constituent sugars by one of the digestive enzymes.

Residual carbohydrates are food for the microorganisms in the ileum and colon. This community of bacteria and yeasts is now commonly called the *intestinal microbiome*. The kind and number of these microorganisms are different for each individual human host, depending on genetics, exposure at birth, environment, lifestyle, and the most recent meals. They use many of the carbohydrates coming their way, including many of the dietary fibers and incompletely digested starches. Some bacteria produce excess short-chain fatty acids (acetic acid, propionic acid, and butyric acid) from these complex carbohydrates. These short-chain fatty acids are an important fuel for the mucosal cells of the colon and small amounts are used by the liver. Other bacteria generate methane and hydrogen from undigested mono- and disaccharides. When somebody with lactose intolerance has a glass of milk, the bacteria will then gobble up the leftovers and start partying. The extra gas and other breakdown products can cause bloating, cramping, and diarrhea.

Table 4.2	Low Activity of a Carbohydrate-Digesting Enzyme Due to a Genetic Variant is not Very Rare		
Condition	**OMIM**	**Carbohydrate**	**Worldwide Prevalence (%)**
Lactose intolerance	223100	(−) Lactose	60−70 [68]
Congenital sucrase-isomaltase deficiency	222900	(−) Sucrose	0.2 [69]
Trehalose intolerance	612119	(−) Trehalose	0.05 [70]
Congenital glucoamylase deficiency	154360	(−) Starch	Occasional
Congenital amylase deficiency	104650	(−) Starch	Probably rare

The column under OMIM (Online Mendelian Inheritance in Man) contains the unified index number in the catalogue of all known genetic conditions. A minus sign indicates that carriers should have less of the listed carbohydrate.

Many people have genetic variants that prevent them from absorbing one or more of the common sugars in foods and beverages (Table 4.2). Many more have variants that encode enzymes with slightly lower or higher activity, most of them unbeknownst to the carriers or their healthcare providers. These individual predispositions shape many of our daily foods preferences.

Since the sucrase-isomaltase can cover for most of the maltase-glucoamylase activities, inborn deficiency with a mild phenotype may not be recognized in clinical practice. Nonetheless, congenital glucoamylase deficiency is occasionally the root cause of chronic diarrhea. Among a case series of more than 500 children with chronic diarrhea from centers all across the USA, at least 1% was found to have congenital glucoamylase deficiency [67]. Avoidance of starch will stop the diarrhea within days. This defect may have subtle but important health implications. Studies on mice with a targeted deletion of the *maltase-glucoamylase* gene indicate that they have diminished capacity for starch digestion, cannot appropriately suppress gluconeogenesis after eating starch, and release less insulin after a meal. Maltase-glucoamylase deficiency may thus just hide behind a metabolic phenotype with mildly altered glucose handling in the liver.

Congenital deficiency of pancreas amylase (*AMY*; OMIM 104650) is rare, but has been observed in a few patients. The number of *amylase* genes that are present and active varies greatly between individuals. On average, there are four separate genes encoding the same amylase enzyme. One of these genes is in reverse orientation to the others. On top of these multiple backup gene versions, there is extensive gene copy number variation [71]. A quarter of all investigated individuals had a large deletion around the region of the amylase genes and almost as many carried a large insertion at this position. This observation mirrors the extensive variation of the number of salivary *amylase* gene copies and has a strong link to historic starch consumption patterns (as discussed in Chapter 3).

4.5.2 Galactose

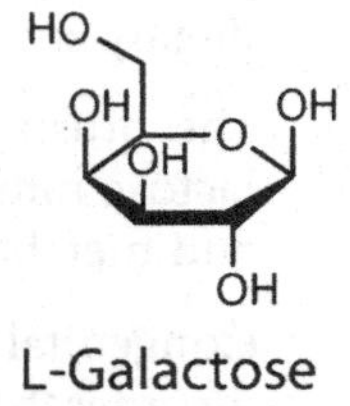

Galactosemia is a rare genetic disorder caused by defective galactose-1-phosphate-uridyltransferase (*GALT*; EC 2.7.7.12; OMIM 606999) or more rarely by mutations in the genes encoding galactokinase (*GALK*; EC604313; OMIM 230200) or UDP-galactose-1,4-epimerase (*GALE*; EC606953; OMIM 5.1.3.2). Symptoms in newborns appear within days of milk feedings. Typical signs include vomiting, diarrhea, and failure to thrive. Unless consumption of all galactose is stopped, the child will rapidly develop cataracts (clouding of the eye lens), enlargement of the liver, kidney damage, and cognitive impairment.

L-Galactose

Lifelong avoidance of galactose is usually effective. The most obvious sources of galactose are milk and all dairy products. Beans, peas, and a few vegetables also often contain enough galactose to cause problems for patients with galactosemia [72]. Women with a severe form of the condition will be harmed by the galactose produced in their own milk ducts [73].

4.5.3 Lactose

Lactose (β-D-galactopyranosyl-(1→4)-D-glucose) is a disaccharide consisting of a glucose and a galactose molecule joined by a β-1→4 link (Figure 4.15).

The nearly exclusive dietary sources of lactose are milk and other dairy products, which are valuable modern sources of calcium, energy, protein, riboflavin, and

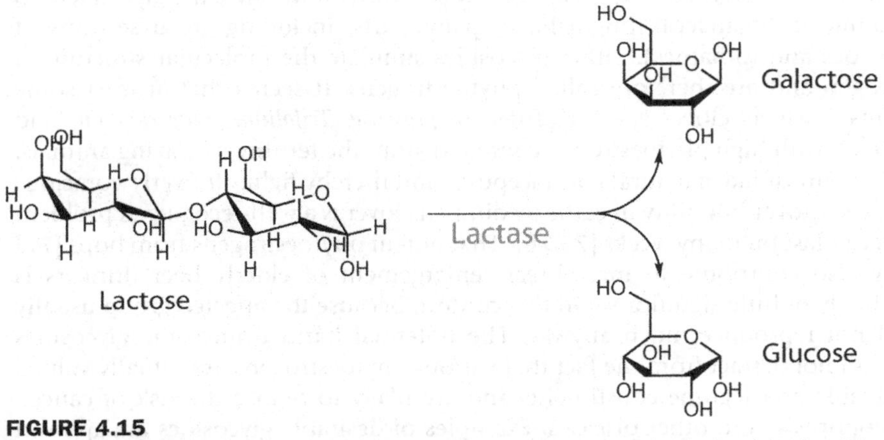

FIGURE 4.15
Lactose consists of a galactose and a glucose part, which can be separated at the brush border of the small intestine by lactase.

other critical nutrients. Lactose is also a filler or binder in many medications (including the placebos in some monthly packages for oral contraceptives) and dietary supplements.

The brush border enzyme lactase (EC 3.2.1.62) in the small intestine splits lactose into its glucose and galactose constituents, which can then be absorbed and metabolized. Lactase is encoded by the *LCT* gene (OMIM 603202).

Congenital lactase deficiency (*CLT*; OMIM 223000), is a very rare recessive disorder that prevents infants from digesting the lactose in milk [74]. They suffer from watery diarrhea with dehydration and acidosis immediately upon the start of feeding. Exclusion of lactose from their diet quickly reverses those symptoms and allows them to grow normally. The disorder is more common in Finland than elsewhere because many Finns have inherited the lactase-inactivating lactase variant Tyr1390Xaa from an early settler [74].

Congenital lactase deficiency must not be confused with lactose intolerance (adult-type hypolactasia; OMIM 223100) in adulthood, which is to be discussed next. Infants obviously need active lactase because without it they would have died in the past. One only has to look at what happens to infants with the rare condition of congenital lactase deficiency described above. Strong selective pressure ensures that most children express an adequately functioning enzyme. However, after weaning there is little reason for humans and other mammals to maintain intestinal lactase activity because milk is not part of their natural food supply and other foods do not contain lactose.

There is a good reason why maintaining lactase activity beyond the time of breastfeeding is a disadvantage. Various complex chemicals in berries and other plant parts have a glucose attached and are therefore called glycosides or glucosides. Intestinal absorption of these glycosides is minimal unless they are digested by lactase. Many of these glycosides can release cyanides (for instance linamarin in cassava roots) or are otherwise toxic (for instance the glycoalkaloid solanine in Solanaceae [nightshade] plant parts, including green sections of potatoes and tomatoes). Other glycosides simulate the molecular structure of estrogen and are therefore called phytoestrogens. It seems that at least some plants, such as clover species (*Trifolium pratense*, *Trifolium subterraneum*, and others), with high phytoestrogen content disrupt the fertility of grazing animals, acting almost like a natural contraceptive, and thereby fight off overly voracious grazers. Clover infertility in cattle feeding on clover is a well-recognized problem that can last for many weeks [75,76]. That potent phytoestrogens from hops [77] may also contribute to male breast enlargement of elderly beer drinkers is probably of little significance in this context, because the affected group usually will not reproduce much anyway. The potential harm from some glycosides should not detract from the fact that various phytoestrogens are actually valued vegetable and legume constituents and are likely to reduce the risk of cancer, osteoporosis, and other diseases. Examples of desirable glycosides are apiin in vegetables, genistin and related isoflavone glycosides in soy, and naringin in oranges.

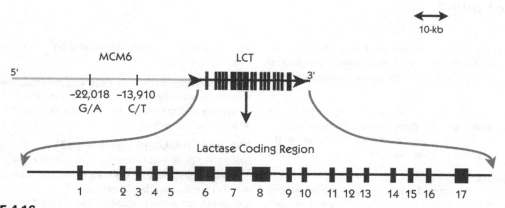

FIGURE 4.16
The lactase gene (LCT) contains 17 exons. Two enhancer elements, around 14,000 and 22,000 bases upstream from the transcribed sequence, control gene expression at the tips of the mucosa brush border villi of the proximal small intestine [79].

This context should explain why most (over 70%) adults around the world do not have significant lactase activity. But how do the others continue to express lactase after childhood? An enhancing factor about 14,000 bases upstream of the transcribed lactase sequence has to be active for expression to occur. The −13910T variant (Figure 4.16), which is common in Northern Europeans, keeps the enhancing region active [78] and maintains intestinal lactase expression throughout life. A second region, even further upstream, also enhances lactase expression. Most people of Northern European ancestry with lactase persistence get an extra boost because they carry the variant −22,018A in this additional enhancer region.

It is important to understand that many people with self-diagnosed lactose intolerance can actually digest the sugar. Healthcare providers also often diagnose the condition based on self-reported symptoms instead of an objective functional or genetic test [80].

A randomized, double-blind study demonstrated that there is no meaningful difference in symptoms when lactose-intolerant individuals drink a cup (250 mL [8 oz.]) of milk with 12 g of lactose or the same amount of lactose-free milk [85]. Symptoms in some people may really be a response to cow milk proteins, not to lactose. Typical symptoms of lactose maldigestion are abdominal pain, bloating, borborygmi (abdominal gurgling), flatulence, and loose stools. Systemic effects may include headache, light-headedness, loss of concentration, and signs of allergy such as asthma; eczema, itching, and running nose. Many people (particularly men) don't mind the lactose-triggered riot in their bowels too much and continue drinking some milk. On the other hand, there is significant overlap [86] between lactose intolerance and IBS. This link indicates that some IBS sufferers who cannot control their symptoms

Got milk?

Many patients do not recognize symptoms or do not have any complaints. The diagnosis of lactose maldigestion/malabsorption can only be made with objective diagnostic tests.

Diagnosis can be based on genotyping, hydrogen in breath after lactose load, detection of ^{13}C-labeled lactose in blood or as $^{13}CO_2$ in breath after test dose (for research purposes), or lactase activity in jejunum biopsy.

Genotyping

In European and some South and North American populations, the absence of an −13910T allele predicts lactose intolerance of adults with reasonable sensitivity (typically better than 90%) and specificity (close to 90%) compared to functional tests [81]. Sensitivity is so good because a majority of people with lactase persistence in these populations carries the −13910T allele. The narrow tailoring of this test to just this one variant limits its use to Caucasians. A more comprehensive assessment of the lactase sequence should eventually capture all the polymorphisms that potentially cause lactase persistence in diverse populations around the world.

Lactose Hydrogen Breath Test [82]

Give 50 g lactose orally and measure hydrogen concentration in breath for 3 hours.

An increase of hydrogen concentration > 20 ppm indicates malabsorption.

Sensitivity and specificity for detecting individuals with no lactase persistence (LCT −13910 C/C) is close to 100%. Sensitivity and specificity are lower for adults with some residual lactase activity.

^{13}C- and ^{2}H-labeled Lactose [83]

Give a 25 g dose of the labeled lactose orally. The ratio of ^{13}C-labeled glucose to ^{2}H-labeled glucose is measured in blood plasma after 45−60 minutes. A low ratio indicates low intestinal lactase activity. The method can detect residual lactase activity in adults with the LCT −13910 C/C genotype, who might have a variant with modest lactase-inducing capacity.

Alternatively, the appearance of $^{13}CO_2$ in breath can be measured at 30-minute intervals for 2.5 hours.

Jejunum Biopsy

Lactase activity in a homogenized biopsy specimen from the small intestine is measured by incubating with lactose and then measuring glucose release [84]. This is considered the reference method and mostly used to rule out an alternative lactose intolerance diagnosis in patients with irritable bowel syndrome (IBS).

KEY CONCEPT

Individuals without intestinal lactase activity tolerate modest amounts of lactose well and do not have to rigorously avoid all dishes with cheese, cream, or milk.

might benefit from a proper assessment of their lactase status and follow-up nutritional guidance.

The typically lower milk consumption means in practice that the normal variant (the one that does not confer lactase persistence) is associated with a reduced intake of calcium [87] and possibly other nutrients. To avoid an increased risk of bone fractures [88] in cultures with strongly engrained patterns of milk use, children and young adolescents in particular need to be provided with lactose-free calcium sources.

Rare defects of the sodium-glucose transporter (SGLT1; OMIM 182380) limit absorption of both glucose and galactose [89]. Normally functioning lactase cleaves the lactose from milk in these patients, but the unabsorbed monosaccharides cause severe and intractable diarrhea, which can quickly become fatal if feeding of milk or glucose persists. Diarrhea stops when the infants get formula without galactose, glucose, or lactose and development proceeds normally as long as they receive a diet with minimal amounts of the offending sugars [90].

4.5.4 Fructose

4.5.4.1 HEREDITARY FRUCTOSE INTOLERANCE

Hereditary fructose intolerance (fructosemia) is one of the more common (about 1:20,000 in Caucasians) inborn errors of metabolism (OMIM 229600), usually due to a mutation in the *aldolase B* gene (*ALDOB*; OMIM 612724), which encodes fructose-1-phosphate aldolase (fructoaldolase; EC 4.1.2.13) in the liver, kidney, and small bowel [91,92]. Ingested fructose gets stuck in these tissues as fructose-1-phosphate, which would normally be split into D-glyceraldehyde and dihydroxyacetone by fructoaldolase. One of the harmful effects of fructose-1-phosphate accumulation is the depletion of phosphate for other reactions. Another set of harmful effects limits gluconeogenesis through inhibition of aldolase A (EC 4.1.2.13) and glycogenolysis through inhibition of phosphorylase B (EC 2.4.1.1). Another metabolic consequence of fructoaldolase deficiency is excessive activation of adenosine deaminase (EC 3.5.4.4), which breaks down purines, increases uric acid production, and depletes magnesium [93].

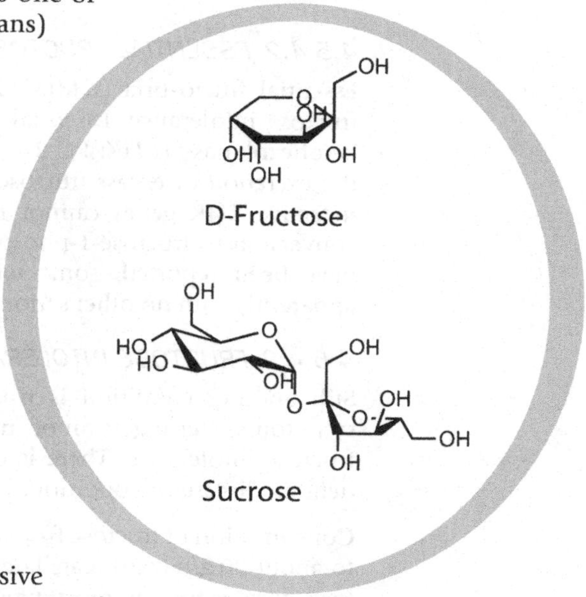

D-Fructose

Sucrose

Consumption of sugar (sucrose), fructose, or fruits will regularly cause vomiting, abdominal cramps, and diarrhea. Such foods can also trigger convulsions and loss of consciousness due to hypoglycemia, which may even be acutely fatal. The infusion of fructose or sorbitol in hospitalized patients must be avoided for those reasons [94]. Because the symptoms start occurring at the earliest age, people with this condition will often develop a strong aversion to the offending foods and maintain good health. There is no need to consume fructose and avoidance may even help maintain a healthy body weight. Persistent fructose intake, on the other hand, will usually promote metabolic acidosis, liver cirrhosis, and kidney damage [93].

About one in 50 adults is heterozygous for a nonfunctional *ALDOB* variant [92]. Because many Americans now consume well in excess of 100 g/day of fructose from sugar, high-fructose corn syrup, and fruits, such a high intake may well pose a risk for heterozygous carriers.

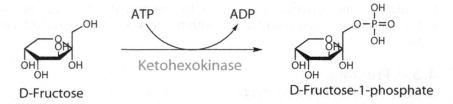

FIGURE 4.17
Ketohexokinase in the liver phosphorylates fructose to fructose-1-phosphate.

D-Fructose

Ketohexokinase

D-Fructose-1-phosphate

4.5.4.2 ESSENTIAL FRUCTOSURIA

Essential fructosuria (OMIM 229800) is completely distinct from hereditary fructose intolerance. Essential fructosuria is caused by diminished activity of ketohexokinase (*KHK*; EC 2.7.1.3; OMIM 614058), which occasionally causes the excretion of excess fructose in urine. Affected individuals with completely defective KHK genes cannot metabolize dietary fructose because the initial conversion to fructose-1-phosphate is blocked (Figure 4.17). Only a few cases have been reported, some with enlargement of the liver and spleen, but apparently with no other serious health consequences related to their condition.

4.5.4.3 FRUCTOSE INTOLERANCE

Still another condition is fructose intolerance, characterized by abdominal symptoms after ingestion of fructose, again not to be confused with hereditary fructose intolerance. There is debate about whether this is even a sufficiently defined disease or condition.

Consumption of fructose from all sources has risen from about 75 g/day in 1970 to about 90 g/day 30 years later [95], largely due to increased use of high-fructose corn syrup in most types of foods and beverages. Possibly the most consequential change is the increased consumption of fructose in its free form. Many people cannot fully absorb large amounts of fructose in the absence of similar amounts of glucose. In a dose-finding series, the absorption capacity for free fructose was less than 25 g in half of the healthy test persons [96]. Each of them readily tolerated twice that amount of fructose when it was ingested with an equal amount of glucose. The 25 g fructose in the challenge dose is approximately the content of a 16-oz. bottle of cola sweetened with high-fructose corn syrup. People who exceed their individual threshold dose may experience symptoms such as abdominal discomfort, bloating, nausea, pain, or diarrhea [97–100]. We should just remind ourselves that people with the completely unrelated hereditary fructose intolerance absorb fructose normally. In some case series, a higher percentage of patients with IBS than controls did not tolerate fructose well, based on a standard hydrogen breath test [101, 102]. In other series, the percentage was similar in both groups [103].

The differences in the amount of fructose people can tolerate may have a genetic link, but no associated genetic variants are known. Any abdominal discomfort is easily avoided by limiting the consumption of beverages and candy containing large amounts of free fructose. This condition is completely self-inflicted.

4.5.5 Sucrose

What most people would call sugar (from the Sanskrit word *sharkara*, which means 'grit') is biochemically sucrose, consisting of one part glucose and one part fructose. There was not much sucrose around in prehistoric times. Arctic and other Northern populations had very little exposure to sucrose in food. Modest amounts were present in some fruits, leaves, and vegetables. Selective plant breeding then produced the sweet-tasting fruits that rapidly gained popularity. The oldest sucrose-rich fruits are probably palm dates, which go back 5000–6000 years. Most modern fruits and vegetables contain much more sucrose than earlier cultivars. Today's mangos contain around 10% sucrose. Bananas have a 6% sugar coating. Peas and carrots weigh in at a sweet 4%.

People on the Indian subcontinent developed the production of refined sugar less than 3000 years ago. Extensive production and use then took off about 1000 years ago during the Arab agricultural revolution and later with the introduction of large-scale sugar cane plantations to the Americas. This means that probably many of us are still not genetically adapted to the present-day intake level (in the USA) of about 30 kg/year [104]. Many people still consume more than 100 g of sucrose on a good day, pushing their sugar digestion much harder than anything their distant ancestors ever experienced.

The sucrase-isomaltase (*SI*; EC 3.2.1.10; OMIM 609845) complex in the small intestinal mucosa releases glucose and fructose from ingested sucrose (Figure 4.18). The sodium-glucose transporter 1 (*SGLT1* or *SLC5A1*; OMIM 182380), which sits right next to the membrane-bound sucrase, moves glucose into mucosal cells. The newly released fructose molecule enters the mucosal cell via the GLUT5 (*SLC2A5*; OMIM 138230) transporter.

A significant number of adults are likely to experience sucrase deficiency symptoms because they cleave and absorb their high sucrose load before it leaves the proximal small intestines and becomes fodder for the ever-hungry trillions of commensal organisms in our ileum and colon. Just to keep this in perspective,

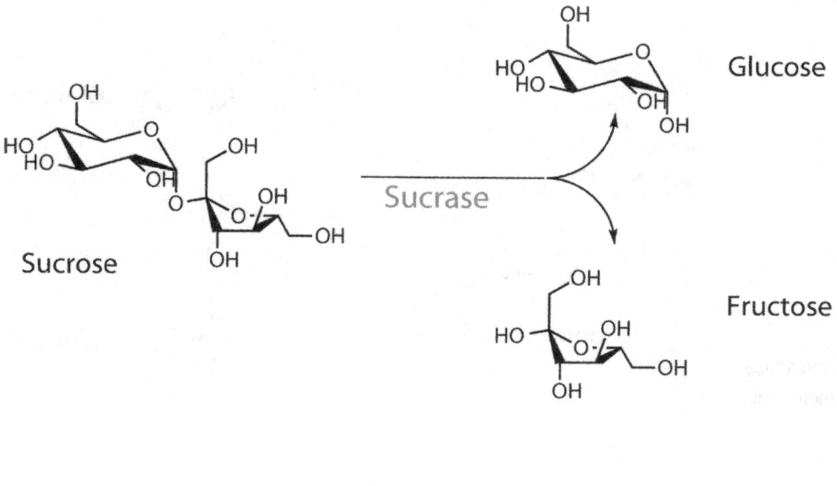

FIGURE 4.18
The intestinal brush border sucrose—isomaltase enzyme complex splits table sugar (sucrose) into glucose and fructose.

> **CONSIDER THIS**
> Some children and adults with chronic diarrhea have low sucrase activity and would do better with less sucrose.

a cup of milk contains about 12 g of lactose. This amount of undigested sugar causes significant abdominal discomfort in many people with lactose intolerance. A similar amount of sucrose is in a single milk chocolate bar. This would be certainly enough to cause discomfort in people with sucrase deficiency. Two bars would do the trick in even more people. Previous estimates suggested that one in 500 Americans has sucrose intolerance [69]. Congenital sucrase deficiency is much more common in Arctic natives, with a prevalence of more than 10% in Greenland Eskimos [105]. The condition is best assessed with a hydrogen breath test after drinking a challenge dose of sucrose (usually 50 g). A more specific breath test uses sucrose labeled with a stable isotope (^{13}C) and measures the isotope label with mass spectrometry in the carbon dioxide of exhaled air [106]. Confirmation can be obtained by measuring enzyme activities in biopsy samples from the jejunum [107].

Genetic variants associated with low sucrase activity are much more common than is generally known and appear to affect several percent of Americans and Europeans [108]. Constitutionally low sucrase activity is depressed even further in people taking the antacid medication ranitidine [109], which could be relevant for individuals with a low-activity variant.

4.5.6 Isomaltose

The amylase enzymes generate both maltose and isomaltose fragments when they digest starch. The difference between the two disaccharides is that an α-(1→4) bond connects the two glucose parts in maltose while in isomaltose they are linked by an α-(1→6) bond. On paper this is just a different number but for the enzyme that has to split the disaccharide it is an entirely different job. As it turns out, the same protein complex that handles sucrose (*SI*; EC 3.2.1.10; OMIM 609845) is also good at splitting isomaltose (Figure 4.19).

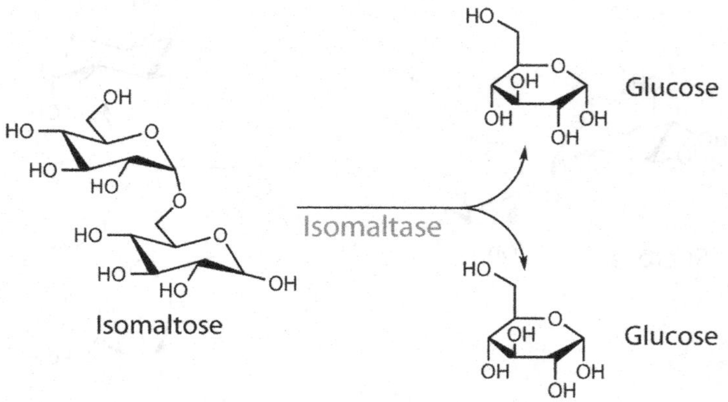

FIGURE 4.19
The intestinal brush-border enzyme isomaltase cleaves isomaltose into two glucose molecules.

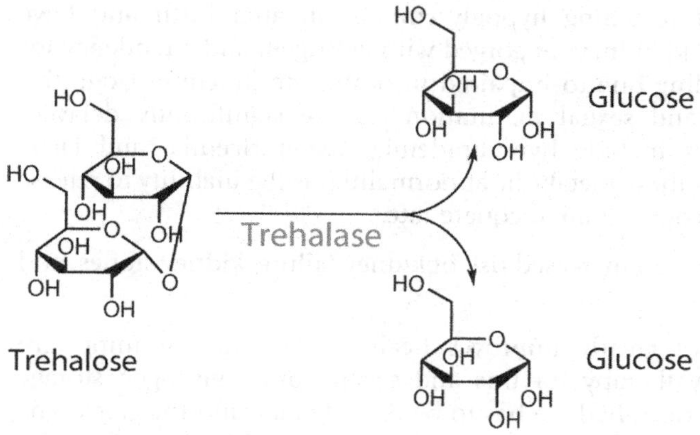

FIGURE 4.20
Trehalase in the brush border of the proximal small intestine cleaves the disaccharide trehalose from bacteria, mushrooms, industrially produced food additives, and yeast in most, but not all, people.

4.5.7 Trehalose

Trehalose (α-D-glucopyranosyl-$(1 \rightarrow 1)$-α-D-glucopyranoside) is a nonreducing disaccharide consisting of two alpha, alpha-linked glucose moieties (Figure 4.20). Mushrooms and yeasts are among the very few natural foods and foods ingredients that contain significant amounts of trehalose. Industrial production from starch has made trehalase available as a food additive.

The disaccharide is readily cleaved by the brush border enzyme trehalase (EC 3.2.1.28; OMIM 275360) in the small intestine, and the two resulting glucose molecules can then be absorbed and metabolized in the usual fashion.

Most people tolerate trehalose without any untoward effects [110]. However, a small minority of otherwise healthy Greenlanders [111], Finns [70], Japanese [112], and presumably other populations [113, 114] have low or absent intestinal trehalase activity and cannot tolerate even modest amounts from mushrooms well. It is likely that this low or absent activity is due to genetic variation but the specific changes that abolish trehalase activity in these individuals are not known. Abdominal symptoms are the same as in other sugar malabsorption syndromes, arising from the osmotic action of undigested sugar and from bacterial gas production.

4.5.8 Glycogen storage diseases

After we are done with digesting and absorbing the carbohydrates in a meal, we have to fall back on our reserves, which is glycogen stored in the liver and skeletal muscles. We need a constant supply of glucose for ongoing metabolic functions and for our glucose-hungry brain even while we sleep. There are several rare glycogen storage disorders that disrupt the availability of glucose reserves.

Let us briefly consider [115] just one of these, glucose-6-phosphatase deficiency (glycogen storage disease type Ia or von Gierke disease; OMIM 232200), which is due to lack of a functional *G6PC* gene (OMIM 613742). Infants born with this very rare condition (less than 1:100,000 generally or 1:20,000 in Ashkenazi

Jews) suffer from life-threatening hypoglycemia soon after birth and have anemia, enlarged liver and kidneys engorged with glycogen, and a tendency for easy bruising and bleeding due to impaired thrombocyte function. Cognitive development, growth, and sexual maturation can be significantly delayed. Metabolic abnormalities include hyperlipidemia, hyperuricemia, and lactic acidemia. The reason for these metabolic abnormalities is the inability to release glucose from stored glycogen at an adequate rate.

During adulthood there is an increased risk of kidney failure, kidney stones, and liver carcinoma.

Good metabolic control greatly improves health outcomes. An important commonality of dietary therapy for this and several other glycogen storage disorders is that dietary carbohydrate has to be provided around the clock. The prevention of accumulation of lactic acid in blood and the blunting of insulin secretion in response to carbohydrate intake also seems important for G6PC deficiency. This was achieved in some cases with the use of low-dose diazoxide [116]. Otherwise, frequent meals and the use of a feeding tube (particularly during the first year of life) are proven strategies. Macronutrient composition is unchanged, but uncooked cornstarch is a staple because it is slow to release glucose and readily available on the go [115]. Citrate supplements may reduce the risk of kidney stone formation.

4.6 ALCOHOLS

4.6.1 Ethanol

Ethanol occurs naturally under some conditions in decaying fruits, honey, and other sugar-rich plant parts. Many animals, including elephants, giraffes, moose, and primates, can get drunk when they gorge on overripe, fermenting fruit. Videos of intoxicated animals are at least as common (and popular) as depictions of the less accidental human equivalent (maybe excluding postings on social media). Humans appear to have learned to brew alcoholic beverages thousands of years ago, as detailed in Chapter 3. More recently, they developed technologies to produce commercially available beverages with up to 80% ethanol, concentrated enough to disinfect wounds, fuel rockets, or cook food. Many people in Western countries consume 20−40 g of ethanol on most days, which is the amount in 2−4 small glasses of wine or beer, and considerably more consume this amount a few times a year. Heavy drinkers often drink more than 100 g in 1 day.

With the consumption of modest amounts, alcohol dehydrogenases (EC1.1.1.1), encoded by several distinct genes, are responsible for most of the conversion of ethanol to acetaldehyde (Figure 4.21). Aldehyde dehydrogenases (EC1.2.1.10), again encoded by several distinct genes, are responsible for the second step of ethanol breakdown. These enzymes convert acetaldehyde to the readily metabolized acetic acid. Other less specific enzymes, particularly cytochrome p450 (CYP2E1; EC 1.14.14.1; one of the mixed function oxidases in the liver), contribute increasingly to overall metabolic activity with higher and

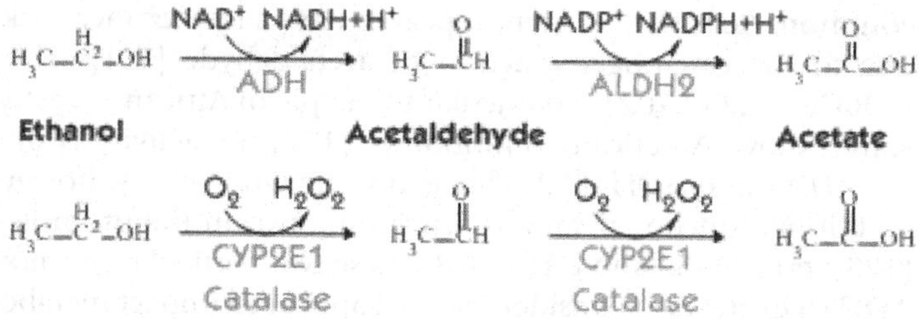

FIGURE 4.21
Most ethanol is converted first to acetaldehyde by alcohol dehydrogenases (ADH) in the cytosol and then to acetate mostly by mitochondrial acetaldehyde dehydrogenase (ALDH2). Excess amounts can be metabolized by cytochromes (mixed function oxidases) and catalase (CYP2E1).

chronic ethanol consumption. Catalase (EC1.11.1.6) is another enzyme that breaks down ethanol in peroxisomes, generating hydrogen peroxide in the process. This metabolic pathway appears to be responsible for some of the ethanol metabolism in brain [117].

The three genes responsible for most ethanol-metabolizing activity are *ADH1A* (OMIM 103700), *ADH1B* (OMIM 103720), and *ADH1C* (OMIM 103730). Some ethanol is metabolized when it passes through the stomach, mostly through the action of ADH3 (OMIM 103730) and ADH4 (OMIM 103740). Great care must be taken when reading older and sometimes even current texts, because the nomenclature has changed [118]. Table 4.3 lists the most relevant isoforms of ethanol-metabolizing alcohol dehydrogenases with both the current and older names.

One of the more fascinating variants is *ADH1B*2* (48His; rs1229984) as explained in Chapter 3. It is found mainly in Asians and occasionally in Sephardic Jews. The allozyme translated from this variant has a much higher activity than the *ADH1B*1* version and maintains that high activity at very low ethanol concentration [119]. This means that ethanol will be cleared very rapidly from the body. Carriers do not tolerate ethanol consumption well, basically because this overactive enzyme converts the alcohol almost instantly to acetaldehyde with all its unpleasant, hangover-inducing side effects. The most visible symptom is flushing of face (Oriental flush). But there may be health benefits, most likely due to lower alcohol consumption. In Korea, for example, where this variant is

Table 4.3	Nomenclature and Kinetic Characteristics of the Major Ethanol-Metabolizing Alcohol Dehydrogenases				
Gene	**OMIM**	**Variants**	**Old Name**	**Old Variant Names**	**Activity**
ADH1A	103700	None	ADH1	None	—
ADH1B	103720	*ADH1B*1* reference	ADH2	*ADH2*1*	Low
		*ADH1B*2* (rs1229984)		*ADH2*2*	Very high
		*ADH1B*3* (rs2066702)		*ADH2*3*	High
ADH1C	103730	*ADH1C*1* reference	ADH3	*ADH3*1*	High
		*ADH1C*2* (rs1693482)		*ADH3*2*	Low

common, *ADH1B*2* carriers appear to have a lower cancer risk [120], presumably due to the carcinogenic action of acetaldehyde [121]. The variant *ADH1B*3* (396Cys; rs2066702) is particular to people of African ancestry and also occurs in some Native American communities [122]. Its activity is intermediate between *ADH1B*1* and *ADH1B*2*. This is not the whole story, however. There are many additional functional variants, including two in the promoter region of *ADH1B* [123]: rs1229982 and rs1159918. These two particular promoter variants increase ADH1B expression considerably and appear to impact metabolic effects, drinking habits, and dependency patterns, in addition to the more commonly investigated variants. The two commonly studied versions of *ADH1C* are *ADH1C*1* (272Arg) with low activity and *ADH1C*2* (272Gln; rs1693482) with high activity. *ADH1C*1* and *ADH1C*2* are about equally common in Caucasians, while *ADH1C*2*, with the higher enzyme activity, is dominant in Africans, Asians, and other groups. ADH4, which is expressed in the stomach lining, also has several common variants with functional differences [124].

Mitochondrial *ALDH2* (OMIM 100650) and, to a lesser extent, cytosolic *ALDH1A1* (OMIM 100640) and *ALDH1B1* (OMIM 100670) are the main genes encoding acetaldehyde-metabolizing enzymes in the brain, liver, and stomach. As with alcohol dehydrogenases, people differ greatly in their ability to detoxify acetaldehyde. Carriers of the common *ALDH2* variant *ALDH2*2* (Glu504Lys; rs671) have a much diminished ALDH activity. This variant is present in about half of all Asians but is rare in most other populations. Alcohol intake appears to depend strongly on *ALDH2* genotype. The average weekly alcohol consumption in Japanese men with the genotypes 1/1, 1/2, and 2/2 in one study was 213, 79, and 26 g, respectively [125]. That means that the small (7%) minority of Japanese men with the homozygous *ALDH2* 2/2 genotype have little more than one drink a week on average, whereas the majority (about 58%) with the 1/1 genotype love to partake at an average level of about ten drinks a week.

The impact of genetic variation in these two groups of genes encoding ethanol-metabolizing enzymes on drinking behavior can be reasonably simplified to this: higher ADH activity and lower ALDH activity tend to increase the spontaneous aversion to alcohol consumption. Alcohol abuse and addiction tend to follow the same rule. However, this simple rule has its limitations because actual enzyme activities are not always predicted well by the usual variants and cultural and other influences can override the genetic cues.

4.7 MINERALS

4.7.1 Sodium

It has been said that the high sodium concentration in our body fluids reflects the marine environment of very distant ancestors. However that may be, this is how our body works now. Much of what happens in our body is driven by the powerful sodium concentration gradient between the inside of our cells and the fluids around them. This is very much like the ever-present electricity that powers our homes and offices and much in between. The high sodium concentration on the outside of our cells and the low concentration on the inside are maintained by a battery of sodium-potassium adenosine triphosphatases (ATPases). We spend more than 20% of our resting energy expenditure just to run this sodium gradient. Complex regulatory mechanisms maintain the concentration of sodium ions in blood within a narrow range, for most populations around 140 mmol/L (same as mEq/L) and inside most cells it is around 4 mmol/L. These mechanisms increase sodium excretion when it is greater than the amount needed to balance losses with sweat, urine, and feces. But with low intake, the adjustment can only go so far. Continued net sodium losses lead to hyponatremia and volume contraction. When the depletion occurs rapidly, as it does with vomiting and diarrhea due to cholera and other gastrointestinal infections, cerebral edema and heart failure may result. Infectious diseases with massive salt loss are so deadly, especially for young children, that they can shape the genetic settings of a local population within a short time. Regulatory systems are not equally good on both ends of the spectrum and the one that keeps our sodium in balance is not an exception. The downside to holding on to sodium very tightly comes from the fact that regulation of sodium balance and regulation of blood pressure use many of the same genes and mechanisms. The genetic variants that are good for surviving acute diarrhea or a low-sodium environment tend to increase blood pressure when sodium intake is high. This is the evolutionary basis of the salt-sensitive genotype.

We should expect that each population becomes adapted to the local environment and evolves its own genetic setting. This includes adaptation at all levels: salt tasting acuity, salt preference, and the metabolic and regulatory response to high and low salt intake. The observed genetic diversity is consistent with a tension between high and low sodium intakes under different circumstances. All of this probably would not matter much today, if it were not for the very high salt intake of many people who are not adapted to these amounts. Americans consume on average more than 3200 mg sodium (that is more than 8 g of salt)/day [126]. Since about half of all people with high blood pressure are sensitive to salt intake in excess of 1600 mg (4 g salt)/day, current salt intake levels literally kill. Pointing to the many people who can seemingly consume a lot of salt without much harm cannot provide consolation or excuse for the many salt-sensitive people.

4.7.1.1 THE GENSALT STUDY

The question is how to identify the expected genetic variants that make some people more sensitive than others. The concept is quite simple: expose a lot of otherwise healthy people to different amounts of salt, measure their blood pressure, and then compare the genetic variants of salt-sensitive and -nonsensitive individuals. The GenSalt study did just that, with some extra sophistication added in. Close to 2000 adults were asked to use a low-sodium diet (1180 mg sodium/day) for 7 days and then a high-sodium diet (7080 mg sodium/day) for another 7 days. Blood pressure measurements were taken in triplicate initially and after each of the 7-day periods. And from this pile of data, many reports of gene–sodium interactions continue to emerge [127]. The limitation of this design is, of course, that it only captures the short-term effects of low or high salt intake.

Since blood pressure deregulation and essential hypertension develop over the course of many years, these observations tend to underestimate the extent of the response. However, this type of investigation will point us in the right direction and help with the identification of the major players in salt-induced hypertension. Several studies of genetic variants with effect on salt sensitivity have been completed or are in progress, including the European Project on Genes in Hypertension (EPOGH) and the International Hypertensive Pathotype (HyperPATH) cohort. Many more will be needed to resolve the genetic intricacies of salt sensitivity.

Numerous genes contribute to the maintenance of blood osmolality, pressure, and volume, and of sodium concentration. Variants in many of them are known to affect the response to salt intake. We will now review a few of the better understood regulatory systems and their contributory genes. When particular genes and gene variants are mentioned, they should only be taken as illustrative examples, since the number of functionally important ones is far too large for inclusion. Many of these variants are included in many studies because they happened to be investigated early on and later research just continued with the same selection of variants. Unbiased studies will certainly uncover many more and may find that the early candidate variants have much less impact than suggested by the initial data. This has a lot to do with selection bias, as will be discussed in Chapter 6.

4.7.1.2 RENAL SODIUM RECOVERY

The nephrons in the kidneys remove from blood as much as 1400 g of salt/day and then whittle this amount down to a few grams for excretion in urine. They do so by selectively removing sodium and chloride ions in the proximal and distal tubules and the collecting ducts.

The renal epithelial sodium channel (*SCNN1B* or ENaC; OMIM 600760) recovers sodium from the distal tubule. Significantly increased activity is the cause of salt-sensitive hypertension in people with Liddle syndrome (also called pseudo-aldosteronism; OMIM 177200). Since a single copy of variant ENaC is enough for manifestation of the condition, inheritance is dominant. Treatment is based primarily on sodium restriction, with the addition of ENaC blockers (amiloride, benzamil, and triamterene) as needed. How common are ENaC variants with increased activity (gain-of-function variants)? The answer is that they are apparently quite common. The abnormally responsive variants Gly589Ser and i12−17CT were found in about 9% of hypertensive patients in Finland [128]. The GenSalt study found that more than 10% of the investigated Chinese adults have ENaC variants associated with increased blood pressure response to high sodium intake [129]. Among them are two (rs4073930 and rs4299163) that may be intronic enhancers of gene expression. The Thr594Met variant (rs1799979) is found in more than 5% of individuals with African ancestry [130], but not in Caucasians or Asians. This variant is very often associated with hypertension and appears to be a marker for salt sensitivity. Several other ENaC variants may contribute to hypertension in Caucasians [131]. The quantitative sodium channel activities of these numerous ENaC variants need to be studied further and related systematically to salt intake and blood pressure. It might be of interest that carriers of these salt-sensitive ENaC variants are also much more likely to get high blood pressure when they are eating natural licorice [132].

Two more genes should be mentioned in this context because the interaction of their products with ENaC can explain why their variants show up in genetic studies of hypertension. These are the ubiquitin protein ligases, NEDD-4 (*NEDD4*; OMIM 602278) and NEDD4.2 (*NEDD4L*; OMIM 606384), that guide the degradation of ENaC.

Adducin is important for renal sodium recovery through a very different mechanism. This component of the cytoskeleton consists of alpha-adducin (*ADD1*; OMIM 102680) and either beta-adducin (*ADD2*; OMIM 102681) or gamma-adducin (*ADD3*; OMIM 601568). Through a signaling cascade, adducin boosts the activity of an ATPase that pumps sodium from tubular epithelial cells into blood and by this means increases sodium reabsorption in the kidney. *ADD1* variants, such as Gly460Trp (rs4961), are associated with hypertension [133]. Reducing sodium intake is definitely worth a try for patients with such *ADD1* variants.

4.7.1.3 THE RENIN/ANGIOTENSIN SYSTEM

Blood pressure regulation depends very significantly on the modulation of aldosterone concentration by an integrated system of checks and balances. A simplified presentation (Figure 4.22) may start in the kidney with the action of

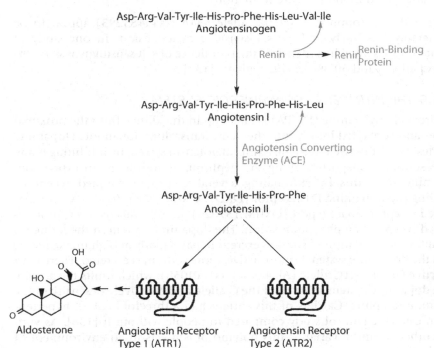

FIGURE 4.22

Six different genes contribute to the renin/angiotensin system, a primary regulator of blood pressure and blood solute concentration.

renin (*REN*; EC 3.4.23.15; OMIM 179820) on angiotensinogen (*AGT*; OMIM 106150), which generates angiotensinogen I. This intermediate is then short-ened by the angiotensin-converting enzyme (*ACE*; EC 3.4.15.1; OMIM 106180) to angiotensin II, which acts on angiotensin receptors type 1 (*AGT1*; OMIM 106165) and type 2 (*AGT2*; OMIM 300034).

Renin is inhibited by binding to the renin-binding protein (*RENBP*; OMIM 312420), which is also an *N*-acetyl-D-glucosamine 2-epimerase (EC 5.1.3.8).

Among the genetic variants in the renin/angiotensin system, angiotensinogen variants have been linked in some studies to sodium sensitivity. Plasma AGT levels are highly heritable, though the link to the risk of hypertension is elusive [134]. Individuals with the rs5051G, rs699 T, or rs943580 T alleles tend to have the lowest AGT concentrations, which make them less likely to be salt-sensitive. Contrary to initial reports, the common ACE insertion/deletion polymorphism does not appear to contribute to salt sensitivity in a major way [135].

4.7.1.4 CORTICOSTEROIDS

Another group of steroid hormones with potent effects on blood pressure, the corticosteroids, are produced in the outer layer of the adrenal gland. The enzyme 11-beta-hydroxysteroid dehydrogenase (EC 1.1.1.146) converts active cortisol into inactive cortisone. The *HSD11B2* gene (OMIM 614232) encodes the isoenzyme that is expressed in the kidney and which plays a particularly important role in blood pressure regulation.

People with the common *HSD11B2* allele 534A (rs45483293) appear to be considerably less likely to have salt-sensitive hypertension. In one study of patients with essential hypertension, the prevalence of salt sensitivity was 14.3% in G/A patients and 50.8% in G/G patients [137].

4.7.1.5 THE INTRARENAL DOPAMINERGIC SYSTEM

Dihydroxyphenylalanine (DOPA) is taken up in the kidney from the proximal tubules and converted locally into the neurotransmitter dopamine. Dopamine provides the counterweight to the renin/angiotensin system by inhibiting many of the sodium transporters along the nephron, as well as renin expression. Dopamine does this by stimulating several G protein-coupled receptors, including the dopamine D1 receptor (*DRD1*; OMIM 126449). The G protein-coupled receptor kinase type 4 (OMIM 137026), in turn, inactivates G protein-coupled receptors by phosphorylation. The dopamine system in the kidney is critical for the regulation of sodium excretion, particularly at high intake levels. When the dopamine system is more active, more sodium is excreted. There is an overactive GRK4 1711C allele (Ala486Val; rs1801058), which limits the activity of the dopamine system more than the C allele and effectively desensitizes the dopamine receptors. Carriers of this variant had a threefold risk of hyperten-sion in one case-control study compared to people without it [138]. As with many other common variants, this polymorphism acts in an environment of several other variants, each of which has its own effect on GRK4 activity. Two

Variation at the AGT locus

What many investigators ignore is the contribution of all variants on the same strand (haplotype) to gene product quality and quantity. Let's take the example of the AGT −6G>A variant (rs5051), which was shown in several studies to increase salt sensitivity. In a cohort of Caucasian families in Utah (one of the reference cohorts of the Hap-Map project for determining common haplotypes around the world), the attributable variance in AGT concentration in plasma was 0.64 for the rs5051 variant, indicating a very high degree of heritability. But there are many other common variants in the same region, many of which also have a strong association with the AGT concentration. Most of the individuals in this population carry just one or two of five haplotypes (variant combinations on the same DNA strand). The illustration below uses a period to indicate that the base is the same as in the H1 haplotype in the top line. What we see is that 60% of these Caucasian individuals (the ones with haplotypes H1 or H2) share an ancestor. We know this because they have mostly the same bases in the 25 examined positions across this region of more than 30,000 bases in length. The one difference between haplotypes H1 and H2 in position 11,535 (rs7079) appears to be due to a later mutation and this difference has little impact on AGT concentrations. The other three haplotypes differ more extensively. Because 94% have the same combination of alleles in the promoter region (rs5051) and in the 3′ untranslated region (rs943580), it makes little difference which one is measured. This is what is referred to as linkage. In one study investigating the impact of AGT variants on the risk of preeclampsia, it was found that the Met235Thr (rs699) polymorphism influenced outcomes in Caucasian women [136]. If the investigated patients had the same haplotype structure as the Utah cohort, it should not have mattered whether the rs5051 or rs943580 variants had been measured instead because they are shared within the two haplotype groups (H1-H2 vs. H3-H4-H5).

Individuals with the H1 or H2 haplotypes tend to have the lowest AGT concentrations; carriers of the H4 haplotype the highest concentrations. Since quite a few bases vary between the H1/H2 and the H4 haplotypes it is difficult to know which ones are responsible for the difference in AGT concentration.

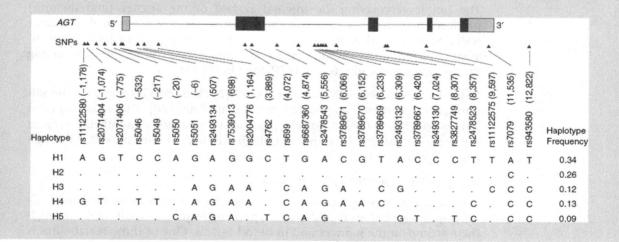

Haplotype	rs11122580 (−1,178)	rs2071404 (−1,074)	rs2071406 (−775)	rs5046 (−532)	rs5049 (−217)	rs5050 (−20)	rs5051 (−6)	rs2493134 (507)	rs7539013 (698)	rs2004776 (1,164)	rs4762 (3,889)	rs699 (4,072)	rs6687360 (4,874)	rs2478543 (5,556)	rs3789671 (6,066)	rs3789670 (6,152)	rs3789669 (6,233)	rs2493132 (6,309)	rs3789667 (6,420)	rs2493130 (7,024)	rs3827749 (8,307)	rs2478523 (8,357)	rs11122575 (9,597)	rs7079 (11,535)	rs943580 (12,822)	Haplotype Frequency
H1	A	G	T	C	C	A	G	A	G	G	C	T	G	A	C	G	T	A	C	C	C	T	T	A	T	0.34
H2	.	.	.	.	.	.	.	.	.	.	.	.	.	.	.	.	.	.	.	.	.	.	.	C	.	0.26
H3	.	.	.	.	.	.	A	G	A	A	.	C	A	G	A	.	C	G	.	.	.	.	C	C	C	0.12
H4	G	T	.	T	T	.	A	G	A	A	.	C	A	G	A	A	C	.	.	.	.	C	.	C	C	0.13
H5	.	.	.	.	.	.	C	A	G	A	.	T	C	A	G	.	.	.	.	G	T	.	T	C	C	0.09

additional exonic variants were looked at more closely: 448G>T (Arg65Leu; rs2960306) and 679C>T (Ala142Val; rs1024323). As can be seen in Table 4.4, people with the 1711C allele had the highest likelihood to have high blood pressure, and those with the 1711T allele had the lowest likelihood. However,

Table 4.4	Odds Ratios of Having Hypertension in Chinese Adults with Common GRK4 Haplotypes [138]	
GRK4 Haplotype	Frequency	Odds Ratio
448G-679T-1711T	0.028	0.29
448T-679T-1711T	0.124	0.65
448G-679C-1711T	0.36	0.86
448G-679T-1711C	0.054	1.17
448G-679C-1711C	0.46	1.39
448T-679T-1711C	0.026	5.71

depending on which of the other two variants were on the same DNA strand (haplotype), there were still several-fold differences between the two groups. The relative contribution of these three polymorphisms to salt-induced hypertension was very similar in another cohort, this one of Japanese adults [139]. The presence of other common GRK4 polymorphisms made no difference. These investigators confirmed that the high salt sensitivity of people with multiple low-activity GRK4 variants was due to the lack of response to dopamine. Once you look at it from the haplotype perspective, someone with the 448G-679T-1711T haplotype has a low risk, while the risk of someone with the 448T-679T-1711C haplotype is almost 20 times higher. And all of this is probably because this high-risk haplotype just cannot take that extra pinch of salt.

4.7.1.6 VASOCONSTRICTION

The cell layer covering the internal surface of the arteries (endothelium) produces peptides that cause contraction of the muscle fibers in the vascular walls. Sodium load is an important trigger for the release of these vasoconstrictive peptides. The endothelins act on a group of specific receptors, including the endothelin receptor, type B (*EDNRB*; OMIM 131244).

Individuals with the *EDNRB* allele 1065G (277Leu; rs5351) appear to be salt sensitive more often than people without that allele [140]. This variant in exon 5 does not change the amino acid sequence in the mature endothelin receptor B protein, but affects the response to high salt intake indirectly, either through altered splicing or linkage to another variant.

4.7.1.7 EICOSANOID METABOLISM

Various metabolites (eicosanoids) of long-chain polyunsaturated fatty acids are important mediators of the blood pressure response to salt intake through their actions in the kidneys and in blood vessels. One of these metabolites is 20-hydroxyeicosatetraenoic (20-HETE), which inhibits tubular sodium transporters and promotes sodium excretion [141]. Cytochrome p450 4A11 (*CYP4A11*; EC 1.14.15.3; OMIM 601310) oxidizes arachidonic acid at the u end (the carbon that is the farthest away from the carboxyl group) to 20-HETE.

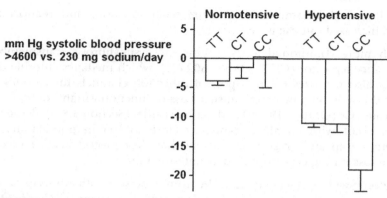

FIGURE 4.23

The impact of genetic variants, such as *CYP4A11* 8590C, often depend on the existing health status, usually because other genetic or exposure factors have to be present. Shown here are findings from the HyperPATH cohort, investigating the short-term response to diets with very little sodium (230 mg/day) and high sodium content (>4600 mg/day), each for 7 days [142]. The bars show the difference in mean arterial pressure with the two sodium intake levels in participants with initially normal (on the left) and high blood pressure (on the right).

The common CYP4A11 variant T8590C has diminished activity, producing less 20-HETE and allowing increased sodium retention. Increasing salt intake raises blood pressure in people with the C/C genotype if their blood pressure is already high [142]. Apparently, it takes two copies of this loss-of-function variant and the additional impairment of other regulatory factors to induce a failure of blood pressure control when challenged by high salt intake (Figure 4.23).

4.7.2 Chloride

Chloride is the main anion in blood and tissues. It is also used as hydrochloric acid for the digestion of foods in the stomach and as hydrochlorous acid for the antibacterial burst of macrophages. Deficiency occurs mainly in response to persistent vomiting, particularly in children. Loss of both sodium and chloride is much more common and usually involves extensive sweating and inadequate salt intake. Health outcomes of chloride deficiency include low blood pressure, alkalosis, hypokalemia, lethargy, loss of muscle and cognitive control, and ultimately death. In such instances, individual genetic vulnerabilities may become important.

The various transporters and channels for absorption, distribution and excretion are too numerous to list here in detail.

Congenital chloride diarrhea (OMIM 214700) is a very rare condition caused by mutations in the *solute carrier family 26 member 3* (*SLC26A3*; OMIM 126650) gene. The defect disrupts normal chloride transport in the ileum and colon, causes intestinal inflammation, impairs renal function, and reduces male fertility. Increased salt use is beneficial.

A much more common defect concerns the cystic fibrosis transmembrane conductance regulator (*CFTR*; OMIM 602421), which mediates chloride transport but also regulates the transport of other ion channels for chloride and sodium through unknown mechanisms. Loss-of-function variants of both CFTR copies cause CF (OMIM 219700) as described earlier (Section 3.8.2). Because the variants occur with an allele frequency close to 5% in populations with Northern European ancestry, it is likely that they protect against excessive chloride loss during catastrophic gastrointestinal infections.

The genome-wide sequencing of DNA from a person with Khoisan ancestry found a variant of the *chloride channel, voltage-sensitive Kb* gene (*CLCNKB; OMIM 602023*) that enhances chloride recovery in the kidney. This variation may be an adaptation to the extremely dry environment of the Kalahari Desert in the African south, where Khoisans have lived as hunter-gatherers for millennia [143]. Loss-of-function variants in this gene are more widely known, because they are often responsible [144] for Bartter syndrome, a condition of excessive renal salt losses with hypokalemia and metabolic alkalosis. Another common *CLCNKB* gain-of-function variant, where replacement of a threonine at position 481 by serine has been linked to essential hypertension in both children and adults in some populations 145. This is an important finding because it demonstrates that hypertension can be related to chloride balance, not just to changes in sodium homeostasis.

4.7.3 Potassium

A large team of ion channels, transporters, and regulators comes together to maintain potassium homeostasis. As if the potassium system was not already complex enough, many of the team members also work on sodium and chloride sensing and transport. It is easy to see that this gene set will be finely tuned to typical amount and mixture of electrolytes in the food supply. An African rain forest nutritope with fruits and leaves provides lots of potassium but sodium may be in short supply. The move north to a frigid tundra many thousands of years ago and the heavy reliance there on meat consumption constitutes a dramatic shift toward higher sodium intake and more limited potassium supplies.

Blood pressure response to potassium intake appears to differ depending on genetic predisposition. A genome-wide association study (GWAS) of Han Chinese adults found that carriers of the *angiotensin II receptor, type 1* (*AGTR1*) C/C genotype of rs16860760 responded to an extra 60 mmol potassium (1140 mg)/day with slightly lower systolic and diastolic blood pressure (-3.7 and -3.1 mm mercury, respectively). Carriers of the T/T genotype, in contrast, had a 1 mm Hg increase [146]. The *AGTR1* gene encodes a receptor for the vasopressor hormone angiotensin II, which rises in response to low renal perfusion and low blood sodium concentration. Increased angiotensin II action in the kidneys (partially by stimulating aldosterone release) promotes reabsorption of sodium and water, and excretion of potassium.

4.7.4 Calcium

The concentration of calcium ions in blood and tissues is tightly controlled because the various calcium-mediated signaling functions would be disrupted by excessively low or high concentrations. Deviations in calcium ion concentrations, for instance after meals, are detected by the calcium sensing receptor (*CASR*; OMIM 601199), which responds by changing the parathyroid hormone secretion rate and modifying other regulatory elements. The interaction of this and other calcium-regulating genes with calcium intake levels is necessarily indirect and usually difficult to ascertain.

A key example for such convoluted nutrient–gene interactions is the impact that variation of the *CASR* gene might have. The S allele of the *CASR* Ala986Ser polymorphism (rs1801725) increases the set point for ionized calcium (Ca^{2+}) concentration [147]. This effect, in conjunction with habitual calcium intake levels, may have consequences for nephrolithiasis [148], colorectal cancer risk [149], and other chronic diseases.

4.7.5 Magnesium

We need magnesium as a cofactor of many enzyme reactions, to stabilize DNA and ribonucleic acid (RNA), and as a key constituent of bone. Of equal importance is the role of magnesium in cell signaling, where it often opposes the action of calcium. Low magnesium consumption, particularly against a background of high calcium intakes, worsens the risk of cancer and cardiovascular disease. We get magnesium mostly from fruits, potatoes, vegetables, and whole grains. The more refined and animal-based foods, on the other hand, tend to be low in magnesium and often have more of its white partner, calcium.

Magnesium from food enters the lining of the intestinal wall through an ATP-powered ion channel formed as a heterodimer by the subunits TRPM6 (OMIM 607009) and TRPM7 (OMIM 605692).

So far, so good. But then we realize that this channel is also the main port of entry for calcium, and both of these pushy ions compete for going through the door inside. That should be not a big problem for magnesium because the channel actually prefers this ion slightly over calcium. But the more calcium we eat, the harder it is for magnesium to make it into the intestinal cell. Large amounts of calcium can just crowd out dietary magnesium. This has great significance for carriers of the common *TRPM7* allele 4727G>A (Thr1482Ile; rs8042919) because low availability of magnesium unbalances cell signaling and promotes malignant cell growth. Even a very modest increase in magnesium intake lowers colon cancer risk in people without the variant. In contrast, people with a 4727A allele have to get a high dose for any benefit (Figure 4.24, left-hand

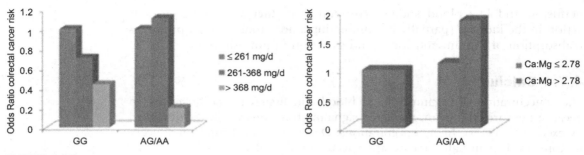

FIGURE 4.24
Carriers of the *TRPM7* 4727A allele can reduce their colon cancer risk [150] by maintaining a high magnesium (Mg) intake (left panel), but only if their calcium (Ca) intake is not also very high (right panel).

side). When they consume much more calcium than magnesium (Figure 4.24, right-hand side), the excess risk is also very apparent [150].

Practice case

Your patient is a woman of African ancestry in her mid-fifties and in good health. She gets 800 mg calcium and 340 mg magnesium from food. A dietary supplement gives her another 600 mg calcium.

If you did not know anything else about her, you might consider her calcium and magnesium intakes in a good range (the calcium recommended dietary allowance [RDA] for women over 50 is 1200 mg/day and the magnesium RDA is 320 mg/day).

But you know that she has the *TRPM7* 1482A/A genotype. This changes your evaluation, since her magnesium intake is too low and her calcium:magnesium intake ratio is too high for her genotype. Judging by the currently available data, her colon cancer risk is almost twice as high with the calcium supplement as it would be without the supplement (approximately 1% during the next 10 years instead of 0.5%). If she had no A allele, she would have no such increased risk with the supplement she uses.

But you can help. You tell her that she will be better off with slightly less calcium, say about 1200 mg in total, as long as she gets an additional 120 mg magnesium from food and maybe a moderately dosed supplement. This would take her calcium:magnesium intake ratio down to 2.61. She should try to eat more magnesium-rich fruits, nuts, seeds, and vegetables. A handful of nuts contain 120 mg magnesium, a cup of beans provides 90 mg, and a medium-sized banana gives her 32 mg. To the extent that she eats bread and other grain-based products, she should look for whole-grain alternatives. Replacing four slices of white bread with the same amount of whole-wheat bread adds about 64 mg magnesium.

It is easy to see that your nutrition plan reflects current nutritional recommendations for healthy adults, which is given heightened importance by her personal genetic predisposition.

4.8 TRACE ELEMENTS
4.8.1 Iron
4.8.1.1 TROUBLESOME METAL
Iron is one of those trace elements which we are all familiar with. Iron deficiency is immediately suspected when someone looks pale and feels tired and in many

instances a lack of the element is actually the cause. After all, we are talking about the most common micronutrient deficiency worldwide. Microcytic anemia (indicating iron deficiency) and low ferritin concentration (signaling diminished iron stores) would confirm the suspicion. About 30% of American women have iron deficiency during the last trimester of their pregnancy [151] and may not be able to provide adequate amounts to the fetus. Iron deficiency in the mother increases risk of delayed fetal growth, preterm delivery, spontaneous abortion, and stillbirth [152]. Given these risks and the relatively low cost of iron, one should expect very aggressive use of iron supplements to make iron deficiency very rare. But there is a dark side to iron nutrition, which is harm from excess. Iron is actually a fairly toxic transition metal that in a free, unbound form catalyzes the generation of free radicals. Another dangerous effect of unbound iron is the promotion of bacterial growth. The body's resources for sequestering free iron and preventing harmful effects are limited and can be overwhelmed by excessive amounts. The need for a plentiful iron supply on one hand and the risk of oxidation, infection, and liver damage, on the other hand, explain the extensive genetic variation in the relevant genes. We are all living records of the tug-of-war that has played out in our ancestors' nutritopes. We also need to recognize that this battle is still ongoing.

4.8.1.2 ABSORPTION AND TRANSPORT

The understanding of the regulation of iron absorption, utilization, and storage has recently shifted significantly [153]. The divalent metal transporter 1 (*DMT1* or *SLC11A2*; OMIM 600523) moves iron from food into the enterocytes of the small intestine. The only iron exporter in the body, ferroportin (*FPN1* or *SLC40A1*; OMIM 604653), moves iron from the enterocytes to transferrin (*TF*; OMIM 190000) in blood. The same ferroportin-dependent mechanism helps macrophages to get rid of the iron recycled from broken down red blood cells. Cells take up iron via the transferrin receptor 1 (*TFRC*; OMIM 190010). The human hemochromatosis protein (HFE; OMIM 613609) competes with transferrin for binding to TFRC and at the same time greatly reduces the affinity of TFRC to iron-laden transferrin. Increased HFE activity reduces the amount of iron that gets into a cell. Iron is stored in the liver enclosed in a cage of ferritin, which is composed of numerous light ferritin (*FTL*; OMIM 134790) and heavy ferritin (*FTH1*; OMIM 134770) subunits.

4.8.1.3 HORMONAL CONTROL

We now know that the hormone hepcidin (*HAMP*; OMIM 606464) coordinates absorption, storage, and utilization of iron (Figure 4.25). High blood levels of hepcidin slow iron export from enterocytes in the small intestine by inactivating ferroportin and inhibiting expression of DMT1 and TFRC. If there is no ferroportin for releasing iron from the enterocytes into blood, the newly absorbed iron returns into the duodenal lumen as the enterocytes are shed at the end of their 2–3-day life span. Ferroportin is also essential for releasing iron from macrophages, which mop up stray heme and hemoglobin iron.

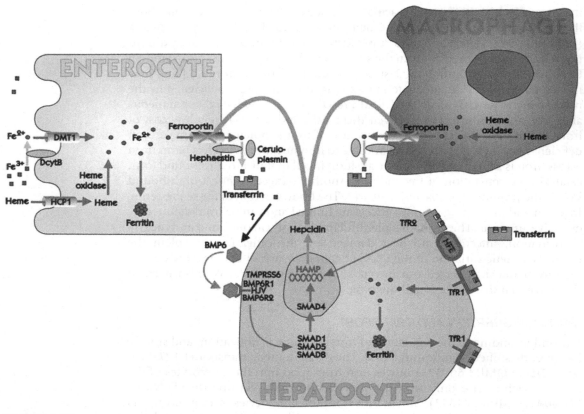

FIGURE 4.25

Iron absorption, delivery to target tissues, storage, and removal is regulated in a tight net of interactions, where hepcidin functions as the central regulatory hormone. Fe, iron.

Both large amounts of iron in blood and inflammation increase the production of hepcidin and its release from liver cells. The iron concentration in blood is sensed by a complex of transferrin receptor 1, HFE, and transferrin receptor 2 (*TFR2*; OMIM 604720). If transferrin receptor 1 binds diferric transferrin (transferrin with two iron atoms), it pushes the associated HFE molecule away. HFE is then more likely to talk to transferrin receptor 2, which triggers increased production and release of hepcidin. Bone morphogenic protein 6 (*BMP6*; OMIM 112266) is the master regulator for hepcidin in response to iron status. BMP6 communicates with the cell nucleus through a membrane complex consisting of hemojuvelin (*HJV*; OMIM 608374) and several other proteins. Signaling through this membrane protein complex is inhibited by the action of a membrane-associated protease, matriptase-2 (*TMPRSS6*; OMIM 609862). The signal transduction cascade from the membrane to the nucleus requires additional proteins, including several of the whimsically named SMAD proteins (SMAD stands for son of mothers against decapentaplegic; giving quirky names to genes is a signature perk of the productive genetic researcher). This is a lot of genes just for the regulation of a single hormone, hepcidin. But that is the way the system works—except when it doesn't.

4.8.1.4 ABNORMAL IRON RETENTION

Healthy, iron-replete individuals usually absorb about 1 mg of iron from a varied diet. Some individuals tend to absorb much more of the available iron in foods, largely due to a genetic predisposition. The extra iron tends to accumulate slowly, eventually causing harm. Hemochromatosis is an inherited condition with a greatly relaxed control of iron excess and an increased tendency to retain excess iron in storage. By definition, more than 45% of the iron-binding capacity of their circulating transferrin is saturated, which is much more than the 25% to 30% in people without the condition. Their blood concentration of free iron (not bound to transferrin) is increased above average by several orders of magnitude.

Genetic variants of *HAMP, HFE, HJV,* and *TFR2* all diminish hepcidin release and thereby increase iron absorption from the intestine and promote its storage in liver and other tissues. Such proabsorption variants probably represent adaptions to nutritopes with low iron availability because they shift the balance toward retaining more iron and safeguarding less against iron overload. This should be expected in light of the great risks of maternal iron deficiency for the fetus. The *HFE* gene, in particular, has common variants that promote iron absorption and retention. One of them has a tyrosine instead of the regular cysteine in position 282 (Cys282Tyr; rs1800562). This variant goes back to a mutation about 6000 years ago in an early ancestor of the Celtic population [154]. *His63Asp* (rs1799945) is a second iron-retaining variant, which probably arose several thousand years ago in today's Basque region in the north of Spain. A third one is Ser65Cys (rs1800730). These variants promote less iron retention than the allele encoding HFE 282Tyr and have much lower penetrance in terms of clinically relevant hemochromatosis.

Ferroportin disease is a more recently discovered condition, which also confers a tendency for iron accumulation. It is caused by variants in the *HAMP* gene (encoding hepcidin) that alter ferroportin responsiveness to hepcidin-mediated inactivation. The defining characteristic is iron overload in liver and macrophages as reflected by the ferritin concentrations, which tend to be massively in excess of 300 µg/L [155]. Transferrin saturation is increased in some patients but not in others, probably depending on the nature of the gene variant. Figure 4.26 summarizes average ferritin concentrations in patients with the same variant. The size of the dots corresponds to the number of reported patients with the same variant. As can be seen, the most common variants occur in a cluster between amino acids 64 and 181 and a few more at the tail end of the protein. Basically, all of these variants are associated with a significant loss of function, almost regardless of the position in the protein sequence where the change occurs. Lists of average observed effects for each known variant will eventually help with the prediction of clinical outcomes based on whole-genome information.

We still have to ask about the biological relevance of polymorphisms that promote iron retention. As we have seen in Chapter 1, hemochromatosis is not

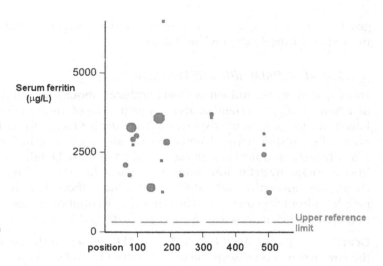

FIGURE 4.26
Several variants have been linked to ferroportin disease. Shown here is the median ferritin concentration for people with variants at the indicated amino acid positions in the mature ferroportin protein [155]. All of them are significantly over the upper ferritin reference limit of 300 µg/L for men (dashed line). The area of each spot corresponds to the number of observed cases.

a harmless condition. The main dangers are inflammation, cirrhosis, and eventually cancer of the liver, heart disease, cognitive decline, and increased vulnerability to some bacterial infectious agents. The occasional episodes of infections in people with hemochromatosis are often unexpected and sometimes even bizarre. One such case was the tragic death of a prominent microbiologist from the plague [156]. Another unusual case was that of an elderly woman with severe hemochromatosis who suffered a life-threatening infection with *Listeria monocytogenes* after eating cheese [157]. Common ferroportin variants also undermine resistance to tuberculosis [158].

Two copies of the HFE 282Tyr variant, a combination of HFE 282Tyr and 63Asp variants (compound heterozygotes), the aforementioned single copies of ferroportin variants, and a few less common *HAMP*, *HJV*, and *TFR2* variants lead to the progressive accumulation of iron in the liver and other tissues and a high risk of the eventual development of hemochromatosis. Homozygous carriers of the *HFE* 282Tyr allele constitute the majority of individuals with hemochromatosis symptoms [159]. Their risk of developing hemochromatosis is more than 4000 times higher than the risk of people with the common *HFE* sequence. Iron accumulation is usually slowed in women while they lose blood regularly during menstruation. Even among men, fewer than half of all *HFE* 282Tyr homozygotes will develop clinically relevant disease. This means that the penetrance of the homozygous *HFE* 282Tyr genotype is less than 50%. Only a small number of them will die from hemochromatosis, but the risk is still significant and explains the need for interventions. Large population studies indicate that hemochromatosis increases risk of premature death by 30–50% and accounts for nearly 1% of mortality for the entire population [160].

Another type of risk is a slowed wound healing and increased surgical complication rate. One large case series found that patients with two or more copies of

HFE variants suffered more complications after open or laparoscopic Roux-en-Y gastric bypass surgery and had a greater length of stay in hospital [161].

A single copy of the *HFE* 282Tyr, 63Asp, or 65Cys alleles (carried by 20–30% of many populations) will increase ferritin concentration slightly in men [162], particularly in conjunction with alcohol abuse [163] or high iron intake.

A few common variants appear to be related to the risk of iron deficiency in normal populations. There is quite clear evidence that rs3811647 is related to differences in transferrin concentration, though without affecting body iron status. More important is the fact that the common hemochromatosis risk allele *HFE* rs1800562 A (*HFE* 282Tyr) provides significant protection against iron deficiency at the same level of iron intake [164]. Two exonic variants in TMPRSS6, rs4820268 and rs855791, however, do seem to be important. In one study, they were related to lower hemoglobin concentration and smaller red blood cell volume, indicators of lower iron availability [165].

Hemochromatosis interventions

The goal is to maintain adequate but not excessive iron stores. Assessment is most practical by monitoring ferritin concentration in blood, with the aim of keeping it in the midrange. Anybody with one of the known variants causing iron retention or excessive ferritin concentration, or with high transferrin saturation should be considered at risk. Typically, 15–30% of the general public is at moderate risk and slightly less than 1% is at high risk.

The most effective course of action for people with existing hemochromatosis or a high risk of developing hemochromatosis is bloodletting with proper medical supervision. This is initially done at frequent (e.g., weekly) intervals and then less often. Donating blood several times a year is helpful for many people, particularly those with one or more of the iron-retaining *HFE* alleles. Blood from hemochromatosis donors can be used for general purposes but has to be labeled to state that the donor has hemochromatosis (21 CFR 640.3[d]). The Food and Drug Administration can provide blood banks upon application with an exemption from this labeling requirement.

At-risk individuals should never eat raw seafood (clams, fish, or oysters), because they are not infrequently contaminated with *Vibrio vulnificus* or other Gram-negative bacteria, to which individuals with iron-retaining conditions are exceptionally vulnerable. Handling raw shellfish should also be avoided.

Iron intake must be tailored to individual needs, particularly taking into account the situation of young women with low ferritin concentration. Dietary measures do not replace the need for iron removal by phlebotomy but have a preventive and supportive value.

Consumption of red meat (beef, lamb, pork, or venison) and processed meat should be tightly limited. Of course, iron-fortified cereals or iron-containing dietary supplements are rarely a good idea for men and older women with the genetic variants known to increase iron retention. The use in young women with genetic risk factors is less concerning, if their iron status is very low.

Consumption of black tea with foods can help to limit the absorption of unwanted non-heme iron but not of heme iron (in fish, meats, and poultry).

Alcohol intake should be kept to a minimum. Individuals with no signs of liver damage may consume up to one serving a day on infrequent occasions.

Dietary supplements with high doses of vitamin C or vitamin A should be avoided. These nutrients should come mainly from vegetables.

4.8.2 Iodine

The only known function of iodine in humans is as a part of thyroid hormones. If your mother had not gotten a good dose of iodine during her pregnancy, you would not be able to read this. Severe mental retardation (cretinism) due to maternal iodine deficiency during pregnancy was once much too common in mountainous regions with low iodine content in the soil and no access to seafood and sea salt. Iodine-fortified salt, better access to iodine-rich foods, and altogether improved nutrition have made this manifestation of prenatal iodine deficiency less common.

4.8.2.1 HYPOTHYROIDISM

Nonfunctioning variants in the genes needed for synthesis and regulation of thyroid hormone occur and neonatal screening usually detects congenital hypothyroidism (OMIM 275200) in about 1:3000 newborn infants [166]. Most often the cause is incomplete formation of the thyroid gland. In children with a fully formed gland, *thyroid peroxidase* (*TPO*; OMIM 606765) is the gene most often affected by variants causing congenital hypothyroidism due to defective hormone synthesis. TPO uses hydrogen peroxide (generated by enzymes encoded by *DUOX1* and *DUOX2*) to iodinate the hormone precursor thyroglobulin (TG). Obviously, TPO can link iodine to thyroglobulin only if there is enough of it. The worst combination will be low TPO activity along with scarce iodine supply. The clinical consequences of such combinations have not been well explored. Other affected genes include *SLC26A4* (encodes pendrin), *DUOX1* and *DUOX2* (maturation factors), *DEHAL1* (a dehalogenase), and *TG* (encodes the hormone precursor thyroglobulin).

4.8.2.2 GOITER

Diffuse enlargement of the thyroid gland (goiter) is a common response to endemic iodine deficiency. However, not everybody will develop goiter in areas with iodine-depleted soil and familial clustering of cases suggests a genetic component [167,168].

Pendrin, the product of the *SLC26A4* (OMIM 605646) gene, moves iodine from thyroid follicle cells into the colloid (Figure 4.27), where it can be used for thyroid hormone synthesis [169]. This anion transporter also helps to maintain the acid–base balance of the fluid (endolymph) in the inner ear by exchanging chloride for bicarbonate. Loss-of-function variants of *SLC26A4* are the cause of Pendred syndrome (OMIM 274600), with congenital deafness, loss of balance (due to vestibular dysfunction), and goiter. As many as one in 13,000 Britons are reported to have Pendred syndrome but the variability of the symptoms make prevalence estimates difficult. Goiter is not a necessary condition in patients with Pendred syndrome. About 90% of them will not have significant thyroid abnormalities as long as they get enough iodine. This means that most low-function alleles of *SLC26A4* just make carriers more vulnerable to iodine deficiency. It is quite likely that the relatively common *SLC26A4* variants increase the tendency of carriers to develop goiter in response to low iodine intake. At the

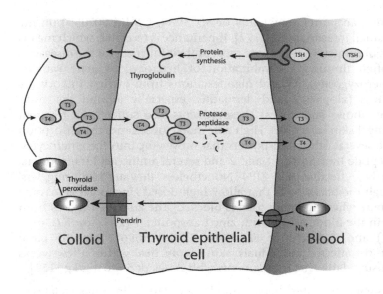

FIGURE 4.27
Function of the pendrin and thyroid peroxidase genes, which are critical for the synthesis of thyroid hormones, are changed in many people by variants. These variants alter iodine requirements and influence the risk of low thyroid hormone production and goiter. I, iodine; Na, sodium; T3/T4, thyroid hormones.

same time, however, they are also likely to be less vulnerable to the harmful effects of high iodine intake, whatever those might be. So the variants are again an echo of the prevailing nutritope of some of the allele carriers' ancestors that shaped them genetically.

4.8.3 Zinc

Humans, like all mammals, use the divalent trace mineral zinc as a cofactor for many enzymes, such as alkaline phosphatase and other proteins [170]. Zinc plays a particularly remarkable role in the regulation of gene function. There are more than a thousand *zinc finger* proteins and related proteins with a tightly coordinated zinc ion that bind DNA and RNA sequences, sometimes also specific proteins. Many of these zinc metalloproteins are transcription factors. Take for example the Ikaros protein family (Aiolos, Eos, Helios, and Ikaros; the more prosaic designations are IKZF1 through IKZF4), which has key functions in the regulation of hematopoiesis and immunity [171]. These transcription factors bind with high specificity to the functional motif (nucleotide sequence) GGAAA in gene promoters and enhancers. They guide the very early development of immune cells in the embryo and later continue to steer maturation of B and T lymphocytes, osteoclasts, and other cells. It is easy to imagine the consequences of zinc deficiency just by looking at this small subset of the zinc finger protein superfamily. Tissues with a rapid turnover are particularly sensitive to impaired zinc availability because the zinc metalloproteins are so intimately involved in cell proliferation. Canker sores in the mouth and sore throat are common manifestations of zinc deficiency because the mucosal cells cannot hold up to the usual wear and tear.

Take the very rare genetic condition acrodermatitis enteropathica (OMIM 201100), which causes an inborn form of zinc deficiency due to loss-of-function variants of a zinc transporter gene (*ZIP4* or *SLC39A4*; OMIM 607059) on the luminal side of small intestinal cells [172]. Inheritance is recessive, which means that both copies of the gene have to lose most of their function for the disease to be manifest, often through a combination of heterozygous gene variants (compound heterozygosity). Typical manifestations from earliest infancy are diarrhea, hair loss (alopecia), and dermatitis, especially around the anus, mouth, and the elbows and knees. These patients also suffer from frequent bacterial and fungal skin infections. High zinc intake overcomes the absorption defect. There are several other transporters that move zinc into the enterocytes, including the peptide transporters 1 and 2 and several amino acid transporters, but they are far less effective than ZIP4. Nonetheless, they are like smugglers' foot trails through the backwoods that allow high-dosed zinc to get across the luminal membrane when the main road is blocked. The zinc then moves from the enterocytes in the bloodstream with zinc transporter 1 (*ZnT1* or *SLC30A1*; OMIM 609521) and 2 (*ZnT2* or *SLC30A2*; OMIM 609617) in the basal membrane. In most affected individuals, skin lesions heal within a few weeks after they start using daily zinc supplements with doses of 5–10 mg/kg [173].

As pointed out, such remarkable therapeutic efficacy of high zinc intake relates to very specific genetic conditions and should not be extrapolated to other patients with severe skin conditions. On the other hand, many variants in this and other zinc transporter genes may increase zinc requirements and predispose to dermatitis without coming to the level of the full-blown acrodermatitis enteropathica syndrome. This suggests that some patients with dermatitis may benefit from increased zinc intake because their zinc transport is genetically inefficient.

IMPORTANT NOTE
High-dosed zinc supplements should be used in combination with copper.

This is also a good opportunity for a reminder that high zinc intake depletes copper stores [174]. Anybody using zinc supplements must make sure to also get additional copper (about 2 mg copper with 50 mg zinc). The rare exceptions to this rule are people with copper retention disorders such as Wilson disease.

Variants in at least some of the more than 20 known zinc transporters affect zinc availability in a range of tissues. For example, a rare missense mutation (His54Arg) in ZnT2 disrupts zinc transfer into milk, which causes zinc deficiency in exclusively breast-fed infants [175].

Zinc plays also a central role in the storage and secretion of insulin from the pancreas. The secretory granules in the islet beta cells contain proinsulin hexamers coordinated with two zinc ions. A special zinc transporter, ZnT8 (*SLC30A8*; OMIM 611145), pumps the necessary zinc into the granules. The ability to store crystalized insulin in the secretory granules is largely abolished without the transporter, as studies in a knockout mouse strain demonstrate [176]. The same effect is provided by autoantibodies against ZnT8, which many patients with type 1 diabetes have [177]. In adults with diabetes, a non-synonymous variant that produces a protein with tryptophan instead of arginine in position 325 (*SLC30A8* Arg325Trp; rs13266634) appears to increase the likelihood that they have antibodies against ZnT8. People with a common *ZnT8* variant (rs11558471) appear to need higher zinc intake than their peers without the variant, at least with regard to their blood glucose control [178]. Meta-analysis of the data from 14 large cohort studies indicated that otherwise similar healthy adults need to get about 4–5 mg extra zinc for each A allele to achieve comparable glucose control (reflected by fasting blood glucose concentration). This may not seem very much, but is quite difficult to get from foods alone without overeating. The nearly 50% of Americans with two A alleles would need two extra servings of meat or fish, or 4–5 servings of beans, lentils, or tofu just to catch up with noncarriers. Since most people do not get the extra zinc, their regulation is not optimal and they have an increased long-term risk of developing type 2 diabetes [178].

Metallothioneins act as zinc storage buffers in cells. The common promoter variant −209A of *metallothionein 2A* (*MT2A*; OMIM 156360) appears to have a somewhat similar effect. Carriers of the A/A genotype had poorer zinc status and were more likely to have fasting hyperglycemia in one study [179].

4.8.4 Copper

Like the other transition metals, copper is both vitally important for cellular metabolism and a direct threat to the integrity of cellular constituents. Both value and danger come from the oxidant properties of the metal ion. Copper is a cofactor of enzymes involved in antioxidant defense, blood coagulation, connective tissue maturation, energy generation, hormone synthesis, iron metabolism, and other functions.

Most copper in blood is carried in ceruloplasmin (CP). The ATP-dependent Menkes disease protein (*ATP7A*, OMIM 300011) moves copper within cells toward newly synthesized proteins, usually with the assistance of specific chaperones. The ATP-dependent Wilson protein (*ATP7B*; OMIM 606882) protects against copper accumulation by helping to secrete it into bile.

Rare mutations that disrupt the normal production of CP (OMIM 117700) cause progressive degeneration of the basal ganglia and the retina. Serum iron is usually low because copper is not available for ferroxidase, which helps with the transfer of iron to transferrin by oxidizing the intracellular diferric iron. Accumulation of iron and copper in the brain, liver, and pancreas are responsible for cirrhosis, diabetes, and dementia in midlife.

Defective Wilson protein leads to the accumulation of copper in brain, cornea, and liver (Kayser–Fleischer rings). Wilson's disease, which occurs with a frequency of about one in 30,000, should be considered in any individual with liver abnormalities or neurological movement disorders of uncertain cause, regardless of age [180]. The diagnosis can usually be made based on low CP concentration in blood, copper concentration in liver, and genetic analyses. Symptoms may not appear until adulthood or even late adulthood. Neurological symptoms include uncoordinated gait (ataxia), low muscle tension (dystonia), and tremor. Liver symptoms are often mild and unspecific but are occasionally of sudden onset with rapid progression to liver failure. Another acute and life-threatening manifestation can be hemolysis. Premature death from liver failure or other complications is the rule without treatment. Chelation therapy with penicillamine or the copper-binding agent triethylenetetramine dihydrochloride (trientine) and eventually orthotopic liver transplantation are often the best courses of action.

One effective form of dietary treatment relies on the use of high-dosed zinc (typically 150 mg/day for adults and about 75 mg in children, taken three times a day, half an hour before meals) to induce metallothionein expression. Binding to metallothionein can safely sequester excess copper and protect against liver damage. Curcumin is thought to be helpful as an antioxidant and to supplement copper-chelating activity [181].

About 1% of adults can be presumed to carry one copy of a low-function *ATP7B* variant. They tend to have slightly lower copper and CP concentrations in blood and some may have mild neurological signs of copper accumulation in brain ganglia [182]. Optimal preventive measures for heterozygous carriers, such as use of zinc supplements or limiting copper intake, will be an important question in the near future, when widespread whole-genome sequencing will identify large numbers of affected individuals.

In a few infants, a combination of an unknown inherited vulnerability and excessive exposure with drinking water delivered through copper piping has caused a syndrome of toxic copper accumulation with irritability, loss of appetite, a slight rise of temperature, and enlarged liver [183]. Eventually, such cases have often progressed to liver cirrhosis and eventually liver failure. Exposure to large amounts of copper has also occurred with the use of copper cookware. These unfortunate cases provide further examples of avoidable health harm due to the combination of nutritional exposure and genetic predisposition.

4.8.5 Selenium

Selenium protects against free radicals, assists with the production of thyroid hormones and insulin, and sustains normal cell growth and fertility. It is not always obvious when we do not get enough. Many consequences of suboptimal selenium status are insidious and indirect, such as increased cancer risk, osteoarthritis, and the potential for more virulent infections.

We all rely on the same food supply to maintain adequate blood and tissue concentrations without realizing that some of us may be more sensitive to low selenium availability than others. The *glutathione peroxidase 1* gene (GPX1; EC 1.11.1.9; OMIM 138320) appears to influence the selenium concentration in blood and thus bioavailability at presumably similar intake levels. People with two copies of the variant 679T (198Leu; rs1050450) have 7% lower selenium concentration in blood than people with two 679C (198Pro) copies [184]. Glutathione peroxidase activities are about 15% lower in people with the 679T variant than without it, at both low and high intake [185]. This means that carriers of the 679T variant can achieve the same enzyme activity, as long as they get enough selenium from food.

The consequences of such differences are uncertain, but previous studies have linked important health outcomes (e.g., cancer) to selenium concentration in blood. We have to assume for now, therefore, that the approximately 10% of people with the 679T/T genotype need slightly higher selenium intakes than individuals with the 679C/C genotype, and that this difference will have health implications for some individuals with marginal selenium intake. In a meta-analysis of case-control studies with cancer patients, the 679T allele (the one associated with lower selenium concentration) appeared to increase overall cancer risk by 12% [186]. Susceptibility to the typical osteoarthritis due to selenium deficiency (Kashin-Beck disease) in a region of China with low-selenium soil is nearly doubled in people with two 679T alleles compared to individuals without this allele [187].

The same 679T/T genotype, on the other hand, may protect some individuals with intake at the high end from harmful effects of selenium excess. This may sound unlikely, but dermatitis, diarrhea, peripheral neuropathy, and other harmful effects may occur when long-term daily intakes exceed 400 μg. High-selenium supplements, Brazil nuts, clams, liver, and some other foods taken on top of a typical diet can easily push intakes above the critical intake threshold. The 290 μg of selenium in a single Brazil nut can do that during a 30-second commercial break in front of the TV.

Such findings cannot be used for individual counseling but improve our understanding of the causes for variability in selenium requirements.

4.8.6 Molybdenum

Only a small number of human enzymes require molybdenum: aldehyde oxidase (EC 1.2.3.1), and mitochondrial amidoxime reducing components 1 and 2 (mARC1 and mARC2), sulfite oxidase (EC 1.8.3.1), and xanthine oxidase (EC 1.17.3.2). All of these enzymes contain molybdenum cofactor (MoCo), which consists of endogenously synthesized molybdopterin and molybdate (MoO_4^{2-}).

Molybdopterin

We need so little of this trace element, usually not much more than 45 μg/day, that our typical intakes (50–100 μg) should be more than enough. Unfortunately, that is not true for everybody.

Very rare inborn errors of molybdenum metabolism are usually caused by the inability to produce the MoCo from guanosine triphosphate [188]. The symptoms of these genetic defects are seizures that usually start a few days after birth, mental retardation, brain abnormalities, and dislocated eye lenses, as well as high concentrations of sulfite, uric acid, and xanthine in blood and urine. Symptoms rapidly get worse in cases with classical molybdenum deficiency and the condition ends with death within months. A report has described a favorable response to parenteral treatment with pyranopterin monophosphate as a substitute for the missing molybdopterin [189].

However, not all cases are as seriously afflicted. Milder forms have been observed, presumably because they have some residual ability to produce MoCo [190]. This is important because dietary treatment in such situations can be effective, particularly with a reduction of the sulfur-containing amino acids methionine and cysteine [191]. Indeed, there is probably a continuum of MoCo synthesis capacity and more individuals than currently known may have some degree of molybdenum deficiency and be vulnerable to high sulfur intake (from cysteine, methionine, and sulfites).

4.9 VITAMINS

4.9.1 Vitamin C

Ascorbic acid (vitamin C) is the key antioxidant in the aqueous compartments of human tissues. It protects against free radicals both directly and by reactivating tocopherols (vitamin E). Adequate vitamin C status reduces the risk of several major chronic diseases, including cancer, cardiovascular disease, and diabetes. Ascorbic acid is also an essential cofactor of a few enzymes involved in the synthesis of carnitine, collagen, hormones, and neurotransmitters. Humans and other primates have lost the ability to produce their own ascorbic acid from glucose and are entirely dependent on dietary intakes. The main food sources are fruits and vegetables and synthetic ascorbic acid in supplements and fortified foods.

The most obvious symptoms of severe vitamin C deficiency include painful swelling and bleeding into gums (including the *scorbutic tongue* shown in Figure 4.28), joints, and extremities, poor wound healing, fatigue, and

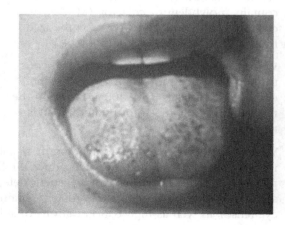

FIGURE 4.28
The many small spots of submucosal bleeding create the typical pattern of the *scorbut tongue* shown here.
Photo courtesy of Public Health Image Library of the Centers of Disease Control.

confusion. Increased risk of heart disease and other chronic disease is more insidious.

Contrary to popular myths, there is no evidence that humans can make ascorbic acid. However, there are several common genetic variants that either decrease or increase vitamin C requirements. We will consider them by function rather than by importance or impact.

Ascorbic acid

4.9.1.1 *TRANSPORTERS*

Vitamin C gets into cells either as ascorbic acid through sodium cotransporters or as dehydroascorbic acid (the oxidized form) via glucose transporters (GLUT1, GLUT3, and GLUT4). Two sodium-dependent transporters, SVCT1 (*SLC23A1*; OMIM 603790) and SVCT2 (*SLC23A2*; OMIM 603791) move ascorbic acid into epithelial cells. It is interesting to note that mammals have two genes for the same transport process. Around 450 million years ago, a fish-like ancestor seems to have had a gene duplication event, and the resulting two genes have distinct patterns of expression and regulation [192]. The fact that the originally identical *SVCT2* gene is now about ten times as large as its sibling *SVCT1* gene should remind us that gene size is not everything. The two transporters are similarly important; they just function differently [193]. For instance, in the small intestine, SVCT1 moves ascorbic acid into the enterocyte and SVCT2 moves it from there into blood.

As an oversimplification, one might say that SVCT1 is particularly important for absorption and protecting against vitamin C losses in the kidney, whereas SVCT2 is important for intestinal absorption and transfer to tissues and the fetus [194]. It should not surprise us, therefore, to find that several *SVCT1* variants influence vitamin C concentrations in the body [194] because less of the ingested amounts get absorbed. Variations in the *sodium-vitamin C transporter 2* (*SVCT2* or *SLC23A2*), on the other hand, affect pregnancy outcome [195]. Women with the common intronic variant *SLC23A2*−08C>T (rs6139591) appear to have a greater risk of premature delivery (at least 3 weeks before full term). With two copies of the variant, the risk is almost three times higher than in women without the variant. The exact nature of this association is awaiting clarification. Consuming four to five servings of fruit and vegetables typically provides about 200 mg vitamin C, which achieves saturation levels of vitamin C in tissues [194]. There is good reason to expect that such generous vitamin C intake compensates for low-functioning polymorphisms in both ascorbic acid transporters and makes their inefficiencies irrelevant. This is really the take-home message again—that seemingly small genetic differences often have significant health differences that could easily be avoided with adequate intakes.

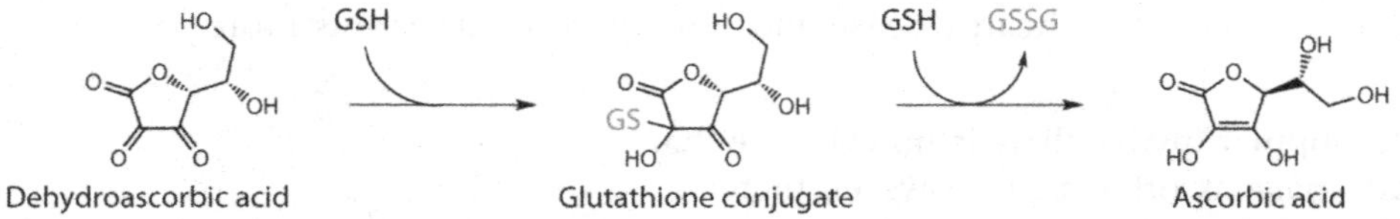

HO   GSH   HO   GSH   GSSG   OH

Dehydroascorbic acid   Glutathione conjugate   Ascorbic acid

FIGURE 4.29
Dehydroascorbic acid is the oxidized metabolite that is produced when ascorbic acid interacts with a free radical with two unpaired electrons (particularly peroxide). Reduced glutathione (GSH) can reactivate dehydroascorbic acid both nonenzymatically and catalyzed by one of several enzymes including omega class glutathione S-transferases 1 and 2.

4.9.1.2 GLUTATHIONE

Several glutathione S-transferase enzymes (EC 2.5.1.18) can considerably draw on our vitamin C budget by competing for the reactivation of the oxidized dehydroascorbic acid form (Figure 4.29). Most will know this class of enzymes better from its ability to promote the excretion of carcinogens and medications by linking them to glutathione.

Glutathione S-transferase M1 (*GSTM1*; OMIM 138350) encodes one of these competing enzymes. About half of all Caucasians have two copies of *GSTM1* null alleles (these are variants with low or no activity due to a deletion). In one American cohort, the ascorbic acid concentration was slightly higher (+8%) in the blood of adults with the *GSTM1* null genotype than in their peers with functional GSTM1[196]. However, no significant difference in ascorbic acid concentration was seen in a smaller Canadian group and the authors reported an interaction with vitamin C intake in which the null genotype increased the risk of deficiency [197]. Such inconsistent findings should not surprise us when we consider the many additional elements that influence vitamin C recycling [198].

4.9.1.3 HAPTOGLOBIN

Another type of not easily anticipated modulator of vitamin C status is the hemoglobin-sequestering protein haptoglobin (*HP*; OMIM 140100). The efficient recycling of iron in free hemoglobin by a haptoglobin-directed process keeps this highly reactive transition metal under wraps. Any hemoglobin spillover has to be mopped up by hemopexin and macrophages. Some of this extra and misdirected iron loads up transferrin and limits its capacity to bind excess iron, particularly after fresh iron comes in from a meal. All of this extra iron that is not safely tucked away into red blood cells acts as an oxidative catalyst for the irreversible breakdown of ascorbic acid.

Two common variants of the *HP* gene encode the Hp1 and Hp2 isoforms. People with only Hp1 isoforms have in their blood Hp1−Hp1 dimers with high hemoglobin-binding efficiency. If the Hp2 isoform is present, either alone or in combination with the Hp1 isoform, several Hp molecules form larger complexes that are relatively poor hemoglobin-binders (Figure 4.30).

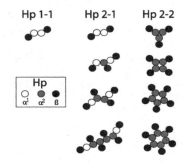

FIGURE 4.30
Haptoglobin (Hp) consists of an alpha and a beta subunit, which then aggregate to larger polymers. The haptoglobin 1 isoform, which contains the alpha1 subunit, forms only dimers in the absence of the haptoglobin 2 isoform. Much larger aggregates form in people with one or two haptoglobin 2 alleles. These large aggregates bind less hemoglobin than the smaller ones.

People with the Hp2 isoform have all the signs of less efficient hemoglobin recycling, including a high iron content of macrophages, increased transferrin saturation, elevated concentration of free iron in plasma, and, most important of all, increased vitamin C requirements due to accelerated oxidative breakdown [199]. As described in Chapter 3, selective pressure has evidently favored the vitamin C-sparing Hp1 isoform. The most dramatic example is given by the Polynesian settlers of Easter Island, who reached this most remote speck of land without knowing it was even there. The shortest travel distance was much further than the first voyage of Columbus from the Canary Islands to the Bahamas. Their presence on Easter Island shows that they survived a sea voyage of many months, possibly even years, without dying from scurvy. The genetic traces of this most extreme example of selection pressure are still there. Almost none of indigenous people carry the Hp2 allele with the higher vitamin C requirement. Presumably, no early voyagers with this allele made it to the island.

The Inuit of the Arctic Circle are another group of people with few carriers of the Hp2 allele. They appear to do well on 10 mg of vitamin C per day. Historically, they could not have survived with a higher requirement because there were no good dietary sources of vitamin C during the long winter months. Many of the early gold diggers did not make it through the winter in the lower Arctic because of scurvy, but the Inuit have survived this hostile nutritope for millennia.

The question remains of why the Hp2 allele is still the more common allele across the world. We have seen in other instances that some nutritopes favor one allele and others favor another one. It is definitely possible with the Hp1/Hp2 polymorphism that we have just not yet identified the circumstances that favor the Hp2 allele. Hp1 may also be associated with its own specific risks and there is some indication that its carriers have more difficulty than Hp2 carriers in fighting off infections, such as with the bacterium *Chlamydia trachomatis*, a major cause of blindness in developing countries [200]. The alternative

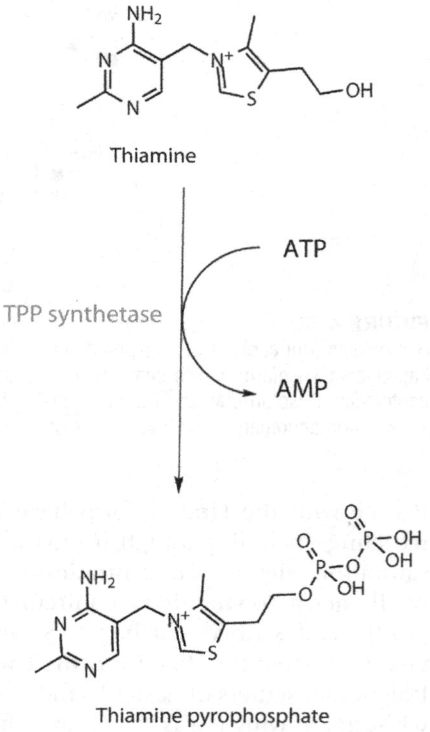

FIGURE 4.31
Thiamine pyrophosphate (TPP) synthetase uses
adenosine triphosphate (ATP) to activate thiamine.
ADP, adenosine diphosphate.

explanation is that the Hp1 allele is a fairly recent development, which has not had enough time and selective pressure to replace the Hp2 allele. The limited geographic distribution of the alleles makes this a reasonable guess.

4.9.2 Thiamine

There are only a handful of thiamine-dependent enzymes, but we just can't do without them. Thiamine pyrophosphate is the cofactor for a small number of vital enzymes, including alpha-ketoglutarate dehydrogenase, BCKD, 2-hydroxyphytanoyl-CoA lyase, pyruvate dehydrogenase, and transketolase.

For illustrative purposes, we shall focus here on transketolase (TKT; EC 2.2.1.1), which needs thiamine pyrophosphate (TPP, also called thiamine diphosphate; Figure 4.31) as a cofactor.

Transketolase works together with transaldolase to convert the pentose phosphate pathway products xylulose-5-phosphate and ribose-5-phosphate into fructose-6-phosphate and glyceraldehyde-3-phosphate for continued oxidative metabolism (Figure 4.32). Just as a reminder, the initial (oxidative) part of the pentose phosphate pathway generates nicotinamide adenine dinucleotide phosphate (NADPH) for the synthesis of fatty acids, steroids and many other compounds, and ribose for DNA and RNA synthesis. With lower transketolase

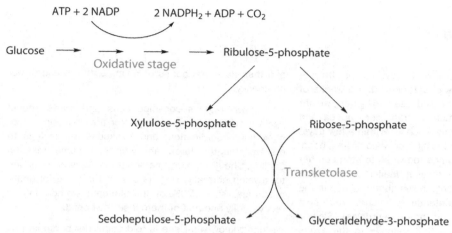

FIGURE 4.32
The initial oxidative steps of the pentose phosphate pathway generate nicotinamide adenine dinucleotide phosphate, reduced NADPH. Transketolase is one of several enzymes that act together to catalyze the successive rearrangement of ribulose-5-phosphate into glyceraldehyde-3-phosphate. ADP, adenosine diphosphate; CO_2, carbon dioxide; NAD, nicotinamide adenine dinucleotide.

activity, less NADPH and nucleotides are available. Evidence for the vital importance of this reaction comes from the fact that at least three distinct genes encode enzyme with transketolase activity.

4.9.2.1 NUTRITIONAL DEFICIENCY

The most direct reason for impaired transketolase activity is inadequate thiamine intake. It comes in fortified flour (in Australia, Canada, and the USA), legumes and some vegetables, the outer layers of grains (therefore with whole grain), pork, and yeast. Deficiency causes eventually irreversible damage to the brain and nerves (corresponds largely to the thiamine deficiency syndrome of dry beriberi) and the heart (wet beriberi). The manifestations of thiamine deficiency in the nervous system include the Wernicke and Korsakoff syndromes (*WKS*; OMIM 277730), which may occur jointly or individually. Wernicke syndrome is characterized by the rapidly manifesting triad of defects in body coordination (ataxia), eye movement (ophthalmoplegia and vertical nystagmus), and mental function (confusion), sometimes also with hearing loss and other neurological pathologies. Korsakoff syndrome relates to slowly disintegrating higher mental functions (loss of reality and psychosis). Very often, the deficiency symptoms are precipitated by high alcohol consumption, but this is not a necessary condition. All it may take in a vulnerable individual is a few weeks of morning sickness [201] or chronic diarrhea [202]. Wernicke encephalopathy is much more common than is generally known. Autopsy series suggest that almost 2% of adults have this severe deficiency disease at the time of death [203]. In alcoholics, the prevalence is over 10%.

Sample case

A 27-year-old pregnant (week 12) woman in Italy is admitted to the obstetrics department after 2 weeks of persistent nausea, vomiting, and loss of 5 kg body weight [201]. She gets medication to treat her nausea and intravenous fluids containing glucose. Within a few days, she progressively starts losing her vision, has headaches, feels drowsy, becomes too weak to stand on her legs, and experiences a tingling feeling in her feet. Magnetic resonance imaging shows abnormalities in the posterior thalamus (hyperintense symmetric areas next to the posterior horn of the lateral ventricles). Slowed nerve conduction indicates that damage to the axons leading to the lower limbs (motor polyradiculopathy) is probably responsible for progressive paralysis of her legs.

With a tentative diagnosis of Wernicke encephalopathy she is given daily 100 mg intravenous thiamine for a week and then daily oral doses of 50 mg until the end of her pregnancy.

She slowly regains consciousness and starts seeing again but it takes many weeks for the confusion, poor movement coordination, and weakness of the legs to resolve. Months later, she still has some residual tingling. She gives birth to a healthy child, who continues to develop normally. Hers is one of the more fortunate outcomes, since 70% of the affected women fail to recover fully and half of them lose their child.

We don't know what she is told about her personal risk and how to guard against another episode. It is reasonable to assume that she has an inherited predisposition that increases her thiamine requirements. Lifelong use of a moderately dosed thiamine supplement (for instance, 5—10 mg/day) seems to be a reasonable precaution.

4.9.2.2 TRANSKETOLASE VARIANTS

A significant heritable vulnerability for thiamine deficiency and symptomatic disease has been suggested for a long time, particularly with an eye toward transketolase. It had been noted that the enzyme from patients with WKS appears to bind TPP less tightly than the enzyme from healthy individuals [204]. It has since emerged that the genes *TKT* (OMIM 606781), *TKTL1* (OMIM 300044), and *TKTL2* (no OMIM entry) all encode enzymes with transketolase activity [205]. Among the variants in these genes, *TKT2* Gln590His (rs11735477) is very common in ethnically diverse populations. The functional impact of these variants remains to be established.

4.9.2.3 TRANSPORT

The thiamine transporters 1 (*SLC19A2*; OMIM 603941) and 2 (*SLC19A3*; OMIM 606152) contribute to absorption of the vitamin from the intestines, its recovery from renal tubules, and its uptake into cells. SLC19A3 sits at the apical side of intestinal and renal cells. Additional transporters help with absorption and renal recovery.

Individuals born with defective SLC19A2 have a syndrome called thiamine-responsive megaloblastic anemia (*TRMA*; OMIM 249270). Characteristic symptoms, in addition to the eponymous megaloblastic anemia, are deafness due to nerve damage and diabetes. Onset can be any time after birth. Sometimes the syndrome is not recognized until adolescence or adulthood. Patients

Neurological signs of thiamine deficiency

Depending on the severity and duration of thiamine deficiency, the condition may present with one or more of the classical triad: impairments in body coordination, eye movement, and mental function. Hearing loss is a less frequent sign. Confabulation (making up facts) and psychosis (Korsakoff syndrome) tend to evolve slowly over an extended period of time.

Mental Function

Assessment of memory with the Mini-Mental State Exam (MMSE) is sufficient for rapid evaluation. The Wechsler Memory Scale (Revised 1987) gives a more reliable quantitative measure.

Confusion becomes apparent, with an impaired ability to correctly answer standard orientation questions (What is the year/season/date/day/month?). This is best asked as part of the full battery of simple questions and memory tasks in the Mini-Mental State Examination [210].

Spontaneous confabulation frequency can be rated on a semiquantitative Likert scale (never, seldom, sometimes, often, or always) based on queries to nursing staff that have contact with the person throughout the day. The Dalla Barba Confabulation Battery [211] quantitates prompted behavior in a standardized manner.

Eye Movements

Ask the test person to keep their head still and use only their eyes to follow your index finger as it moves from eye level to the far right, far left, upward, and downward. Involuntary rapid eye movements against the path of the index finger (nystagmus) indicate abnormal oculomotor (eye muscle) control.

Body Coordination

The Fregly—Graybiel walk-a-line test [212] explores abnormalities of the cerebellum. The test subject is asked to keep their arms folded across the chest. Each test is performed twice with the eyes open and twice with the eyes closed. Ask the person to do the following:

1. Stand with one foot directly behind the other for 60 seconds.
2. Walk setting one foot directly before the other for ten steps.
3. Stand for 30 seconds each on the left and then the right foot.

Healthy men who do not suffer from alcoholism can usually stand with one foot behind the other and their eyes open for nearly the full time; women for more than 45 seconds [213]. Normal standing time is about half as long with the eyes closed. The untroubled number of heel-to-toe steps is normally at least seven with open eyes and at least three with closed eyes. A healthy person is usually able to stand on one leg for more than 22 seconds with open eyes and for more than eight seconds with closed eyes.

respond well to high-dosed thiamine (50 mg/day), but brain damage may continue to progress.

Expression of SLC19A3 is particularly high in the thalamus region of the brain. This is suggestive because the thalamus is most notably affected in Wernicke syndrome. A report described two Japanese brothers with the typical symptoms of Wernicke syndrome and persisting seizures. The use of high-dosed thiamine stopped the seizures within a day and the other symptoms after a few weeks. In both cases, the cause of the syndrome was variants in the SLC19A3 gene [206].

GENETIC JARGON
The term *megaphenic* indicates that a locus has an outsized effect on a trait.

4.9.2.4 ALCOHOL ABUSE

As mentioned above, about 10% of people with serious chronic alcohol abuse will eventually develop WKS. There is every reason to assume that such thiamine-related brain damage is preventable with increased thiamine intake in most cases. People with chronic alcohol abuse often have marginal thiamine intakes due to severe nutritional neglect. In addition, chronic ethanol exposure appears to interfere with both absorption and utilization of thiamine [207]. This reduced thiamine availability puts people with the least effective transporters and thiamine-dependent enzymes at greatest risk.

Few doubt that there are genetic factors that increase WKS risk but efforts to track them down have been insufficient. Very limited data point to several variants in both *SLC19A2* and *SLC19A3* [208]. Loss-of-function variants of the transketolases also have been suggested [207]. *TKTL1* (OMIM 300044) is particularly interesting because it is located on the X chromosome. Men cannot compensate for a defective variant in this gene because they have only one copy. Deletions and other variants do occur but their relevance is still unclear. A significant causative role of *TKTL1* could explain why WKS is much more common in men than in women [209]. It may not only be the fact that severe chronic alcohol abuse with nutritional neglect is more common in men but also that men lack the *TKTL1* backup copy that women have.

4.9.3 Riboflavin

Riboflavin (vitamin B2) is the precursor of flavin mononucleotide (FMN) and flavin adenine dinucleotide (FAD), the cofactors of numerous flavoproteins. A lack of this vitamin causes a few visible symptoms, including cracking and inflammation of the corners of the mouth (cheilosis) and the lips (Figure 4.33), normocytic anemia, seborrheic dermatitis, and impaired growth in childhood. More import may be the increased risk of cancer and other chronic disease.

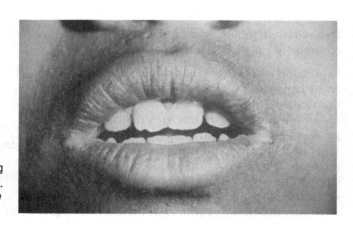

FIGURE 4.33
The most directly observable signs of riboflavin deficiency are dry, scaly, and swollen lips and cracking of the corners of the mouth, as shown in this individual. *Photo courtesy of Public Health Image Library (PHIL) of the Centers of Disease Control.*

The main reason for the apparent absence of common health problems due to riboflavin deficiency in the USA is the existing fortification program and the low requirements compared to the ready availability of riboflavin from dairy products, eggs, meats, and other food sources.

The monocarboxylic acid transporter *SLC16A12* (OMIM 611910), the G protein-coupled receptor 172B (GPR172B or *RFT1*; OMIM 607883), and the riboflavin transporter 2 (*C20orf54* or *RFT2*; OMIM 613350) provide for riboflavin transfer into cells. Riboflavin kinase (RFK; EC 2.7.1.26) generates FMN and FAD synthase (EC 2.7.7.2), encoded by the *FLAD1* gene (OMIM 610595), which then complete the production of FAD.

Riboflavin

The link of the FAD-dependent enzyme, 5,10-methylene tetrahydrofolate reductase (5-mTHF, *MTHFR*; EC 1.5.1.20; OMIM 607093), to riboflavin is explained in more detail in Section 4.9.6 on folate. The allozyme produced from the common *MTHFR* 677T allele retains full activity in the presence of high FAD concentration but is very sensitive to low FAD availability. Carriers of the *MTHFR* 677T/T genotype usually benefit more from high riboflavin intake than from high folate intake.

Less is known about other genetically predicated riboflavin deficiencies or other genetic conditions where the adjustment of riboflavin availability would be mainly beneficial to individuals with a particular genotype. Metabolic screening has detected multiple acyl-CoA dehydrogenase deficiency in a newborn infant, which normalized with regular feeding [214]. The problem was not a metabolic defect in the child but persistent riboflavin deficiency in the adequately fed mother, presumably due to a genetic defect. This otherwise healthy woman had the metabolic signatures of multiple acyl-CoA dehydrogenase deficiency herself due to her riboflavin deficiency. There were no signs of variants in the genes encoding the electron transfer flavoprotein dehydrogenase gene complex (ETFDH; EC 1.5.5.1), which is usually responsible for this metabolic pattern. The responsible genetic variant could not be identified. The woman used a riboflavin supplement during two later pregnancies and none of these children showed signs of deficiency at birth. The good responsiveness of the mother's metabolic disorder to riboflavin supplementation is similar to what is seen in patients with mild forms of multiple acyl-CoA dehydrogenase deficiency (MADD, caused by a defect in any of the three genes encoding the electron transfer flavoprotein dehydrogenase complex (OMIM 231680)).

In another patient, mild riboflavin deficiency was explained by a combination of haploinsufficiency (meaning that she has only one copy) of the *RFT1* gene and increased requirement due to pregnancy [215]. Patients with the rare Fazio-Londe disease (disordered metabolic pattern like in MADD, hypotonia, progressive bulbar palsy, and respiratory insufficiency due to paralysis of the diaphragm; OMIM 211500) and the closely related Brown-Vialetto-van Laere syndrome (similar to Fazio-Londe disease plus deafness; OMIM 211530) all appear to have a defect in the *RFT2* gene. Confirmation of this conclusion comes from the observation that they tend to respond well to riboflavin supplementation [216]. Supplementation with riboflavin (10 mg/kg) resolved the metabolic disorder within days and improved muscle function very slowly over a period of many months. Cessation of the supplementation for a short time immediately brought the metabolic disorder back. The same supplementation treatment was even more effective in a younger sister, where the intervention was started 3 months after birth. The long-term outcome with supplementation in such cases is still uncertain, however.

These examples are intended to illustrate that riboflavin insufficiency due to specific genetic vulnerabilities has serious health consequences and needs to be recognized in a timely manner to serve the affected patients well.

4.9.4 Niacin

Strictly speaking, niacin is not really a vitamin, because most of us can make it from its precursor tryptophan. Much of this has been discussed above in Section 4.2.5 on Hartnup disease, where diminished intestinal absorption and uncontrolled renal loss limits the availability of tryptophan. The symptoms of severe niacin deficiency are typical skin changes after sun exposure (pellagra with patchy spots of excessively pigmented and unpigmented areas), bright-red tongue, gastrointestinal symptoms (constipation, diarrhea, and vomiting), depression, and cognitive decline.

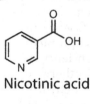

Nicotinic acid

Nicotinamide

The conversion of tryptophan to niacin in the liver is a multistep process that relies on at least six enzymes and an adequate supply of iron [217], pyridoxine, and riboflavin (Figure 4.34). The amount of tryptophan that is available for conversion is around 5 g/day [218]. The standard assumption is that of every available 60 mg about 1 mg is converted into niacin. This would mean that 80–90 mg (1/60 of 5000 mg) can be produced in a day [219]. The conversion rate is not affected much by niacin intake [220].

How to recognize pellagra

Remember that pellagra in affluent countries occurs most often in homeless men with chronic alcohol abuse and that skin changes may be attributed to exposure to the elements and neglect. Cancer, other chronic disease, or medications may lead to the misinterpretation of symptoms [222]. The condition may be confused with lupus erythematosus, photo dermatitis, or porphyria.

The traditional four Ds are a good start to identify individuals with the niacin deficiency disease pellagra: diarrhea, dermatitis, dementia, death. Assuming that you don't want to wait for the fourth D, we can focus on the first three:

Diarrhea

This symptom is not regularly present and can be intermittent. Don't rely on it.

Dermatitis

Photosensitive dermatitis is the most consistent symptom. Circumscribed patches of skin lack pigmentation and others are pigmented much more than nearby skin. Look at the nearly rectangular white area on the arm of this man and contrast it with the very darkly colored patch nearby. The hand on the left is highly pigmented and has a rough appearance (*pelle agra* means rough skin in Italian). Some people have much more dramatic skin lesions than shown here. There may be swelling, erythematous eruptions, intraepidermal blisters and

erosions. A somewhat related symptom is a beefy, red, swollen tongue (glossitis). Hair loss with patchy baldness (alopecia) is another symptom.

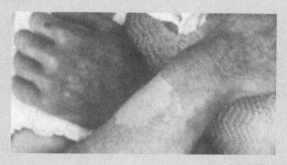

Dementia

Central nervous symptoms leading up to outright dementia include aggressive behavior, confusion, dizziness triggered by rapid movements, emotional instability, insomnia, intolerance of bright light or strong odors, nausea, and restlessness. There are often general neurological symptoms, such as ataxia (uncoordinated gait), muscle weakness, and peripheral neuropathy.

Other

Congestive, dilating heart failure with edema should not be missed when it is present.

What is most interesting in a nutrigenetic context is that individual conversion rates vary considerably [221]. Individual results in the largest study observed conversion rates between 34:1 and 86:1. Due to limitations of the study design, we do not know how much of that reflects true individual predisposition but a twofold range of actual conversion rates in an otherwise homogenous group appears to be plausible. In other words, one might ask whether pellagra is a nutrigenetic disease that affects predominantly individuals with a genetic vulnerability.

High-dose nicotinic acid and closely related derivatives are very effective lipid-lowering agents. Individual responses vary greatly and depend on genetic variants. A more detailed description can be found in Chapter 7.

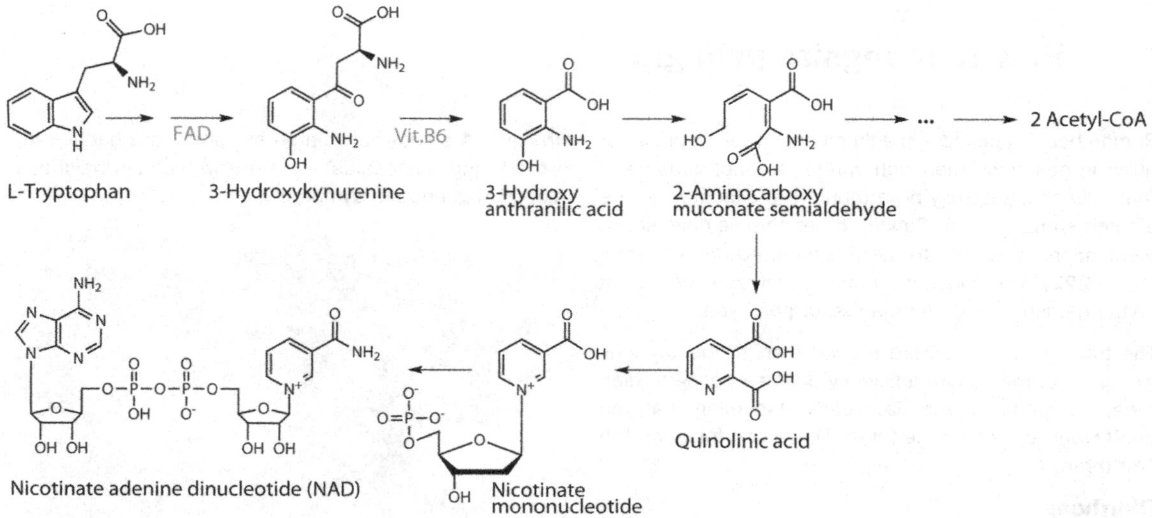

FIGURE 4.34
Humans can produce a limited amount of niacin from tryptophan. FAD, flavin adenine dinucleotide; Vit., vitamin.

4.9.5 Vitamin B6

The large number (more than a hundred) and diversity of enzymes that use it as a cofactor show what a mover and shaker pyridoxine (vitamin B6) is. We get most of it from fish, fruits, meats, and potatoes. This wide range of food sources already suggests that it is not very difficult to get enough, almost regardless of dietary preferences.

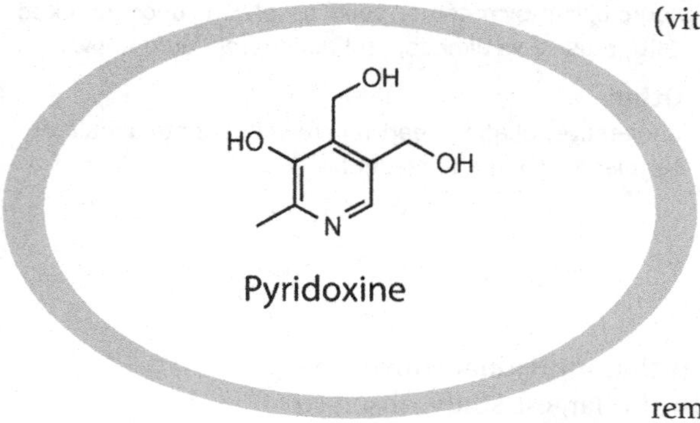

Pyridoxine

The main reactions of pyridoxine metabolism relate to its transport and conversion into active metabolites. Alkaline phosphatase (ALP; EC 3.1.3.1) in the small intestine, encoded by the *ALPI* gene (OMIM 171740) removes the phosphate from phosphorylated metabolites and makes sure that pyridoxine can leave the enterocytes and freely enter cells in tissues that need it. The ubiquitous tissue nonspecific alkaline phosphatase (*ALPL*; OMIM 171760) has a similar function. Pyridoxine kinase (EC 2.7.1.35) phosphorylates pyridoxine, which keeps it from leaving a cell. The flavoprotein pyridoxine-5′-phosphate oxidase (EC 1.4.3.5), encoded by the *PNPO* gene (OMIM 603287), accomplishes most of the other conversions, including the synthesis of pyridoxal phosphate (PLP) and pyridoxamine phosphate (PMP). PLP is the cofactor for transaminases and many other pyridoxine-dependent enzymes. Most of the stored vitamin B6 is bound to the various glycogen phosphorylases (EC 2.4.11).

As might be expected, a genetic deficiency in *PNPO* (OMIM 610090) causes convulsions and severe metabolic disturbances, and rapidly leads to multiorgan failure. Direct administration of the active PLP metabolite was helpful but did not fully restore the health of the single patient for whom this therapy has been attempted. The example of this single gene emphasizes that humans just cannot do without active pyridoxine metabolites. Most of the mutations in the necessary genes are likely to disrupt an affected pregnancy at a very early stage and never come to our attention. However, variants that cause slight modifications of vitamin B6 status do exist. The greatest impact on vitamin B6 concentration in blood is made by the common variant rs4654748 in the 5' UTR of the *ALPL* gene [223,224]. The C allele, which is as common as the T allele, is associated with a 15% lower vitamin B6 concentration. This will hardly be noticeable at the personal level but the lifelong impact and its very high frequency add up to important health effects.

4.9.6 Folate

This vitamin from citrus fruits, dark green vegetables (broccoli, Brussels sprouts, and kale), legumes (beans and lentils), and fortified foods is critically important for good health from the moment of conception. The different metabolic forms serve to provide one-carbon metabolites for the synthesis of carnitine, choline, creatine, DNA, and RNA. Folate also transfers methyl groups for the epigenetic methylation of DNA, a key element in the regulation of genome function. Folate is also important for the detoxification of formate and methanol. Inadequate folate supplies limit the production of cells in blood and tissues. The most directly observable sign is anemia with causative macrocytic red blood cells and hypersegmented granulocytes. Increased risk of birth defects, cancer, cardiovascular disease, and depression are important. Antifolate drugs (including the prototypical methotrexate) are used for the treatment of cancers, leukemias, and autoimmune disease, including psoriatic and rheumatoid arthritis. An excess of folic acid intake (from dietary supplements and fortified foods) is likely to increase the risk of cancer in some susceptible individuals. All of these functions and conditions are thoroughly influenced by the interaction of genetic and environmental factors.

More than a dozen key genes are involved in folate metabolism and important genetic variants have been reported for all of them. Most of the folate in food has a polyglutamate tail attached, which has to be cleaved off by folylpolyglutamate carboxypeptidase (EC 3.4.17.21), encoded by the *glutamate carboxypeptidase II* gene (*GCP2*; OMIM 600934), in the intestinal wall before free folate can be taken up through the reduced folate transporter 1 (*RFC1* or *SLC19A1*; OMIM 600424) and exported into blood by multidrug resistance-associated protein 2 (*MRP2* or *ABCC2*; OMIM 601107). 5-Methyltetrahydrofolate, the nearly exclusive metabolite in blood plasma, enters cells via the bidirectional anion transporter RFC1. Folate stays in the cell, because folylpolyglutamate synthase (EC 6.3.2.17) traps it there by adding a polyglutamate tail. Folate metabolites can leave the cell only when gamma-glutamyl hydrolase (EC 3.4.19.9; OMIM 601509) cleaves the tail off again.

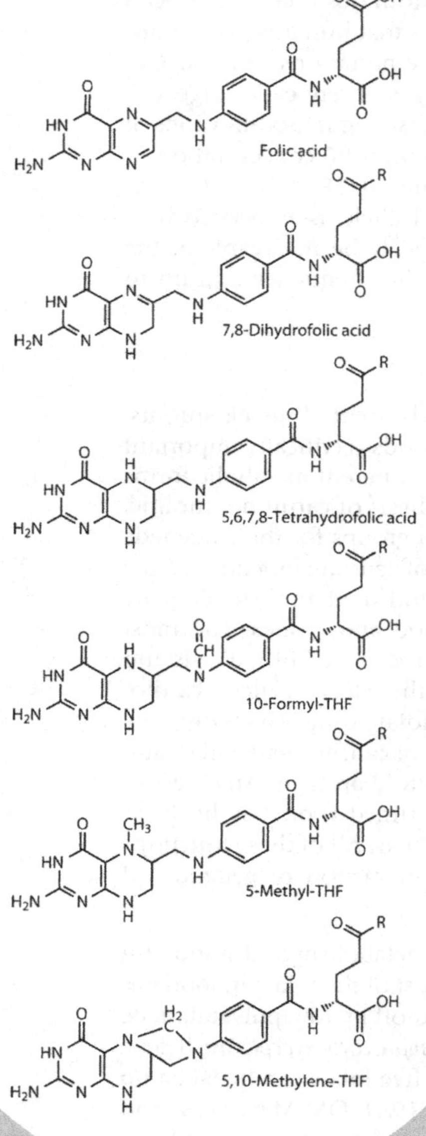

Folic acid

7,8-Dihydrofolic acid

5,6,7,8-Tetrahydrofolic acid

10-Formyl-THF

5-Methyl-THF

5,10-Methylene-THF

Dihydrofolate reductase (EC 1.5.1.3; OMIM 126060) activates dietary folate and reactivates dihydrofolate after it has been oxidized during its use in thymidine synthesis.

5,10-Methylene tetrahydrofolate reductase (*MTHFR*; EC 1.5.1.20; OMIM 607093) and the trifunctional enzyme methylene tetrahydrofolate dehydrogenase 1 (*MTHFD1*; EC 3.5.4.9; OMIM 172460) link one-carbon metabolism (which provides methyl, formyl, and methylene groups for the synthesis of carnitine, choline, creatine, DNA, and RNA, and for DNA methylation) to folate metabolism (Figure 4.35).

Multiple variants in numerous genes affect the bioavailability of dietary folate and the distribution and effectiveness of the diverse folate metabolites in blood and tissues. Health effects of inadequate folate availability include the risk of birth defects, cardiovascular disease, depression, osteoporosis, and possibly cancer due to accumulation of uracil in DNA and the resulting chromosomal instability. We have to limit, therefore, our discussion to just a few examples. Several absorption and transport defects occur in rare instances and cause severe deficiency with grave clinical consequences.

The polymorphic variants have a greater relevance for most purposes because they are, by definition, so much more common than the rare syndromes caused by random mutations. Starting with absorption from the intestine, the common *GCP2* variant His475Tyr (1561C>T) has been reported to promote folate uptake. Carriers of this variant allele were found to have higher folate concentration in blood, which in turn was linked to lower homocysteine concentration [225].

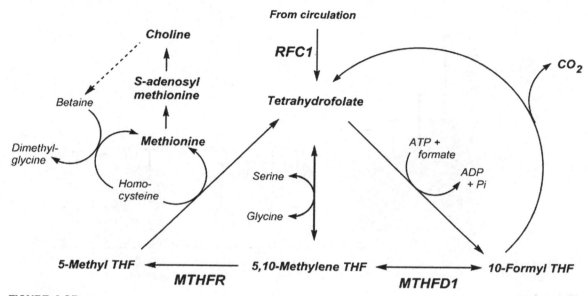

FIGURE 4.35
Key steps of folate-mediated one-carbon metabolism.
ADP, adenosine diphosphate; ATP, adenosine triphosphate; CO$_2$, carbon dioxide

The *MTHFR* variant 677C>T (rs1801131) is more widely known because it has been investigated so much more. There is little doubt that people with the 677T/T genotype generate 5-methyltetrahydrofolate less efficiently than people with the 677C/T or 677C/C genotypes. The reason is that the protein (allozyme) produced from the 677T allele is less stable than the one from the 677C allele. In the end, this means that less active enzyme is available and remethylation of homocysteine is slowed. Generous folate intakes can almost make up for the low activity. Figure 4.36 shows the typical responses of individuals with the three *MTHFR* 677 genotypes to different folate intake levels. Carriers of the *MTHFR* 677T/T genotype needed to consume about four times more folate than individuals with the C/T or C/C genotypes to achieve the same plasma homocysteine concentrations. This does not conclusively resolve the question of optimal intake levels for either group but the relative difference is conclusive.

Some studies in Americans have found a less distinct pattern, most likely because of the higher intake levels due to the food fortification programs in the USA. The 5-mTHF cofactor FAD stabilizes the allozyme produced from the 677T allele and thereby prevents rapid degradation and loss of activity. In one study, for example, higher riboflavin intakes were more important than high folate consumption for the prevention of an unfavorable *MTHFR* 677T/T-related health outcome, such as osteoporosis [226], hypertension [227], or cardiovascular disease [228]. This is a good example of the often multiple nutrient–genotype interactions that make it so hard to unravel the relevance of nutrigenetic variants.

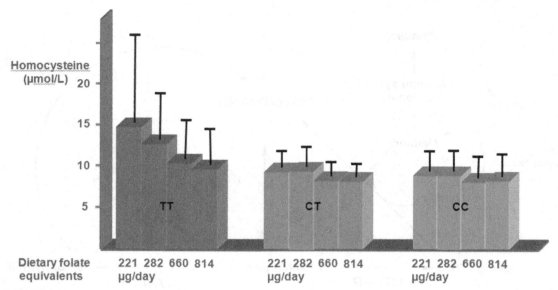

FIGURE 4.36
People with the *MTHFR* 677TT (T/T) genotype have lower plasma homocysteine concentrations with increasing folate intake levels, while people with the other genotypes have a minimal effect at best [225].

But the complexities and complications of the *MTHFR* story do not end here. It is important to recognize that there are dozens of common variants that occur in a gene of average size. This is certainly true for *MTHFR*. Resequencing the *MTHFR* gene in just 240 individuals (taken from the Coriell Institute Cell Repository) identified 62 variants [229]. Many of these variants may not affect expression and function of the mature enzyme, but some of them do. Expression of the major

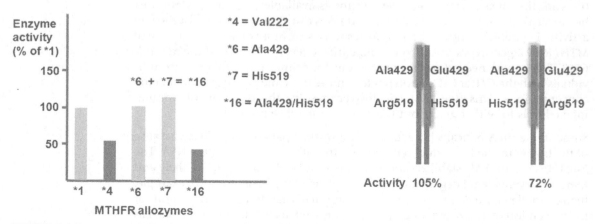

FIGURE 4.37
The combination of genetic variants (haplotypes) determines functional activity of the MTHFR (5-mTHF) enzyme [229]. As is the case for the Ala429 and His519 variants, enzyme activity may be slightly higher than normal, if they are on different strands, but distinctly lower than normal, if they are on the same strand, as illustrated on the right-hand side. Ala, alanine; Arg, arginine; Glu, glutamic acid; His, histidine; Val, valine.

variants and some haplotype combinations in an in vitro vector system (Figure 4.37) showed that several of them had greatly reduced enzyme activity, not just the 677T allele (here marked as allozyme *4) that usually receives the most attention. Another commonly investigated variant is the 1298C variant that encodes an alanine in place of the usual arginine in position 429. Contrary to the common assumption, this allozyme is slightly more active than the normal version. The activity is reduced only when Ala429 occurs in combination with His519 or Gln594. It is remarkable that the His519 on its own has slightly higher activity than the more common Arg519 version. This is an example of why we sometimes need to know the exact haplotype, i.e., the combination of variants on the same DNA strand. If these Ala429 and His519 variants occur on different strands, the activity of the resulting 5-mTHF allozyme is slightly elevated. If they are on the same strand, the activity of the allozyme encoded by this strand is reduced.

Case Study

Nitric Oxide is no Laughing Matter

A Wisconsin couple welcomes the birth of their first-born son. He is healthy, feeds well, and is thriving. When he is 3 months old, his parents find a mass in the calf of his lower right leg and his pediatrician schedules further evaluation [230]. Surgical biopsy reveals that it is an infantile fibrosarcoma (a rare malignant growth of the soft tissue), which is removed 3 days later. The boy is anesthetized for 45 minutes for the biopsy and for 270 minutes for the removal of the tumor. The surgeons use 0.75% sevoflurane and 60% nitrous oxide for anesthesia on both occasions. He is recovering well and discharged home in apparent good health after a week.

Four weeks after the initial surgery, the infant starts having repeated seizures and at one point stops breathing. When he is again admitted to the hospital, the child has little muscle tension, no reflexes, and his breathing is poorly coordinated (ataxic). Imaging of his brain finds generalized atrophy and enlarged cisterns (spaces between the two innermost membranes around the brain). The child dies 11 days later due to respiratory arrest. Autopsy finds that the entire brain has shrunken and many neurons have been replaced by glial cells (asymmetric cerebral atrophy with severe demyelination, astrogliosis, and oligodendroglial-cell depletion in the midbrain, medulla, and cerebellum). What happened here?

Metabolic profiling finds elevated homocysteine concentrations in blood and urine; normal blood concentrations of folate, methyl malonic acid, and vitamin B12; normal folate concentration in cerebrospinal fluid; and no indication of abnormal excretion of organic acids. The elevated homocysteine concentration suggests a problem with folate metabolism, and possibly a genetic disorder because folate concentrations in blood and brain appear normal. And sure enough, the enzyme activity of 5,10-methylene tetrahydrofolate reductase (MTHFR; EC1.5.1.20; OMIM 607093) in cultured fibroblasts is less than one-fifth of the normal value, both with and without addition of FAD, the vitamin B2 (riboflavin) cofactor. Complete analysis of the more than 7000 bases of the MTHFR messenger RNA finds several places where the sequences in both strands differ from the common version with normal activity. One of them on the strand inherited from his father is the relatively common change from cytosine to thymine at position 677 (rs1801133), which replaces an alanine in the enzyme with a valine. More than one-third of all Americans carry this variant. This change not only cuts MTHFR enzyme activity by half [229] but also increases riboflavin requirements [231] and causes the enzyme to become inactive more quickly [232]. A second variation on this strand changes the guanine in position 1755 to an adenine, thus replacing a methionine with an isoleucine in the enzyme produced from this strand. This change takes the already diminished MTHFR enzyme activity down by another 15% or so [229].

The boy inherited from his mother another two variants on his other MTHFR strand. The change at position 1298 is not likely to affect enzyme activity on its own, despite the fact that it changes glutamic acid into alanine in position

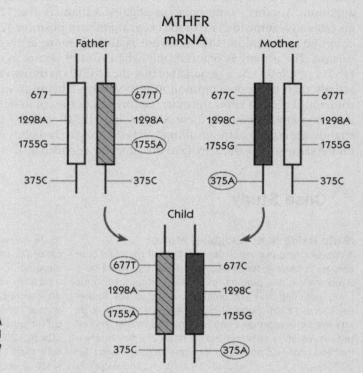

MTHFR mRNA

Father

677	(677T)
1298A	1298A
1755G	(1755A)
375C	375C

Mother

677C	677T
1298C	1298A
1755G	1755G
(375A)	375C

Child

(677T)	677C
1298A	1298C
(1755A)	1755G
375C	(375A)

The sequence variants on the two messenger RNA (mRNA) strands explained the severely depressed 5,10-methylene tetrahydrofolate reductase activity in this patient.

429 of the protein sequence [229]. Of much greater significance is the change at the position located 375 bases after the part of the mRNA that actually encodes amino acids in the mature *MTHFR* enzyme. This 3′ UTR loops back to the start (the 5′ UTR), stabilizing and positioning the mRNA for optimal translation into protein. All kinds of regulatory proteins and small RNA segments (miRNA) bind to this region and control the rate of protein synthesis. This is not a good place to have a variation. In this child it seems to have blocked most protein production from the affected DNA strand. The protein produced from both strands combined probably provided only a quarter of the normal enzyme activity (40% of the paternal half and a very small percentage of the maternal half).

Exposure to nitrous oxide slowly and irreversibly inactivates methylcobalamin (vitamin B12) in 5-methyltetrahydrofolate-homocysteine *S*-methyltransferase (*MTR*; EC 2.1.1.3; OMIM 156570). Every 46 minutes of nitrous

oxide anesthesia cut the available MTR activity in half [233] and recovery of the original activity takes many days because the enzyme has to be newly synthesized and needs a new supply of vitamin B12. All methylation reactions, including the synthesis of the choline-containing phospholipid sphingomyelin, are dependent on the methionine replenished by MTR. There is enough folate, but it gets trapped as 5-methyl folate and cannot be used due to the blocked MTR reaction [234].

The supply of methyl groups for the MTR reaction was already limited by the low-activity *MTHFR* variants in this child. His exposure to nitrous oxide finally throttled phospholipid synthesis to a debilitating trickle. There clearly was not enough material to keep the myelin sheaths around the neurons intact, much less support the growth of new myelinated neurons. A healthy infant's brain nearly doubles in size and adds about 200 g of white matter during the first year [235]. The white matter of the brain consists to a large extent of phospholipids

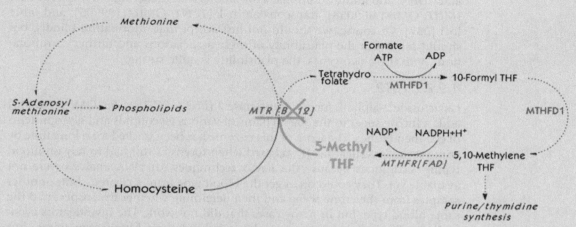

Normal phospholipid synthesis in this child was disrupted by irreversible oxidation of vitamin B12 in the 5-methyltetrahydrofolate-homocysteine *S*-methyltransferase (MTR) enzyme, worsened by genetically low MTHFR activity. In addition, the trapping of most available folate as 5-methyl-tetrahydrofolate slows the folate-dependent production of precursors (purines and thymidine) for DNA synthesis. ADP, adenosine diphosphate; ATP, adenosine triphosphate; THF, tetrahydrofolate.

(sphingomyelin). The catastrophic consequences of blocked phospholipids synthesis came fully to bear in this unfortunate infant with death due to brain atrophy and demyelination.

Modest supplementation with methionine [236], maintaining an optimal supply of riboflavin and vitamin B12, and avoiding nitrous oxide exposure could have kept this infant healthy. This would have been easy to do if the genetic predisposition had been known and understood in time. Among every few thousand otherwise healthy infants one is born with *MTHFR* activity below 25% of the normal value. Would you have appreciated the precarious situation of this child if information about the genetic variants had been available?

4.9.7 Vitamin B12

How important can a vitamin be, if we need it only for three enzymes (5-methyltetrahydrofolate-homocystein *S*-methyl transferase [MTR; EC 2.1.1.13], methylmalonyl-CoA mutase [EC 5.4.99.2], and L-beta-leucine aminomutase [EC 5.4.3.7])? Well, as we know, it is important enough that its lack causes birth defects, dementia, neuropathy, potentially fatal pernicious anemia, and other unpleasantness. Because of the central role of vitamin B12-containing MTR for homocysteine remethylation, suboptimal availability impacts a range of vital functions, such as regulation of DNA expression through epigenetic modifications. We might as well accept, therefore, that vitamin B12 status matters and ask whether genetic variants play a role.

A good first pass, when looking for genetic modulators of nutrient adequacy, is GWAS in large cohorts. One investigation of this kind found that common variants of *FUT2* and *CUBN* are the best genetic predictors of B12 concentration in adults from three Italian regions [222]. Later GWAS also saw links with *FUT2* and *CUBN*, and found additional associations with methylmalonyl-CoA mutase (*MUT*; OMIM 609058), transcobalamin 1 (*TCN1*; OMIM 189905), and other loci [237]. Of course, we should not just accept that information blindly, but should ask about the plausibility of these associations and further corroboration. With these two genes, the plausibility is quite strong.

4.9.7.1 FUT2

Galactoside 2-alpha-L-fucosyl transferase 2 (*FUT2*; EC 2.4.1.69; OMIM 182100) adds a fucose sugar to the sugar chains of various glycolipids and glycoproteins of cell membranes. The work of this enzyme has been tracked for a long time by crime scene investigators. It was used when forensics still had to rely on blood typing techniques because the newer techniques for DNA analysis were not available yet. They could often get the blood type from a cigarette butt or other samples from the crime scene and then determine whether the suspect had the same blood type. But in many cases that did not work. The investigators eventually realized that most people leave behind their blood type traces, but a minority does not. The difference was a genetic trait, which came to be called *secretor status*. Further investigations clarified that common variants of the *FUT2* gene determine blood type secretor status in the ABH and Lewis b antigen systems. For individuals of Caucasian ancestry the *FUT2* 461G allele (rs601338) is most often associated with secretor status, but in other populations other associated variants have a similar effect. So far, so good, but where is the link to vitamin B12 status? More recent work found that some bacteria attach to specific sugars on cell membrane structures. In particular, *Helicobacter pylori*, which infects the stomach lining of just about half of humanity, feels more at home in the stomachs of secretors because it can attach better to the glycan branches with the extra fucosyl residues [238]. The damage that *H. pylori* inflicts on the gastric lining is most severe in secretors and most likely to interfere with the release of intrinsic factor for vitamin B12 absorption. With less intrinsic factor, the secretors absorb vitamin B12 from food and from biliary recirculation less efficiently. They simply have a harder time maintaining good vitamin B12 stores. It should be no surprise then that the many European adults with *H. pylori* infections have a poorer vitamin B12 status than their peers without infection, and many of those infected individuals with diminished vitamin B12 status are secretors with the *FUT2* 461G variant [239]. As with any good story, there is probably a sequel in the making. The same study that showed these neat associations in Europeans came up dry in selected African populations. This suggests the importance of additional lifestyle and genetic factors.

4.7.9.2 CUBN

The second candidate gene identified by GWAS data was *CUBN* (OMIM 602997). Cubilin, the protein encoded by *CUBN*, is the intestinal receptor for vitamin B12 bound to intrinsic factor (*IF*; OMIM 609342). Cubilin needs a partner protein (encoded by *AMN*; OMIM 605799) to facilitate endocytosis of the B12—IF complex. Congenitally disrupted absorption of the B12—IF complex is the cause of the Imerslund-Graesbeck syndrome (*IGS*; OMIM 261100), a rare disease with megaloblastic anemia and renal protein loss. A small fraction of high-dosed vitamin B12 supplementation (typically 1000 µg) can be absorbed without the aid of the IF/cubilin system and rapidly corrects the deficiency symptoms. Injected vitamin B12 is often used to jump-start therapy. There is anecdotal evidence that low-activity variants increase vulnerability to B12 deficiency in some individuals [240,241].

4.9.8 Biotin

We need biotin as cofactor for four different carboxylases (acetyl-CoA carboxylase [EC 6.4.1.2], pyruvate carboxylase [EC 6.4.1.1], propionyl-CoA carboxylase [EC 6.4.1.3], and 3-methylcrotonyl-CoA carboxylase [EC 6.4.1.4]) and for specific interactions with DNA that regulate expression of some genes.

Biotin

In keeping with the consideration that for most nutrients we will find inborn deficiencies, about one in 80,000 infants is born with severe biotin deficiency (OMIM 253260), and about one in 35,000 is born with partial deficiency [242]. We know about them because newborn screening in the USA and many other countries looks for these conditions. It has to be recognized that currently available screening methods (usually with MS/MS) cannot detect all the milder forms and these may present months or even years after birth with clinical problems.

Neonatal screening for biotinidase deficiency is thought to be cost effective despite its rare occurrence because the severe health consequences can be prevented with biotin supplementation in a majority of affected infants. The main threats are metabolic dysfunction with acidosis and neurological abnormalities with variable presentation of ataxia, hearing loss, hypotonia, optic nerve atrophy, seizures, and slowed mental development. Dermatitis (scaly red rash around the anal and genital area, eyes, mouth, and nose), hair loss, and immunological deficiencies with recurrent infections are commonly observed. Without treatment, many of the affected children would become comatose after birth and die. The actual progression of the condition depends on the type of genetic defect and biotin intake.

One of the more common causes of biotin-responsive deficiency is defective biotinidase (EC 3.5.1.12), which is needed for recycling of biotin stores. Excessive losses of biotin with urine occur because the peptide-bound biocytin form of recycled biotin (Figure 4.38) escapes unhindered.

Treatment of biotinidase deficiency with high-dosed biotin (initially intramuscular injections daily with 150 µg biotin or more, then long-term oral

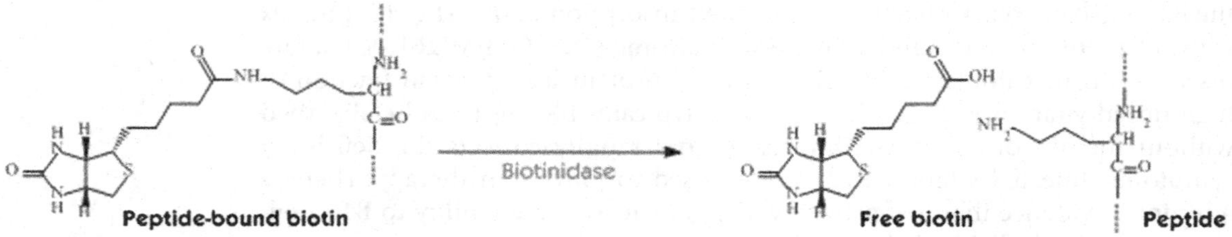

FIGURE 4.38
Biotinidase releases free biotin from degraded biotin-dependent enzymes.

supplementation with 10 mg/day) normalizes symptoms in children treated soon after birth within days or weeks and long-term outcomes are excellent [243]. Children identified several months or years later still benefit from treatment but some of the behavioral, developmental, hearing, and vision deficits will not be fully reversible. Egg white contains avidin, which binds biotin very tightly and limits its absorption. Cooking reduces avidin activity only to about 40% [244]. However, occasional egg consumption is not likely to interfere with high-dose biotin supplementation.

A less common cause for biotin deficiency with viability after birth is defective holocarboxylase synthetase (EC 6.3.4.10), which is the enzyme needed to attach biotin to its apoprotein targets. The enzyme is encoded by the *HLCS* gene (OMIM 609018). Defects of individual carboxylase genes are also known. Only some patients with variants in these genes will benefit significantly from biotin supplementation. Nonetheless, therapeutic trials need to be done to determine effectiveness in individual patients.

The prevalence and significance of very mild gene variants associated with increased biotin requirements have not been investigated systematically. Anecdotal observations [245] make it plausible that individuals with heterozygous gene defects or low-activity gene variants are more vulnerable to low biotin intake, particularly in combination with high egg white consumption or treatment with drugs that impair biotin bioavailability, such as the antiepileptic drugs phenytoin and carbamazepine [246].

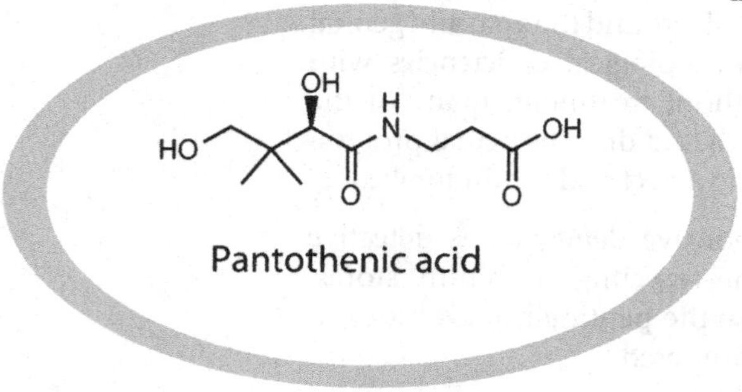

Pantothenic acid

4.9.9 Pantothenic acid

Pantothenic acid (PA) is needed for the synthesis of coenzyme A without which the Krebs cycle, beta-oxidation, or other metabolic pathways would not work. Metabolism would just shut down and that would be the end of it. But this does not happen, because PA is one of those vitamins for which there is almost always enough around, both from intestinal bacteria and from food.

When deficiency occurs, it is usually because a mutation limits coenzyme A synthesis or PA recovery.

One of the critical steps in coenzyme A synthesis depends on pantothenate kinase (EC 2.7.2.33). There are at least four distinct genes encoding this enzyme. No defects of the initially identified gene, *PANK1*, have been reported in humans, presumably because it would prevent cell viability immediately after conception. Loss-of-function mutations in *PANK2* cause a very severe neurodegenerative condition (Hallervorden-Spatz disease; OMIM 234200).

4.9.10 Vitamin A

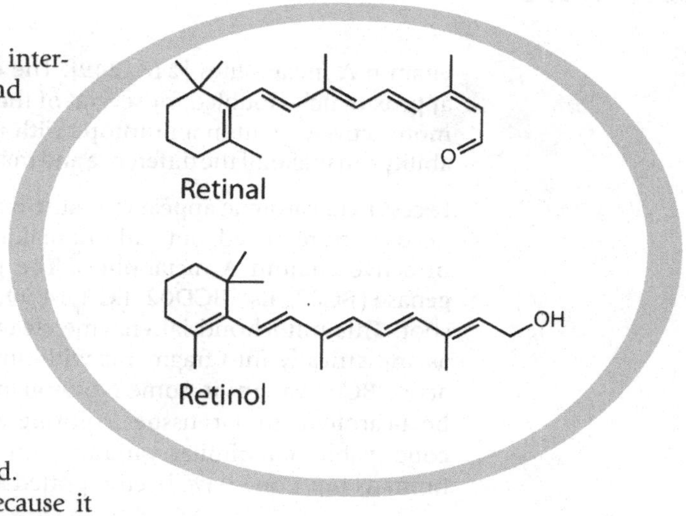

The term *vitamin A* includes mainly the interconvertible compounds retinol and retinal. Both can be converted into several other functionally important metabolites, particularly retinoic acid, but the transformation is irreversible.

Food contains both preformed vitamin A (mostly as retinyl esters) and a variety of provitamin A carotenoids. To qualify as a provitamin A carotenoid, the candidate compound has to contain a retinal that can be released from the parent compound. Beta-carotene presents the ideal case because it generates two retinal molecules when it is split right down the middle (Figure 4.39). Carotenoid 15,15′-monooxygenase 1 (BCM01 or BCD01; EC 1.14.99.36) in the small intestine can do this neat trick [247]. The enzyme also works on other carotenoids, such as alpha-carotene (generating only one retinal molecule). The protein is encoded by the *BCM01* gene (OMIM 605748). Several common *BCM01* variants in humans influence the concentration of beta-carotene, but not always of

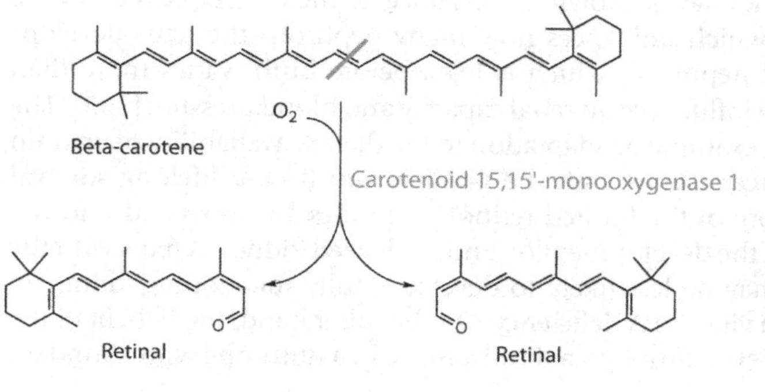

Retinal

Retinol

Beta-carotene

O_2

Carotenoid 15,15′-monooxygenase 1

Retinal

Retinal

FIGURE 4.39
Carotenoid 15,15′-monooxygenase 1 in the small intestine converts beta-carotene into two vitamin A (retinal) molecules. O_2, oxygen molecule.

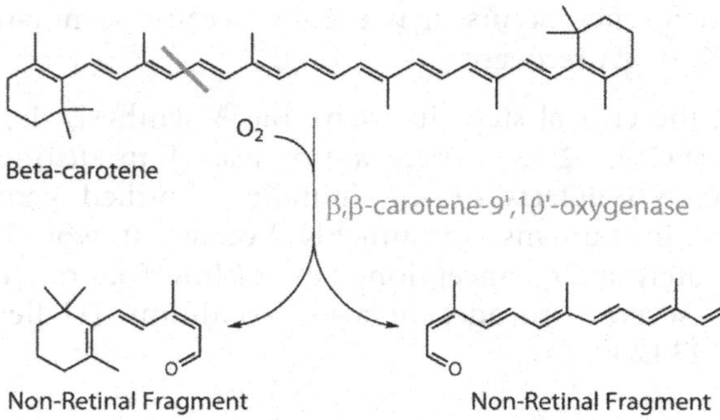

FIGURE 4.40
The mitochondrial enzyme β,β-carotene-9′,10′-oxygenase chops up beta-carotene into pieces of unequal size neither of which has vitamin A activity. O_2, oxygen molecule.

vitamin A metabolites [248, 249]. The efficiency of beta-carotene conversion is approximately doubled for several of the variants. It is easy to see that having the more active variant in a nutritope with marginal beta-carotene and retinol availability can make all the difference and may even be a matter of a child's life or death.

Excess beta-carotene appears to cause oxidative stress in mitochondria [250] and needs to be removed, but without making the situation worse by generating a lot of active vitamin A metabolites. The iron-containing β,β-carotene-9′,10′-oxygenase (BC02, also BCD02; EC 1.14.99.-) takes care of the problem right on the spot. This mitochondrial enzyme cleaves beta-carotene and other carotenoids asymmetrically into fragments without vitamin A activity (Figure 4.40). A less active *BC02* variant in some Norwegian sheep is linked to the accumulation of beta-carotene in fat tissue, coloring it distinctly yellow-orange [251]. It is conceivable that similar variants with impact on carotenoid removal exist in humans but none have been reported so far.

A special form of aldehyde dehydrogenase (EC 1.2.1.36) encoded by *ALDH1A2* (OMIM 603687) converts retinal irreversibly into the potent transcriptional regulator retinoic acid.

The common variant rs7169289 G of the *ALDH1A2* gene appears to increase kidney size at birth by one-fifth [252]. ALDH1A2 in the fetal kidney converts retinol irreversibly into the potent transcriptional regulator retinoic acid, which promotes the formation of nephrons during fetal development [253]. Retinoic acid in the developing kidney is known, for instance, to increase expression of the proto-oncogene *RET*, which influences how many nephrons the fetus develops [254]. The number of nephrons, which is fixed before birth, varies more than threefold, with lifelong influences on renal capacity and blood pressure [255]. The variant may thus be an example of adaptation to the dietary availability of vitamin A. If vitamin A is scarce, a fetus with the G allele will have a lifelong survival advantage because more of the limited retinol molecules are converted into retinoic acid and support the development of a normal-sized kidney. A fetus with the A allele, in contrast, may be less likely to develop a fully functioning kidney in a situation of maternal vitamin A deficiency. On the other hand, the benefit of the G allele may vanish or even turn into a disadvantage in a nutritope with abundant vitamin A because the kidneys grow to a bigger size than needed. This may increase the risk of hypertension and renal failure later in life [256].

A case of haploinsufficiency

What happens when the conversion of provitamin A carotenoids to retinal does not work properly was seen in an otherwise healthy adult with a defective *BCMO1* gene [227]. This middle-aged woman had distinct yellow skin and a fourfold increase in beta-carotene concentration in blood (over 8000 µg/L), which was not caused by excessive beta-carotene consumption. It should be pointed out that even a daily consumption of 2 kg carrots or more will not raise beta-carotene concentrations over 5000 µg/L [258]. Despite the abundant availability of the provitamin A precursor, the retinol concentration in this woman's blood was at the very low end of the normal range, presumably maintained by intake of preformed retinol from meats and eggs [259].

The remarkable finding was that the woman was heterozygous for the defective *BCMO1* Thr170Met variant and that the other strand was fully functional. This means that loss of one gene copy was enough to cause accumulation of beta-carotene in the body and poor vitamin A status. The likely explanation is that the enzyme operates under saturation conditions and increased substrate concentration (from higher intake) cannot increase its activity [257].

This situation where a heterozygous genotype results in clinically significant loss of function is called *haploinsufficiency*.

4.9.11 Vitamin D

Vitamin D is an unusual case among nutrients because we don't normally get it from food. The natural form of vitamin D, cholecalciferol or vitamin D3, is produced in skin when medium-wave UV light (UV-B, 280–320 nm) breaks one of the rings of the cholesterol precursor 7-dehydrocholesterol (Figure 4.41). This

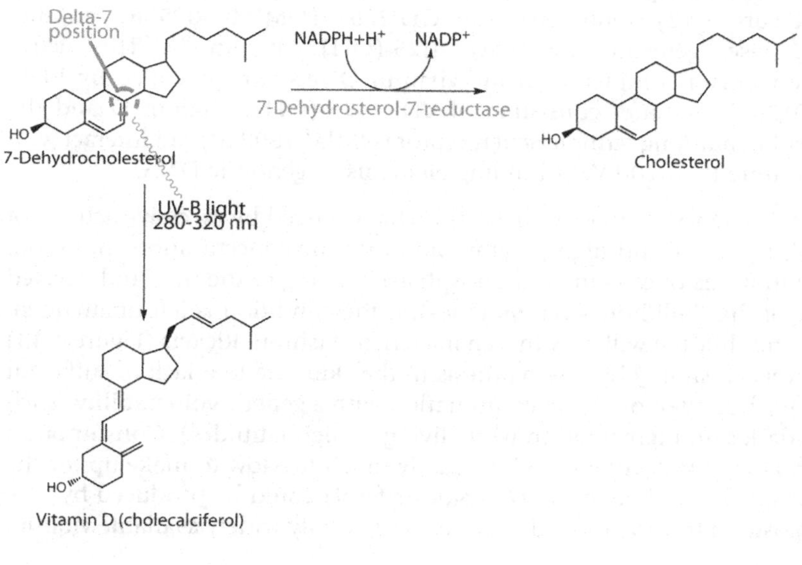

FIGURE 4.41
Most vitamin D in humans is produced when ultraviolet light strikes a cholesterol precursor in the skin. NAD, nicotinamide adenine dinucleotide; UV, ultraviolet.

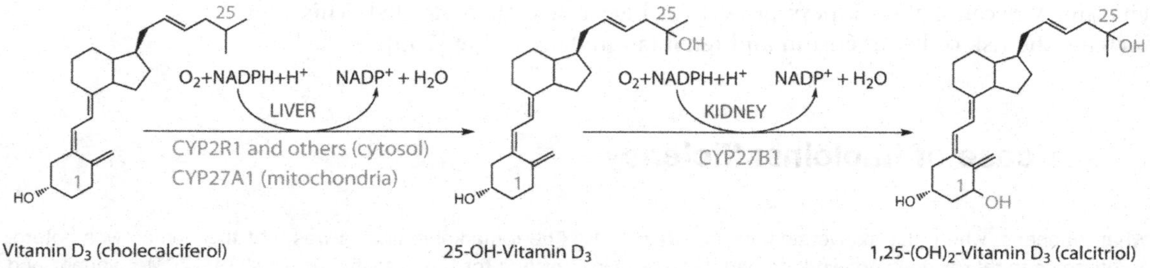

FIGURE 4.42
Whether vitamin D comes from the skin or from food, it needs to be activated in a two-step process before it can do its work.

means that the critical factor is a health behavior, i.e., exposing skin to the sun at the right time. This is somewhat similar to physical exercise, which also greatly impacts nutritional status. Nutrition still comes into the picture because we can get some vitamin D from fatty ocean fish (rarely enough, though). In reality, however, the most important alternative source by far is now synthetic vitamin D, both in the form of fortified foods and dietary supplements.

The classical functions of vitamin D depend on activation of the inert precursor by two consecutive hydroxylation reactions. The initial conversion to 25-OH-vitamin D occurs in the liver. There is still considerable uncertainty about which genes encode the enzymes with vitamin D 25-hydroxylase activity and what their respective properties and roles are [260]. Mitochondrial *CYP27A1* (OMIM 606530) and five different microsomal cytochromes have some activity, but *CYP2R1* (OMIM 608713) appears to be the most important one by far. The others provide some limited backup capacity. The 25-OH-vitamin D must then pass into the primary filtrate of the kidney, where it binds to cubilin (*CUBN*; OMIM 602997) and gets taken up into the epithelia of the proximal renal tubules (Figure 4.42). Only then can *CYP27B1* (OMIM 609506), another monooxygenase, generate the active 1,25-(OH)2-vitamin D. This active metabolite forms a complex with the vitamin D receptor (encoded by *VDR*; OMIM 601769). A dimer consisting of VDR-1,25-(OH)2-vitamin D and the retinoic acid-containing retinoic acid receptor (OMIM 180220) can interact with any of the more than 200 VDR-binding elements in genomic DNA.

Vitamin D deficiency in infancy (rickets) is characterized by excessive activity of alkaline phosphatase and high parathyroid hormone concentrations in blood, low concentrations of calcium and phosphate, beading of the ribs, and delayed hardening of the skull bones (craniotabes). If the condition is left unattended, the legs of the children will bow in a characteristic fashion. Rickets (Figure 4.43) is usually caused simply by low synthesis in the skin due to a lack of sufficient UV exposure, but often occurs in combination with a genetic vulnerability (such as having dark skin pigmentation while living at high latitudes). Consumption of vitamin D with unfortified food is usually much too low to make up for the lack of sunshine. Until vitamin D2 (ergocalciferol) could be produced by irradiating ergosterol from yeast, cod liver oil was the only widely available vitamin

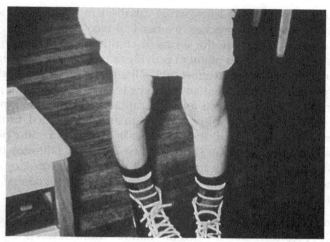

FIGURE 4.43
A lack of active vitamin D is the most common cause for rickets in young children. A genetic predisposition often has increased vulnerability to low vitamin D supplies from ultraviolet light and dietary intake. Physical signs include soft bones of the legs that bend when children start walking and put weight on them. Note the bowed legs and knees in this child.
Photo courtesy of Public Health Image Library of the Centers of Disease Control.

D supplement. Children today are much more fortunate, since they can get taste-free, odorless vitamin D supplements.

As might be expected, rickets can also be due to a genetic disruption of the enzymes for vitamin D activation, but these cases are very rare. A single infant with rickets due to defective *CYP2R1* was reported [261]. This child had extremely low blood concentrations of 25-OH-vitamin D and the biochemical and skeletal abnormalities typically seen with untreated rickets. Another form of congenital rickets is due to defective *CYP27B1* and the inability to produce the active metabolite 1,25-(OH)2-vitamin D. Affected children have normal 25-OH-vitamin D concentration, but very little 1,25-(OH)2-vitamin D. Such children are very successfully treated with oral doses of 1,25-(OH)2-vitamin D (usually called calcitriol).

Multiple sclerosis

Heterozygous loss-of-function variants of *CYP27B1*, which are known to limit 1,25-(OH)2-vitamin D production, have now been firmly linked to familial cases of multiple sclerosis [262]. This hereditary link establishes that vitamin D deficiency causes multiple sclerosis, probably in conjunction with additional factors. We also know now that genetic predisposition acts together with poor vitamin D supply, since multiple sclerosis is most common in Scotland and other regions with too little sun to make vitamin D for many months of the year [263]. Knowing about this nutrient—gene interaction is very important. It establishes a solid basis for the prevention of multiple sclerosis by ensuring adequate vitamin D supplies. The tiny minority of patients with a genetic vitamin D activation defect have unknowingly shown the way for the vast majority, who could activate vitamin D if only they had enough to work with.

It is easy to see the profound impact of variation in vitamin D metabolism. Before we move on to other genes, we need to recognize that the genes that control skin pigmentation are the strongest commonly inherited modifiers of vitamin D synthesis. Melanin pigment provides a highly effective shield against both the harmful and beneficial effects of UV light. Vitamin D synthesis in skin is inversely proportional to pigmentation intensity. The North–South gradient of skin pigmentation density reflects a rapid adaptation to the local availability of UV-B light for vitamin D synthesis. As populations migrated further north, where UV exposure is minimal, their skin got lighter and in some populations lost most of its pigmentation. Those settling down at latitudes in between developed the ability to tan, because they still needed the protection during the peak summer months, but could not get enough UV light for several months in the fall and winter [264].

This happened independently multiple times in different populations. It was the same selective pressure that blanched the skin of Neanderthals [265] when they migrated out of Africa into Europe several hundred thousand years ago that acted much later when modern man followed the same paths. If they could not make enough vitamin D in their skin, their offspring would not survive. It was as simple as that. Skin damage and cancer risk by UV light at high latitudes just don't weigh into this evolutionary balance because the harm is much less than what could happen closer to the equator and cancer usually develops long after reproductive age.

The variants in the *melanocortin 1* gene (*MC1R*; OMIM 155555), a key player among the more than a hundred controllers of skin pigmentation [266], can now tell the story. The Neanderthals in what is today Italy and Spain had a unique low-function variant with a glycine in position 307 of MC1R. Based on what we know, our cousins had light skin and red hair. Dutch redheads came to their flaming hair tint and pale, easily sunburnt skin all on their own with a cysteine replacing the original lysine in position 151 (rs1805007) or an arginine in position 160 (rs1805008) with a tryptophan [267]. Similarly, other light-skinned populations around the world have their own characteristic variants.

GWAS indicate that several polymorphisms, including rs7041 (1296T>G; Asp432Glu) and rs4588 (1307C>A; Thr436Lys), in the gene for a vitamin D-binding protein (*GC*; OMIM 139200) influence vitamin D status as measured by 25-OH-vitamin D concentration [268, 269]. The protein sequence variants encoded by rs7041 and rs4588 are separated by only two amino acids and almost always go together. The 1307A allele (and accordingly the 1296G allele, because the two are tightly linked) boosts the increase in 25-OH-vitamin D concentration (Figure 4.44) in response to supplementation [270]. This could mean either that peripheral cells have more vitamin D available for uptake or that protein binding limits uptake. In other words, we can be certain about the difference, but cannot be sure of its biological implications.

A common variant (rs7944926) of the gene (*DHCR7* or *NADSYN1*; OMIM 602858) encoding delta-7-sterol reductase (EC 1.3.1.21) is another modifier of

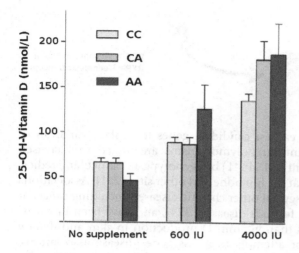

FIGURE 4.44
The response to vitamin D supplementation depends on the common rs4588 polymorphism of the *GC* gene [270]. The vitamin D concentration in blood of people with the 1307AA (A/A) genotype in response to vitamin D supplementation is greater than in people with the CC (C/C) genotype.

25-OH-vitamin D concentration in blood [268]. As pointed out in Chapter 3, this enzyme catalyzes the final step of cholesterol synthesis. The allele encoding a less active enzyme form can be expected to increase the concentration of 7-dehydrocholesterol (the vitamin D precursor in skin) and give sunlight (UV-B) a better opportunity to generate vitamin D. Remember that thinner skin and therefore less precursor is the main reason why vitamin D status declines with age. The function of the *DHCR7* gene cannot decline too much because that would then cause severe disruption of embryonic development. Genetically very low DHCR7 function causes the Smith-Lemli-Opitz syndrome (*SLOS*; OMIM 270400), which is characterized by multiple malformations and mental retardation. Thus, particularly in people in the North living without much sun, the evolutionary balance comes down to more efficient vitamin D production in the large majority of heterozygotes versus severe disability in a smaller number of homozygotes.

The *CYP2R1* gene also contains a common variant; rs10741657, which modifies the concentration of 25-OH-vitamin D in blood, as might be expected from its key position in vitamin D activation [268].

Finally, a word about the vitamin D receptor (*VDR*; OMIM 601769). VDR is a protein that helps 1,25-dihydroxy-vitamin D to activate specific DNA segments. It does so as a heterodimer with the retinoic acid receptor, which, in turn, has retinoic acid bound to it (Figure 4.45). It is not surprising to find that changes in the amount or sequence of VDR can have a profound impact on the effectiveness of vitamin D. A few cases of rickets have been reported to be due to defective VDR. Increasing intake of vitamin D will not help because the active metabolite does not bind well to the DNA targets.

There are several common variants (Table 4.5) that appear to modify VDR expression or alter binding to DNA. Unfortunately, the quantitative molecular consequences of the polymorphisms still need to be clarified.

FIGURE 4.45
A heterodimer consisting of VDR with 1,25-dihydroxy-vitamin D and RXR with retinoic acid binds to specific DNA segments.

VDR = Vitamin D receptor
RXR = Retinoic acid receptor

VDR was actually one of the earliest candidate genes to explain variation in osteoporosis risk. Several common polymorphisms are linked to increased osteoporosis risk. The VDR BsmI (rs154441) b/b genotype, in particular, predicts a slightly lower risk of fracture at the hip bone and other sites [271]. Associations of these polymorphisms with several other chronic diseases, including cancer at various sites [272, 273] and infectious diseases [274], and have been inconsistent and much less important than vitamin D production in skin and dietary intakes. Like with many other attempts to assess gene–disease associations, these studies fall short on reliable long-term measures of vitamin D intake. The actual significance is that these genetic variants modify the amount of vitamin D that an individual needs to prevent one of these chronic diseases. It is often said that known genetic variants do not explain much of our health problems, but this is a crude oversimplification fed by lack of information in a field where knowledge is growing by leaps and bounds. A GWAS of vitamin D insufficiency in Europeans found that carriers of just three common variants (rs2282679 in *GC*; rs12785878 near *DHCR7*, and rs10741657 near *CYP2R1*) were two and a half times more likely to have avoidable vitamin D insufficiency than people without these variants [268]. This is a sizable risk by any measure and is worth addressing.

Too much activated vitamin D would be dangerous because it would constantly promote hypercalcemia. This is exactly what happens when newborn infants with idiopathic infantile hypercalcemia get their usual vitamin D prophylaxis (typically 400 IU/day). In most cases they have a defective version of the *CYP24A1* (OMIM 126065) gene [275], which encodes the enzyme (25-hydroxyvitamin D 24-hydroxylase; EC 1.14.13.126) for inactivating 1,25-(OH)2-vitamin D by conversion to calcitroic acid (Figure 4.46). The common

Table 4.5	Common VDR Variants which Influence Vitamin D Effectiveness	
Name	**Variant**	**rs Number**
ApaI, A>a	A>T, intron 8	rs17879735
BsmI, B>b	A>G, intron 8	rs1544410
TaqI, T>t	C>T, Ile352Ile	rs731236
FokI, F>f	C>T, Met1Thr	rs2228570/rs10735810

rs, reference single nucleotide polymorphism (SNP).

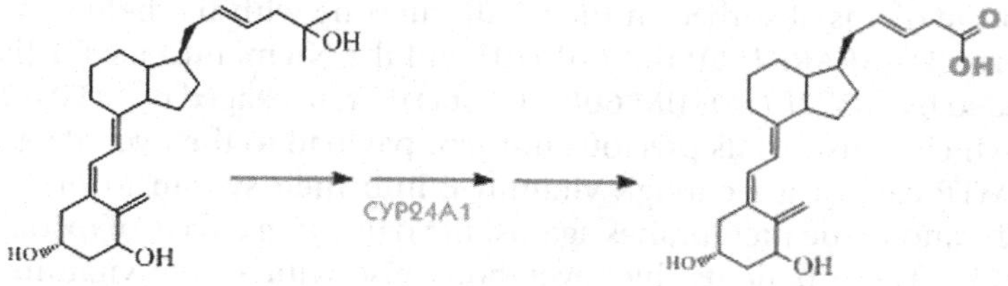

FIGURE 4.46
Inactivation of the functional metabolites completes vitamin D metabolism and keeps this hormone-like system responsive in both directions.

1,25-(OH)$_2$-Vitamin D$_3$ (calcitriol)

Calcitroic acid

CYP24A1 variant G of rs2248137 appears to be associated with reduced 25-hydroxyvitamin D concentration [276].

4.9.12 Vitamin E

We need this antioxidant to protect us against oxygen free radicals, particularly in lipophilic compartments that none of the water-soluble anti-oxidants can reach. When a highly unsaturated fatty acid component of membrane phospholipids becomes oxidized by a free radical, it becomes unnaturally kinked, bending right to where a vitamin E molecule sits (if we have enough of it) and gets cured on contact. Vitamin E should also be there when a polyunsaturated fatty acid in LDL gets attacked by a free radical and is at risk of generating oxidized LDL, a highly atherogenic lipoprotein. It is the only compound that can do that job.

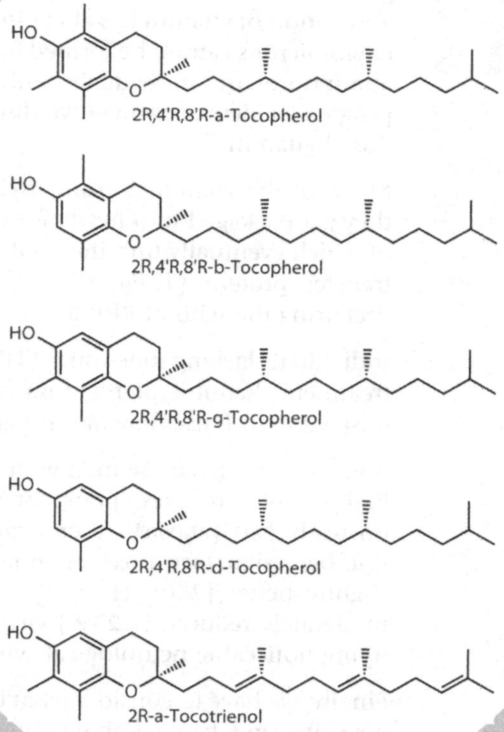

Ten compounds with vitamin E activity are known to occur in nature. The most common ones are alpha-, beta-, gamma-, and delta-tocopherol, occurring mostly in nuts, oils, and seeds. Four analogous tocotrienols are found in some foods.

2R,4'R,8'R-a-Tocopherol

2R,4'R,8'R-b-Tocopherol

2R,4'R,8'R-g-Tocopherol

2R,4'R,8'R-d-Tocopherol

2R-a-Tocotrienol

Vitamin E from the diet is absorbed in the small intestine with the help of the scavenger receptor B1 (*SRARB1*; OMIM 601040) and the Niemann-Pick C1 like protein 1 (encoded by *NPC1L1*; OMIM 608010). It is then packaged by MTP into chylomicrons, which transport its precious nutrient payload to the liver. People with defective MTP cannot get enough vitamin E into their system to protect their lipoproteins and tissue membranes against the damaging effects of oxidant free radicals [277]. These patients, like everybody else with severe vitamin E deficiency, will develop progressive ataxia if they do not get treatment. The vitamin E deficiency also causes the production of misshapen red blood cells with spur-like protrusions (acanthocytosis). Use of a very large daily oral dose of vitamin E (100–200 IU/kg) will halt progression of the neurological abnormalities, and may even lead in some cases to a modest improvement in the symptoms [278]. Production of normal red blood cells resumes quickly. Vitamin E injections, which would bypass the malfunctioning absorption mechanism, are rarely used but may restore plasma vitamin E concentration more effectively than oral intake.

Chylomicron retention disease (*CMRD*; OMIM 246700) is another rare genetic condition that disrupts vitamin E absorption and has similar consequences and guidelines for vitamin E therapy [279]. Another rare condition that prevents the absorption of vitamin E is abetalipoproteinemia (*ABL*; OMIM 200100) in which chylomicrons cannot be formed because of defective synthesis of their structural apolipoprotein. Individuals with abetalipoproteinemia develop very severe progressive neurodegenerative disease that is preventable with the use of high-dosed vitamin E.

Much of the vitamin E from chylomicrons eventually reaches the liver. There they get packaged into newly forming very-low-density proteins (VLDL), many of which eventually turn into LDL. The key packaging agent is alpha-tocopherol transfer protein (*TTPA*; OMIM 600415). TTPA is highly selective, greatly preferring the natural RRR-alpha-tocopherol form over all others.

Individuals lacking functional TTPA develop progressive ataxia without effective treatment. Retinitis pigmentosa is another condition in some patients. High-dose vitamin E use may halt progression in some cases.

A 62-year-old Japanese man with two copies of a loss-of-function TTPA variant had blunted reflexes, poor orientation in space (proprioception), and an unsteady gait (ataxia). After 7 months of treatment with 800 mg vitamin E, Romberg sign (see box) became negative and his orientation in space was slightly better [280]. Heterozygotes for a *TTPA* variant were found to have moderately reduced (<25%) vitamin E concentration in blood without developing noticeable neurological symptoms [280].

Finally, we have to consider what happens to vitamin E once it gets into a cell. If new vitamin E kept coming in, it would get quite crowded in there after a while. Breakdown of vitamin E into inactive, water-soluble metabolites takes care of that little problem. These vitamin E metabolites can readily leave the cell and are excreted in urine and feces. Omega-oxidation by cytochrome P450 4F2 (*CYP4F2*; OMIM 604426) is the main mechanism of vitamin E catabolism [281], but omega-1 and omega-2 oxidation by unknown enzymes also play a role. Now, one might ask why we should bother with such arcane minutiae. An important reason is that common *CYP4F2* variants make the encoded enzyme more or less active. This means that the speed of vitamin E degradation differs, which ultimately influences the concentration of active vitamin E in blood and tissues [281]. Replacement of tryptophan in position 12 of the mature CYP4F2 protein by glycine (Trp12Gly; rs3093105) increases the activity of the variant enzyme more than twofold. Another variant, V433M (rs2108622) reduces enzyme activity by almost a half. The high-activity variant Trp12Gly is about twice as common in people with African ancestry than in those with Caucasian ancestry (21% vs. 11%); for the low-activity variant it is the other way around (9% vs. 17%). This means that, based on these variants alone, at the same intake levels we have to expect higher vitamin E tissue concentrations associated with Caucasian ancestry than with African ancestry.

Neurological signs of vitamin E deficiency

One or more of a group of neurological symptoms and signs can indicate significant damage to the cerebellum, spinal cord, or nerves in people with severe vitamin E deficiency.

Ataxia

Course intention tremor when reaching for objects;
Gait abnormalities and other symptoms of impaired movement coordination indicate cerebellar dysfunction;
Inability to perform rapidly alternating movements (dysdiadochokinesia);
Misjudging the range of movements (dysmetria);
Poorly coordinated and unsteady gait (gait ataxia);
Postural instability;
Slurred speech and poor articulation (dysarthria).

Abnormal Reflexes

Babinski sign (large toe curves upwards when the lateral side of the sole is rubbed);
Loss of deep tendon reflexes (muscle contraction when the tendon is tapped).

Impaired Proprioception

Romberg test evaluates the sense of positioning in space (proprioception). Damage to the dorsal column of the spinal cord may diminish proper proprioception.

Ask the subject to stand freely in a safe place with feet together and eyes closed. Be ready to steady or catch the subject. Look for swaying motions or a tendency to fall and interpret this as positive Romberg sign. A well-functioning vestibular system may be sufficient to maintain balance despite loss of proprioception. The diagnostic accuracy may be greatly improved by having the test subjects make head movements following a sound signal to diminish vestibular function [282].

Other Signs

Impaired vibratory sense;
Muscle weakness.

4.9.13 Vitamin K

The body uses the compounds grouped under the name of vitamin K as cofactors for the gamma-carboxylation of specific glutamates in a small number of proteins,

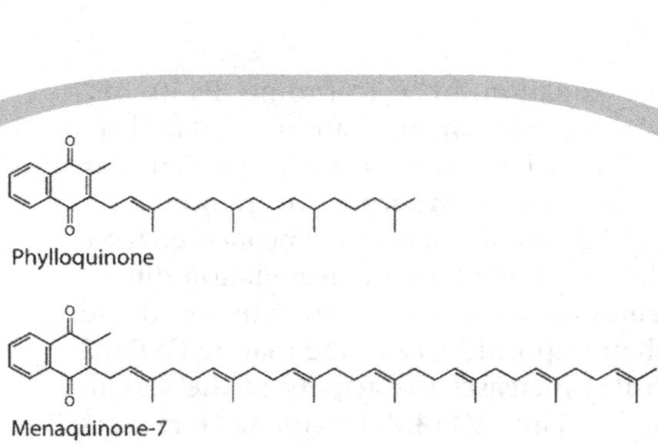

Phylloquinone

Menaquinone-7

most of which are vitally important at some time during our life and we could not survive without them. They can do this because gamma-carboxyl glutamate in proteins binds calcium with high specificity. Some of these vitamin K-dependent proteins sustain blood coagulation (factors II, VII, IX, and X), while others put a break on coagulation or otherwise help with its control (proteins C, S, and Z). Other proteins of this class (matrix Gla protein and osteocalcin) control calcification in bones (prevent-ing osteoporosis) and tissues (preventing atherosclerosis and ectopic soft tissue calcification). Still others (including Gas6) participate in cell signaling, control of cell differentiation, embryonic development, and prevention of excessive cell growth. Since the body cannot make vitamin K, we have to get it from other sources. Green leafy plants (basil pesto, broccoli, Brussels sprouts, kale, and spinach) supply us with phyllo-quinone and bacteria give us the menaquinones, a group of related compounds with vitamin K activity. Much of the bacterial vitamin K (as menaquinones) actually comes from our friends in the ileum and colon. Smaller amounts come from fermented foods, such as smelly cheeses and fermented tofu. The menaqui-nones have between four and 13 isoprenyl units in their side chain. Shown here is the common menaquinone-7 with seven isoprenyl units.

Absorption of vitamin K relies largely on chylomicrons and parallels the previously described uptake and transport of cholesterol, carotenoids, and other fat-soluble substances. Vitamin K stores in the body are minimal and last only a few days. Since the amounts of vitamin K available to us are very small (in the microgram range), we have to rely on recycling reactions (Figure 4.47). Vitamin K activation depends largely on warfarin-sensitive vitamin K-epoxide reductase (*VKOR1*; EC 1.1.4.1; OMIM 608547) and to a lesser extent on warfarin-insensitive vitamin K-epoxide reductase (EC 1.1.4.2). Additional enzymes with this activity are NADPH dehy-drogenase, quinone 1 (*NQ01*; EC 1.6.5.2; OMIM 125860), and NADPH:quinine oxidoreductases 1 (*NDOR1*; EC 1.6.99.2; OMIM 606073); these enzymes are also called diaphorases. Each time the activated vitamin K hydroquinone is used for a carboxylation reaction, it is oxidized and has to be reduced first to its vitamin K form and then to vitamin hydroquinone before it can be used again. The first step uses the same vitamin K-epoxide reductases as the initial activation.

The amount of available vitamin K in the liver and other tissues and its activation are subject to considerable genetic variation. Absorption efficiency and distri-bution of the chylomicron remnants with newly absorbed vitamin K to specific

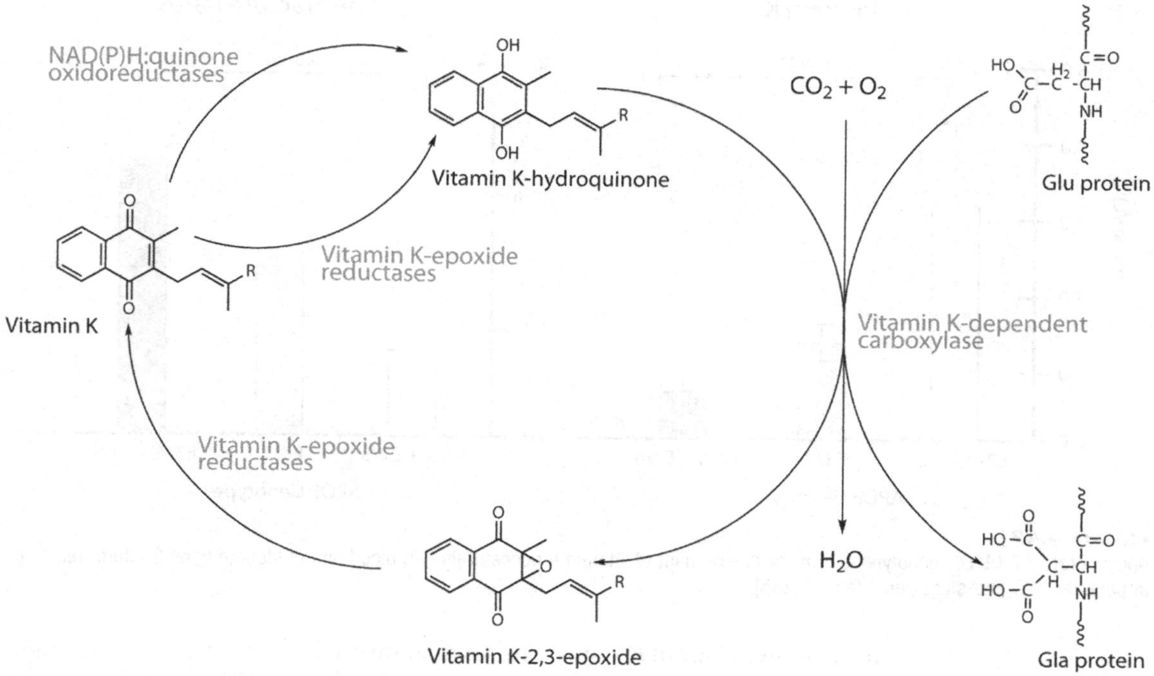

FIGURE 4.47
Vitamin K is oxidized when it participates in a carboxylation reaction as a cofactor. The reactivation proceeds in two distinct steps. CO_2, carbon dioxide; H_2O, water; O_2, oxygen molecule.

tissue is strongly influenced by lipoprotein metabolism and its determinants. Apo-E is such a modulator of vitamin K absorption and transport because it binds to several lipoprotein receptors in the bone, liver, and other tissues. The relatively common *E4* allele (haplotype combination of rs429358 and rs7412) is more likely to steer vitamin K in the chylomicron remnants to the liver because the encoded Apo-E isoform is a more effective LDL receptor ligand than the E3 and E2 isoforms [283]. Once the remnants have delivered vitamin K to a target tissue, there is no efficient mechanism for transporting from there to another tissue. Compare the difference of this situation with the transport of vitamin E described earlier. Once vitamin E is delivered to the liver by chylomicron remnants, it can be transferred into newly produced VLDL and transported to all kinds of other tissues. There is no such luck with vitamin K. If the liver in somebody with the *APOE*E4* allele is getting hold of vitamin K, then vitamin K pretty much stays there.

Genetic variants with influence on vitamin K have the potential to impact vitamin K-related conditions or diseases. The risk of osteoporosis and bone fractures increases with declining vitamin K status, not least because the vitamin K-dependent bone proteins require a particularly high concentration of this nutrient [284]. This known relationship raised the question of whether individuals with the *APOE*E4* variant tend to be at increased risk because they tend

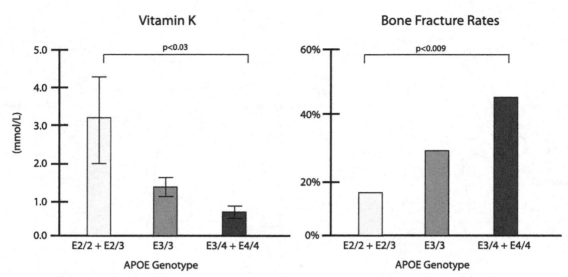

FIGURE 4.48
Apolipoprotein E (*APOE*) genotype is a major determinant of vitamin K concentration in blood and of lifetime bone fracture frequency in patients with end-stage renal failure [285].

to have poorer vitamin K status. The original investigations were done in a high-risk population: patients with end-stage renal failure. *APOE* genotype was the most important determinant of vitamin K concentration in blood and was associated with a distinctly increased risk of bone fractures [285]. Patients with the *E4* allele were almost four times more likely than patients with the *E2* allele to have suffered a bone fracture (Figure 4.48), which paralleled a similar difference in vitamin K concentration.

Later investigations of healthy people indicate a modest effect on women's bone health, specifically on the bone mineral density at the neck of the hip bone (trochanteric region), where fractures are most likely to occur [286]. This relatively weak effect should not surprise us because additional genetic variants are known to influence vitamin K availability and intake levels vary so greatly. The fact that this genotype-specific nutrient effect keeps showing up tells us that we must not ignore it in future studies.

Another well-known source of variation in vitamin K availability is the *VKORC1* gene, which contains several functionally important polymorphisms. This variation has great practical importance, since a common method of anticoagulation (blood thinning) uses medications that inhibit the vitamin K-epoxide reductase VKOR1, as shown in Figure 4.47. Patients with high-activity VKOR1 variants need a higher dose of the anticoagulant than patients with low-activity variants. Genetic testing before the start of anticoagulation therapy accelerates the search for the right dose and reduces the risk of excessive bleeding. More details on the practical use of such testing schemes are given in Chapter 7.

4.9.14 Choline

The amine choline is an important building block for the human body. Most of the phospholipids in cell membranes and the myelin sheath of neurons contain choline. The compound is also the essential precursor of the neurotransmitter acetylcholine.

We can produce choline on our own through a reaction catalyzed by phosphoethanolamine methyltransferase (*PEMT*; EC 2.1.1.17; OMIM 602391). This reaction (Figure 4.49) consumes three methyl groups from the methyl donor *S*-adenosylmethionine (SAM).

Most of these methyl groups come ultimately from the glucose metabolite serine and are funneled to SAM in a series of folate- and vitamin B12-dependent reactions (Figure 4.50).

It took until 1998 before choline was declared an essential nutrient by the Institute of Medicine's Food and Nutrition Board. This belated recognition has a lot to do with the great variability in choline requirements. It is quite likely that a minority of humans usually do not need to get any choline from food because they make enough on their own. This is particularly true of young women because estrogen strongly promotes choline production. Only half of them show any signs of deficiency if they go without choline for a month. This should not surprise us, since the key gene for choline synthesis, *PEMT*, is strongly induced by estrogen. Postmenopausal women and men, who do not have a lot of estrogen, usually need to get choline from food to avoid significant health risks. We know now that a lack of dietary choline often causes the retention of fat (triglycerides) in the liver because choline is needed for packaging fat into VLDL and exporting them into the bloodstream. This can result in increased accumulation of fat in the liver in the absence of excessive alcohol consumption or other pathology. Persistence of this nonalcoholic steatohepatitis (NASH) may transition to liver cirrhosis and eventually in some cases to liver cancer. Another known sign of choline deficiency is the increased fragility of muscle. The detection of increased creatine kinase (CK) activity in blood after choline depletion is a sign of damaged muscle cells.

4.9.14.1 DIETARY NEEDS

Several genes have been linked to increased choline requirements. The first one to be reported was the *MTHFD1* 1958A (rs2236225) variant. This variation changes the ancestral lysine in position 653 into a glutamine, making the protein less stable. Men with one or two 1958A alleles need to get about twice as much choline from food as people without this allele [287]. This means that the 60% of American men with a 1958A allele need about 8 mg choline per kilogram body weight instead of the 4 mg/kg needed

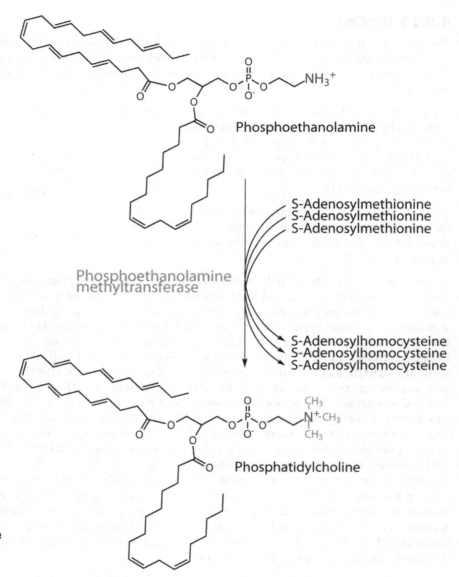

Phosphoethanolamine

Phosphoethanolamine
methyltransferase

S-Adenosylmethionine
S-Adenosylmethionine
S-Adenosylmethionine

S-Adenosylhomocysteine
S-Adenosylhomocysteine
S-Adenosylhomocysteine

Phosphatidylcholine

FIGURE 4.49
The body can produce
choline by methylation of the
precursor ethanolamine in
a phospholipid.

for men with the G/G genotype. For a 70 kg man, the extra choline would
be the amount in two eggs with three slices of bacon. Alternatively, he could
add half a pound of cooked soybeans to his menu or maybe a combination
of choline-rich foods and a moderately dosed dietary supplement. And why
might he want to do that? Because deficiency leads to fatty liver and muscle
damage in men.

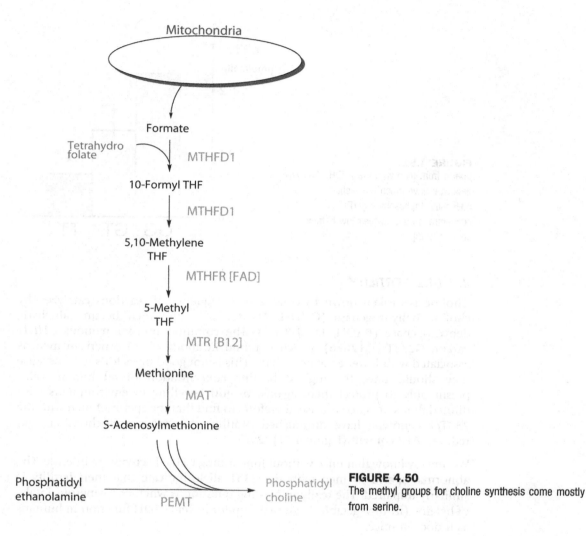

FIGURE 4.50
The methyl groups for choline synthesis come mostly from serine.

Common variants in the estrogen-inducible *PEMT* gene similarly increase dependency on external choline [288]. A variant near one of the estrogen response elements (rs12325817) increases the likelihood that carriers develop signs of choline deficiency with a low-choline diet. Another common *PEMT* variant, V175M, diminishes function of the gene. Carriers of two 175M copies, that is, those without much PEMT activity, are about twice as likely to have nonalcoholic fatty liver disease (NAFLD) than those without this variant [289].

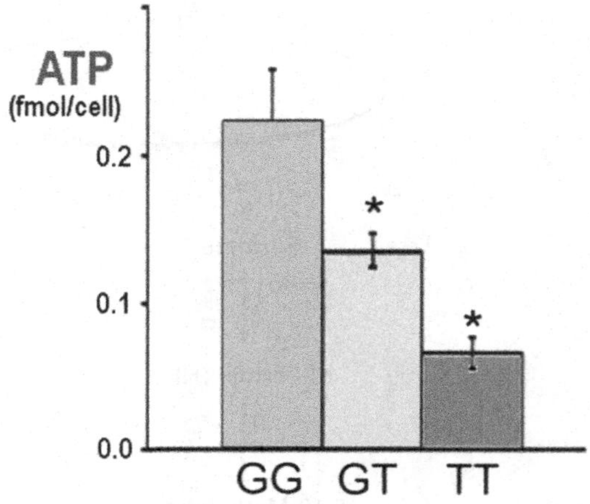

FIGURE 4.51
Sperm from men with the CHDH 233 T/T genotype have much diminished adenosine triphosphate (ATP) concentration due to their low betaine content [290].

4.9.14.2 FERTILITY

Choline is broken down to betaine in a sequence of reactions catalyzed by choline dehydrogenase (CHDH; EC 1.1.99.1) and then betaine aldehyde dehydrogenase (BADH; EC 1.2.1.8). The common nonsynonymous *CHDH* variant G233T (rs12676), which is found in about 7% of American men, is associated with lower enzyme activity. This is not good news for sperm because they should have the highest betaine concentration of all human cells, presumably to protect them against osmotic swelling in environments with diluted fluids. It is no surprise, therefore, to find that the sperm of men with the 233T/T genotype have diminished motility, misshapen mitochondria, and reduced ATP content (Figure 4.51) [290].

We already know that mice without functioning CHDH activity are infertile. The abnormal sperm in men with the 233T allele indicate that their fertility is similarly impaired. The really interesting question is whether a generous intake of betaine (from vegetables) can make up for lower CHDH function in humans as it does in mice.

4.9.14.3 FISH-ODOR SYNDROME

One of the unpleasant side effects of high choline intake is the development of a fishy body odor. This aroma does not come from choline itself, but the metabolite trimethylamine (TMA), which is generated by gut bacteria. In most people, breakdown by flavin-dependent monooxygenase 3 (*FM03*; EC 1.14.13.8; OMIM 136132) limits the accumulation of TMA. In some people, this gene is less active, significant amounts of TMA are excreted in urine, and the odor breaks through (OMIM 602079), particularly after consumption of choline-rich foods, such as eggs, fish, and meat. Patients should be guided to limit their choline intake, but this can be problematic in people with genetically high choline requirements.

Practice questions

What is the advantage of newborn screening by tandem mass spectrometry compared to the classical Guthrie test?

How can you optimize the initial warfarin dosing for an elderly Asian female with persistent atrial fibrillation?

Are goiter, pellagra, or scurvy genetic diseases?

How can lactase persistence be associated with an increased risk of colorectal cancer in some and who would that most likely be?

Which nutrients should be supplemented in people with defective phenylalanine hydroxylase?

What are the specific dietary treatment options for a month-old infant with newly diagnosed accumulation of phytanic acid due to Refsum disease?

Which nutrient should be of concern in someone who is heterozygous for defective aldolase B and how common are such defects in your country?

What is a likely nutrient of concern for a pregnant woman with a bad case of morning sickness, who is increasingly confused, has started losing her vision and hearing, and is now too weak to stand on her legs?

Why do many men need much more choline than others and what can they do about it?

Variants in which genes are most likely to affect vitamin D requirements in people around you?

Deficiency of which nutrient is likely to interfere with proprioception, how would you test this, and what genetic defect might be responsible for the deficiency?

What kind of diet is appropriate for a young woman who is compound heterozygous for the *HFE* 282Tyr and 63Asp alleles?

What molecular and genetic mechanisms keep many people of Asian ancestry from drinking alcohol?

How can you know whether someone is salt sensitive and which genes might be involved?

Identify from published reports a genetic variant (other than those mentioned in this book) that makes only carriers of one allele responsive to a specific nutritional factor.

SUMMARY AND SEGUE TO THE NEXT CHAPTER

Some of the differences in response to nutritional circumstances are readily apparent, but most nutrigenetic variations with serious health consequences are not. Common genetic variants influence absorption, metabolism, use, and elimination of most nutrients. These variants exist because there are trade-offs between the risks of deficiency and oversupply. We each have a unique set of optimal nutrient intake levels, with multiple mechanisms that set our individual response to each nutrient and nutritional condition. We already know many of the mechanisms that determine individual responses to all these nutritional factors but more remain to be explored. We also know the functional impact of many specific genetic variants but know little about their combined impact.

There can be little doubt that much of our current knowledge can and should be used in clinical practice. Many genotype-specific nutritional responses are robust and strongly supported by multiple lines of evidence. The challenge for the practitioner is to know enough about the underlying nutritional mechanisms for the educated weighing of implementation approaches.

The next chapter will explore how individual genetic traits determine whether particular food patterns prevent common conditions and disease.

References

[1] Garrod AE. The incidence of alkaptonuria: a study in chemical individuality. 1902 [classical article]. Yale J Biol Med 2002;75(4):221—31.

[2] Garrod Archibald. Inborn Errors of Metabolism. The Croonian Lectures delivered before the Royal College of Physicians of London, in June. Lancet 1908;172(4427):3.

[3] Laxon S, Ranganath L, Timmis O. Living with alkaptonuria. BMJ 2011;343:d5155.

[4] Mayatepek E, Kallas K, Anninos A, Muller E. Effects of ascorbic acid and low-protein diet in alkaptonuria. Eur J Pediatr 1998;157(10):867—8.

[5] de Haas V, Carbasius Weber EC, de Klerk JB, Bakker HD, Smit GP, Huijbers WA, et al. The success of dietary protein restriction in alkaptonuria patients is age-dependent. J Inherit Metab Dis 1998;21(8):791—8.

[6] Fernandez-Canon JM, Granadino B, Beltran-Valero de Bernabe D, Renedo M, Fernandez-Ruiz E, Penalva MA, et al. The molecular basis of alkaptonuria. Nat Genet 1996; 14(1):19—24.

[7] Howell RR. We need expanded newborn screening. Pediatrics 2006;117(5):1800—5.

[8] Prado-Lima PS, Cruz IB, Schwanke CH, Netto CA, Licinio J. Human food preferences are associated with a 5-HT(2A) serotonergic receptor polymorphism. Mol Psychiatry 2006;11(10):889—91.

[9] Nakamura Y, Ito Y, Aleksic B, Kushima I, Yasui-Furukori N, Inada T, et al. Influence of HTR2A polymorphisms and parental rearing on personality traits in healthy Japanese subjects. Journal of Human Genetics 2010;55(12):838—41.

[10] White MJ, Young RM, Morris CP, Lawford BR. Cigarette smoking in young adults: the influence of the HTR2A T102C polymorphism and punishment sensitivity. Drug and Alcohol Dependence 2011;114(2—3):140—6.

[11] Cavicchi C, Malvagia S, la Marca G, Gasperini S, Donati MA, Zammarchi E, et al. Hypoci-trullinemia in expanded newborn screening by LC-MS/MS is not a reliable marker for orni-thine transcarbamylase deficiency. J Pharm Biomed Anal 2009;49(5):1292—5.

[12] Sathe MN, Patel AS. Update in pediatrics: focus on fat-soluble vitamins. Nutr Clin Pract 2010;25(4):340—6.

[13] Dunn GA, Morgan CP, Bale TL. Sex-specificity in transgenerational epigenetic programming. Horm Behav 2011;59(3):290—5.

[14] Ivanov PL, Wadhams MJ, Roby RK, Holland MM, Weedn VW, Parsons TJ. Mitochondrial DNA sequence heteroplasmy in the Grand Duke of Russia Georgij Romanov establishes the authenticity of the remains of Tsar Nicholas II. Nat Genet 1996;12(4):417—20.

[15] Grunewald S, Fairbanks L, Genet S, Cranston T, Husing J, Leonard JV, et al. How reliable is the allopurinol load in detecting carriers for ornithine transcarbamylase deficiency? J Inherit Metab Dis 2004;27(2):179—86.

[16] Yazbek SN, Spiezio SH, Nadeau JH, Buchner DA. Ancestral paternal genotype controls body weight and food intake for multiple generations. Hum Mol Genet 2010;19(21):4134—44.

[17] Irwin JA, Saunier JL, Niederstatter H, Strouss KM, Sturk KA, Diegoli TM, et al. Investigation of heteroplasmy in the human mitochondrial DNA control region: a synthesis of observations from more than 5000 global population samples. J Mol Evol 2009;68(5):516—27.

[18] Gibson NR, Jahoor F, Ware L, Jackson AA. Endogenous glycine and tyrosine production is maintained in adults consuming a marginal-protein diet. Am J Clin Nutr 2002;75(3):511—18.

[19] Mitchell JJ, Trakadis YJ, Scriver CR. Phenylalanine hydroxylase deficiency. Genet Med 2011;13(8):697—707.

[20] Hanley WB. Non-PKU mild hyperphenylalaninemia (MHP)—the dilemma. Mol Genet Metab 2011;104(1—2):23—6.

[21] Humphrey M, Nation J, Francis I, Boneh A. Effect of tetrahydrobiopterin on Phe/Tyr ratios and variation in Phe levels in tetrahydrobiopterin responsive PKU patients. Mol Genet Metab 2011;104(1—2):89—92.

[22] Staudigl M, Gersting SW, Danecka MK, Messing DD, Woidy M, Pinkas D, et al. The interplay between genotype, metabolic state and cofactor treatment governs phenylalanine hydroxylase function and drug response. Hum Mol Genet 2011;20(13):2628—41.

[23] Waisbren SE, Noel K, Fahrbach K, Cella C, Frame D, Dorenbaum A, et al. Phenylalanine blood levels and clinical outcomes in phenylketonuria: a systematic literature review and meta-analysis. Mol Genet Metab 2007;92(1−2):63−70.

[24] Prick BW, Hop WC, Duvekot JJ. Maternal phenylketonuria and hyperphenylalaninemia in pregnancy: pregnancy complications and neonatal sequelae in untreated and treated pregnancies. Am J Clin Nutr 2012;95(2):374−82.

[25] Webster D, Wildgoose J. Tyrosine supplementation for phenylketonuria. Cochrane Database Syst Rev 2010;(8)::CD001507.

[26] Kalkanoglu HS, Ahring KK, Sertkaya D, Moller LB, Romstad A, Mikkelsen I, et al. Behavioural effects of phenylalanine-free amino acid tablet supplementation in intellectually disabled adults with untreated phenylketonuria. Acta Paediatr 2005;94(9):1218−22.

[27] Broer S. The role of the neutral amino acid transporter B0AT1 (SLC6A19) in Hartnup disorder and protein nutrition. IUBMB life 2009;61(6):591−9.

[28] Edelmann L, Wasserstein MP, Kornreich R, Sansaricq C, Snyderman SE, Diaz GA. Maple syrup urine disease: identification and carrier-frequency determination of a novel founder mutation in the Ashkenazi Jewish population. Am J Hum Genet 2001; 69(4):863−8.

[29] Puckett RL, Lorey F, Rinaldo P, Lipson MH, Matern D, Sowa ME, et al. Maple syrup urine disease: further evidence that newborn screening may fail to identify variant forms. Mol Genet Metab 2010;100(2):136−42.

[30] Bouchard-Mercier A, Paradis AM, Perusse L, Vohl MC. Associations between polymorphisms in genes involved in fatty acid metabolism and dietary fat intakes. Journal of Nutrigenetics and Nutrigenomics 2012;5(1):1−12.

[31] Keller KL, Liang LC, Sakimura J, May D, van Belle C, Breen C, et al. Common variants in the CD36 gene are associated with oral fat perception, fat preferences, and obesity in African Americans. Obesity (Silver Spring) 2012 May;20(5):1066−73.

[32] Ichimura A, Hirasawa A, Poulain-Godefroy O, Bonnefond A, Hara T, Yengo L, et al. Dysfunction of lipid sensor GPR120 leads to obesity in both mouse and human. Nature 2012;483(7389):350−4.

[33] Oh DY, Talukdar S, Bae EJ, Imamura T, Morinaga H, Fan W, et al. GPR120 is an omega-3 fatty acid receptor mediating potent anti-inflammatory and insulin-sensitizing effects. Cell 2010;142(5):687−98.

[34] Goossens GH, Petersen L, Blaak EE, Hul G, Arner P, Astrup A, et al. Several obesity- and nutrient-related gene polymorphisms but not FTO and UCP variants modulate post-absorptive resting energy expenditure and fat-induced thermogenesis in obese individuals: the NUGENOB study. Int J Obes (Lond) 2009;33(6):669−79.

[35] Uibo R, Lernmark A. GAD65 autoimmunity-clinical studies. Advances in Immunology 2008;100:39−78.

[36] Suhre K, Shin SY, Petersen AK, Mohney RP, Meredith D, Wagele B, et al. Human metabolic individuality in biomedical and pharmaceutical research. Nature 2011;477(7362):54−60.

[37] Matsuzaka T, Shimano H. Elov16: a new player in fatty acid metabolism and insulin sensitivity. J Mol Med (Berl) 2009;87(4):379−84.

[38] Morcillo S, Martin-Nunez GM, Rojo-Martinez G, Almaraz MC, Garcia-Escobar E, Mansego ML, et al. ELOVL6 genetic variation is related to insulin sensitivity: a new candidate gene in energy metabolism. PLoS One 2011;6(6):e21198.

[39] Corella D, Peloso G, Arnett DK, Demissie S, Cupples LA, Tucker K, et al. APOA2, dietary fat, and body mass index: replication of a gene−diet interaction in 3 independent populations. Arch Intern Med 2009;169(20):1897−906.

[40] Garaulet M, Lee YC, Shen J, Parnell LD, Arnett DK, Tsai MY, et al. CLOCK genetic variation and metabolic syndrome risk: modulation by monounsaturated fatty acids. Am J Clin Nutr 2009;90(6):1466−75.

[41] Tein I, Elpeleg O, Ben-Zeev B, Korman SH, Lossos A, Lev D, et al. Short-chain acyl-CoA dehydrogenase gene mutation (c.319C>T) presents with clinical heterogeneity and is candidate founder mutation in individuals of Ashkenazi Jewish origin. Mol Genet Metab 2008;93(2):179−89.

[42] van Maldegem BT, Duran M, Wanders RJ, Waterham HR, Wijburg FA. Flavin adenine dinu-cleotide status and the effects of high-dose riboflavin treatment in short-chain acyl-CoA dehydrogenase deficiency. Pediatric Research 2010;67(3):304—8.

[43] Nakamura MT, Nara TY. Structure, function, and dietary regulation of delta6, delta5, and delta9 desaturases. Annu Rev Nutr 2004;24:345—76.

[44] Ameur A, Enroth S, Johansson A, Zaboli G, Igl W, Johansson AC, et al. Genetic adaptation of fatty-acid metabolism: a human-specific haplotype increasing the biosynthesis of long-chain omega-3 and omega-6 fatty acids. Am J Hum Genet 2012;90(5):809—20.

[45] Gregory MK, Gibson RA, Cook-Johnson RJ, Cleland LG, James MJ. Elongase reactions as control points in long-chain polyunsaturated fatty acid synthesis. PLoS One 2011;6(12):e29662.

[46] Ordovas JM, Corella D, Cupples LA, Demissie S, Kelleher A, Coltell O, et al. Polyunsaturated fatty acids modulate the effects of the APOA1 G-A polymorphism on HDL-cholesterol concentrations in a sex-specific manner: the Framingham Study. Am J Clin Nutr 2002;75(1):38—46.

[47] Van Duyn MA, Moser AE, Brown 3rd FR, Sacktor N, Liu A, Moser HW. The design of a diet restricted in saturated very long-chain fatty acids: therapeutic application in adrenoleuko-dystrophy. Am J Clin Nutr 1984;40(2):277—84.

[48] Deon M, Garcia MP, Sitta A, Barschak AG, Coelho DM, Schimit GO, et al. Hexacosanoic and docosanoic acids plasma levels in patients with cerebral childhood and asymptomatic X-linked adrenoleukodystrophy: Lorenzo's oil effect. Metabolic Brain Disease 2008;23(1):43—9.

[49] Hargrove JL, Greenspan P, Hartle DK. Nutritional significance and metabolism of very long chain fatty alcohols and acids from dietary waxes. Exp Biol Med (Maywood) 2004;229(3):215—26.

[50] Terre'Blanche G, van der Walt MM, Bergh JJ, Mienie LJ. Treatment of an adrenomyeloneur-opathy patient with Lorenzo's oil and supplementation with docosahexaenoic acid—a case report. Lipids Health Dis 2011;10:152.

[51] Cappa M, Bizzarri C, Petroni A, Carta G, Cordeddu L, Valeriani M, et al. A mixture of oleic, erucic and conjugated linoleic acids modulates cerebrospinal fluid inflammatory markers and improve somatosensorial evoked potential in X-linked adrenoleukodystrophy female carriers. J Inherit Metab Dis 2012;35(5):899—907.

[52] Lemaitre RN, Tanaka T, Tang W, Manichaikul A, Foy M, Kabagambe EK, et al. Genetic loci associated with plasma phospholipid n-3 fatty acids: a meta-analysis of genome-wide association studies from the CHARGE Consortium. PLoS Genet 2011;7(7):e1002193.

[53] Merino DM, Johnston H, Clarke S, Roke K, Nielsen D, Badawi A, et al. Polymorphisms in FADS1 and FADS2 alter desaturase activity in young Caucasian and Asian adults. Mol Genet Metab 2011;103(2):171—8.

[54] Dumont J, Huybrechts I, Spinneker A, Gottrand F, Grammatikaki E, Bevilacqua N, et al. FADS1 genetic variability interacts with dietary alpha-linolenic acid intake to affect serum non-HDL-cholesterol concentrations in European adolescents. J Nutr 2011;141(7):1247—53.

[55] Dwyer JH, Allayee H, Dwyer KM, Fan J, Wu H, Mar R, et al. Arachidonate 5-lipoxygenase promoter genotype, dietary arachidonic acid, and atherosclerosis. N Engl J Med 2004;350(1):29—37.

[56] Kohlschutter A, Santer R, Lukacs Z, Altenburg C, Kemper MJ, Ruther K. A child with night blindness: Preventing serious symptoms of refsum disease. Journal of Child Neurology 2012 May;27(5):654—6.

[57] Yang C, Yu L, Li W, Xu F, Cohen JC, Hobbs HH. Disruption of cholesterol homeostasis by plant sterols. J Clin Invest 2004;114(6):813—22.

[58] Yamanashi Y, Takada T, Suzuki H. Niemann-Pick C1-like 1 overexpression facilitates ezeti-mibe-sensitive cholesterol and beta-sitosterol uptake in CaCo-2 cells. J Pharmacol Exp Ther 2007;320(2):559—64.

[59] Rios J, Stein E, Shendure J, Hobbs HH, Cohen JC. Identification by whole-genome rese-quencing of gene defect responsible for severe hypercholesterolemia. Hum Mol Genet 2010;19(22):4313—18.

[60] Herron KL, Vega-Lopez S, Conde K, Ramjiganesh T, Shachter NS, Fernandez ML. Men classified as hypo- or hyperresponders to dietary cholesterol feeding exhibit differences in lipoprotein metabolism. J Nutr 2003;133(4):1036–42.

[61] Chakrabarty G, Manjunatha S, Bijlani RL, Ray RB, Mahapatra SC, Mehta N, et al. The effect of ingestion of egg on the serum lipid profile of healthy young Indians. Indian J Physiol Pharmacol 2004;48(3):286–92.

[62] Wolff E, Vergnes MF, Defoort C, Planells R, Portugal H, Nicolay A, et al. Cholesterol absorption status and fasting plasma cholesterol are modulated by the microsomal triacylglycerol transfer protein -493 G/T polymorphism and the usual diet in women. Genes Nutr 2011;6(1):71–9.

[63] Masson LF, McNeill G, Avenell A. Genetic variation and the lipid response to dietary intervention: a systematic review. Am J Clin Nutr 2003;77(5):1098–111.

[64] Sarkkinen E, Korhonen M, Erkkila A, Ebeling T, Uusitupa M. Effect of apolipoprotein E polymorphism on serum lipid response to the separate modification of dietary fat and dietary cholesterol. Am J Clin Nutr 1998;68(6):1215–22.

[65] Herron KL, McGrane MM, Waters D, Lofgren IE, Clark RM, Ordovas JM, et al. The ABCG5 polymorphism contributes to individual responses to dietary cholesterol and carotenoids in eggs. J Nutr 2006;136(5):1161–5.

[66] Clark RM, Herron KL, Waters D, Fernandez ML. Hypo- and hyperresponse to egg cholesterol predicts plasma lutein and beta-carotene concentrations in men and women. J Nutr 2006;136(3):601–7.

[67] Lebenthal E, Khin Maung U, Zheng BY, Lu RB, Lerner A. Small intestinal glucoamylase deficiency and starch malabsorption: a newly recognized alpha-glucosidase deficiency in children. J Pediatr 1994;124(4):541–6.

[68] Itan Y, Jones BL, Ingram CJ, Swallow DM, Thomas MG. A worldwide correlation of lactase persistence phenotype and genotypes. BMC Evol Biol 2010;10:36.

[69] Peterson ML, Herber R. Intestinal sucrase deficiency. Trans Assoc Am Physicians 1967;80:275–83.

[70] Arola H, Koivula T, Karvonen AL, Jokela H, Ahola T, Isokoski M. Low trehalase activity is associated with abdominal symptoms caused by edible mushrooms. Scand J Gastroenterol 1999;34(9):898–903.

[71] Iafrate AJ, Feuk L, Rivera MN, Listewnik ML, Donahoe PK, Qi Y, et al. Detection of large-scale variation in the human genome. Nat Genet 2004;36(9):949–51.

[72] Acosta PB, Gross KC. Hidden sources of galactose in the environment. Eur J Pediatr 1995;154(7 Suppl. 2):S87–92.

[73] Brivet M, Raymond JP, Konopka P, Odievre M, Lemonnier A. Effect of lactation in a mother with galactosemia. J Pediatr 1989;115(2):280–2.

[74] Kuokkanen M, Kokkonen J, Enattah NS, Ylisaukko-Oja T, Komu H, Varilo T, et al. Mutations in the translated region of the lactase gene (LCT) underlie congenital lactase deficiency. Am J Hum Genet 2006;78(2):339–44.

[75] Thain RI. Bovine infertility possibly caused by subterranean clover. Further report and herd histories. Aust Vet J 1966;42(6):199–203.

[76] Kallela K, Heinonen K, Saloniemi H. Plant oestrogens; the cause of decreased fertility in cows. A case report. Nord Vet Med 1984;36(3–4):124–9.

[77] Stevens JF, Page JE. Xanthohumol and related prenylflavonoids from hops and beer: to your good health! Phytochemistry 2004;65(10):1317–30.

[78] Fang L, Ahn JK, Wodziak D, Sibley E. The human lactase persistence-associated SNP -13910*T enables in vivo functional persistence of lactase promoter-reporter transgene expression. Hum Genet 2012 Jul;131(7):1153–9.

[79] Harris EE, Meyer D. The molecular signature of selection underlying human adaptations. Am J Phys Anthropol 2006;(Suppl. 43)::89–130.

[80] Lovelace HY, Barr SI. Diagnosis, symptoms, and calcium intakes of individuals with self-reported lactose intolerance. Journal of the American College of Nutrition 2005;24(1):51–7.

[81] Marton A, Xue X, Szilagyi A. Meta-analysis: the diagnostic accuracy of lactose breath hydrogen or lactose tolerance tests for predicting the North European lactase polymorphism C/T-13910. Aliment Pharmacol Ther 2012;35(4):429–40.

[82] Matthews SB, Waud JP, Roberts AG, Campbell AK. Systemic lactose intolerance: a new perspective on an old problem. Postgrad Med J 2005;81(953):167–73.

[83] Vonk RJ, Stellaard F, Priebe MG, Koetse HA, Hagedoorn RE, De Bruijn S, et al. The 13C/2H-glucose test for determination of small intestinal lactase activity. Eur J Clin Invest 2001;31(3):226–33.

[84] Dahlqvist A. Assay of intestinal disaccharidases. Scand J Clin Lab Invest 1984;44(2):169–72.

[85] Suarez FL, Savaiano DA, Levitt MD. A comparison of symptoms after the consumption of milk or lactose-hydrolyzed milk by people with self-reported severe lactose intolerance. N Engl J Med 1995;333(1):1–4.

[86] Vernia P, Di Camillo M, Marinaro V. Lactose malabsorption, irritable bowel syndrome and self-reported milk intolerance. Digestive and Liver Disease: official journal of the Italian Society of Gastroenterology and the Italian Association for the Study of the Liver 2001;33(3):234–9.

[87] Obermayer-Pietsch BM, Bonelli CM, Walter DE, Kuhn RJ, Fahrleitner-Pammer A, Berghold A, et al. Genetic predisposition for adult lactose intolerance and relation to diet, bone density, and bone fractures. Journal of Bone and Mineral Research: the official journal of the American Society for Bone and Mineral Research 2004;19(1):42–7.

[88] Honkanen R, Kroger H, Alhava E, Turpeinen P, Tuppurainen M, Saarikoski S. Lactose intolerance associated with fractures of weight-bearing bones in Finnish women aged 38–57 years. Bone 1997;21(6):473–7.

[89] Martin MG, Turk E, Lostao MP, Kerner C, Wright EM. Defects in Na+/glucose cotransporter (SGLT1) trafficking and function cause glucose-galactose malabsorption. Nat Genet 1996;12(2):216–20.

[90] Martin MG, Turk E, Kerner C, Zabel B, Wirth S, Wright EM. Prenatal identification of a heterozygous status in two fetuses at risk for glucose-galactose malabsorption. Prenat Diagn 1996;16(5):458–62.

[91] Kriegshauser G, Halsall D, Rauscher B, Oberkanins C. Semi-automated, reverse-hybridization detection of multiple mutations causing hereditary fructose intolerance. Mol Cell Probes 2007;21(3):226–8.

[92] Coffee EM, Tolan DR. Mutations in the promoter region of the aldolase B gene that cause hereditary fructose intolerance. J Inherit Metab Dis 2010;33(6):715–25.

[93] Cox TM. Fructose intolerance: diet and inheritance. Proc Nutr Soc 1991;50(2):305–9.

[94] Yasawy MI, Folsch UR, Schmidt WE, Schwend M. Adult hereditary fructose intolerance. World J Gastroenterol 2009;15(19):2412–3.

[95] Bray GA, Nielsen SJ, Popkin BM. Consumption of high-fructose corn syrup in beverages may play a role in the epidemic of obesity. Am J Clin Nutr 2004;79(4):537–43.

[96] Rumessen JJ, Gudmand-Hoyer E. Absorption capacity of fructose in healthy adults. Comparison with sucrose and its constituent monosaccharides. Gut 1986;27(10):1161–8.

[97] Beyer PL, Caviar EM, McCallum RW. Fructose intake at current levels in the United States may cause gastrointestinal distress in normal adults. J Am Diet Assoc 2005;105(10):1559–66.

[98] Szilagyi A, Malolepszy P, Yesovitch S, Vinokuroff C, Nathwani U, Cohen A, et al. Fructose malabsorption may be gender dependent and fails to show compensation by colonic adaptation. Dig Dis Sci 2007;52(11):2999–3004.

[99] Born P, Sekatcheva M, Rosch T, Classen M. Carbohydrate malabsorption in clinical routine: a prospective observational study. Hepatogastroenterology 2006;53(71):673–7.

[100] Tsampalieros A, Beauchamp J, Boland M, Mack DR. Dietary fructose intolerance in children and adolescents. Arch Dis Child 2008;93(12):1078.

[101] Reyes-Huerta JU, de la Cruz-Patino E, Ramirez-Gutierrez de Velasco A, Zamudio C, Remes-Troche JM. Fructose intolerance in patients with irritable bowel syndrome: a case-control study. Rev Gastroenterol Mex 2010;75(4):405–11.

[102] Choi YK, Kraft N, Zimmerman B, Jackson M, Rao SS. Fructose intolerance in IBS and utility of fructose-restricted diet. J Clin Gastroenterol 2008;42(3):233–8.

[103] Skoog SM, Bharucha AE, Zinsmeister AR. Comparison of breath testing with fructose and high fructose corn syrups in health and IBS. Neurogastroenterol Motil 2008;20(5):505–11.

[104] Cordain L, Eaton SB, Sebastian A, Mann N, Lindeberg S, Watkins BA, et al. Origins and evolution of the Western diet: health implications for the 21st century. Am J Clin Nutr 2005;81(2):341–54.

[105] Asp NG, Berg NO, Dahlqvist A, Gudmand-Hoyer E, Jarnum S, McNair A. Intestinal disaccharidases in Greenland Eskimos. Scand J Gastroenterol 1975;10(5):513–19.

[106] Robayo-Torres CC, Opekun AR, Quezada-Calvillo R, Villa X, Smith EO, Navarrete M, et al. 13C-breath tests for sucrose digestion in congenital sucrase isomaltase-deficient and sacrosidase-supplemented patients. J Pediatr Gastroenterol Nutr 2009;48(4):412–18.

[107] Mones RL, Yankah A, Duelfer D, Bustami R, Mercer G. Disaccharidase deficiency in pediatric patients with celiac disease and intact villi. Scand J Gastroenterol 2011;46(12): 1429–34.

[108] Treem WR. Congenital sucrase-isomaltase deficiency. J Pediatr Gastroenterol Nutr 1995;21(1): 1–14.

[109] Minai-Tehrani D, Ghaffari M, Sobhani-Damavandifar Z, Minoui S, Alavi S, Osmani R, et al. Ranitidine induces inhibition and structural changes in sucrase. J Enzyme Inhib Med Chem 2011 Aug 18 [Epub ahead of print].

[110] van Can JG, Ijzerman TH, van Loon LJ, Brouns F, Blaak EE. Reduced glycaemic and insulinaemic responses following trehalose ingestion: implications for postprandial substrate use. Br J Nutr 2009;102(10):1395–9.

[111] Gudmand-Hoyer E, Fenger HJ, Skovbjerg H, Kern-Hansen P, Madsen PR. Trehalase deficiency in Greenland. Scand J Gastroenterol 1988;23(7):775–8.

[112] Oku T, Nakamura S. Estimation of intestinal trehalase activity from a laxative threshold of trehalose and lactulose on healthy female subjects. Eur J Clin Nutr 2000;54(10):783–8.

[113] Bergoz R, Vallotton MC, Loizeau E. Trehalase deficiency. Prevalence and relation to single-cell protein food. Ann Nutr Metab 1982;26(5):291–5.

[114] Buts JP, Stilmant C, Bernasconi P, Neirinck C, De Keyser N. Characterization of alpha, alpha-trehalase released in the intestinal lumen by the probiotic Saccharomyces boulardii. Scand J Gastroenterol 2008;43(12):1489–96.

[115] Froissart R, Piraud M, Boudjemline AM, Vianey-Saban C, Petit F, Hubert-Buron A, et al. Glucose-6-phosphatase deficiency. Orphanet J Rare Dis 2011;6:27.

[116] Nuoffer JM, Mullis PE, Wiesmann UN. Treatment with low-dose diazoxide in two growth-retarded prepubertal girls with glycogen storage disease type Ia resulted in catch-up growth. J Inherit Metab Dis 1997;20(6):790–8.

[117] Zimatkin SM, Liopo AV, Deitrich RA. Distribution and kinetics of ethanol metabolism in rat brain. Alcohol Clin Exp Res 1998;22(8):1623–7.

[118] Quertemont E. Genetic polymorphism in ethanol metabolism: acetaldehyde contribution to alcohol abuse and alcoholism. Mol Psychiatry 2004;9(6):570–81.

[119] Thomasson HR, Beard JD, Li TK. ADH2 gene polymorphisms are determinants of alcohol pharmacokinetics. Alcohol Clin Exp Res 1995;19(6):1494–9.

[120] Ji YB, Tae K, Ahn TH, Lee SH, Kim KR, Park CW, et al. ADH1B and ALDH2 polymorphisms and their associations with increased risk of squamous cell carcinoma of the head and neck in the Korean population. Oral Oncol 2011;47(7):583–7.

[121] Seitz HK, Stickel F. Molecular mechanisms of alcohol-mediated carcinogenesis. Nat Rev Cancer 2007;7(8):599–612.

[122] Gizer IR, Edenberg HJ, Gilder DA, Wilhelmsen KC, Ehlers CL. Association of alcohol dehydrogenase genes with alcohol-related phenotypes in a native american community sample. Alcohol Clin Exp Res 2011;35(11):2008–18.

[123] Pochareddy S, Edenberg HJ. Variation in the ADH1B proximal promoter affects expression. Chem Biol Interact 2011;191(1–3):38–41.

[124] Preuss UW, Ridinger M, Rujescu D, Samochowiec J, Fehr C, Wurst FM, et al. Association of ADH4 genetic variants with alcohol dependence risk and related phenotypes: results from a larger multicenter association study. Addict Biol 2011;16(2):323–33.

[125] Higuchi S, Matsushita S, Muramatsu T, Murayama M, Hayashida M. Alcohol and aldehyde dehydrogenase genotypes and drinking behavior in Japanese. Alcohol Clin Exp Res 1996; 20(3):493–7.

[126] Centers for Disease Control and Prevention (CDC). CDC Grand Rounds: Dietary Sodium Reduction—Time for Choice. MMWR Morb Mort Weekly Rep 2012;61:89—91.

[127] GenSalt Collaborative Research Group. GenSalt: rationale, design, methods and baseline characteristics of study participants. J Human Hypertension 2007;21(8):639—46.

[128] Hannila-Handelberg T, Kontula K, Tikkanen I, Tikkanen T, Fyhrquist F, Helin K, et al. Common variants of the beta and gamma subunits of the epithelial sodium channel and their relation to plasma renin and aldosterone levels in essential hypertension. BMC Med Genet 2005;6:4.

[129] Zhao Q, Gu D, Hixson JE, Liu DP, Rao DC, Jaquish CE, et al. Common variants in epithelial sodium channel genes contribute to salt sensitivity of blood pressure: The GenSalt study. Circ Cardiovasc Genet 2011;4(4):375—80.

[130] Su YR, Rutkowski MP, Klanke CA, Wu X, Cui Y, Pun RY, et al. A novel variant of the beta-subunit of the amiloride-sensitive sodium channel in African Americans. J Am Soc Nephrol 1996;7(12):2543—9.

[131] Dong YB, Zhu HD, Baker EH, Sagnella GA, MacGregor GA, Carter ND, et al. T594M and G442V polymorphisms of the sodium channel beta subunit and hypertension in a black population. J Human Hypertension 2001;15(6):425—30.

[132] Miettinen HE, Piippo K, Hannila-Handelberg T, Paukku K, Hiltunen TP, Gautschi I, et al. Licorice-induced hypertension and common variants of genes regulating renal sodium reabsorption. Annals Med 2010;42(6):465—74.

[133] Lanzani C, Citterio L, Jankaricova M, Sciarrone MT, Barlassina C, Fattori S, et al. Role of the adducin family genes in human essential hypertension. J Hypertens 2005;23(3):543—9.

[134] Watkins WS, Rohrwasser A, Peiffer A, Leppert MF, Lalouel JM, Jorde LB. AGT genetic variation, plasma AGT, and blood pressure: An analysis of the Utah Genetic Reference Project pedigrees. Am J Hypertension 2010;23(8):917—23.

[135] Beeks E, Kessels AG, Kroon AA, van der Klauw MM, de Leeuw PW. Genetic predisposition to salt-sensitivity: a systematic review. J Hypertens 2004;22(7):1243—9.

[136] Ni S, Zhang Y, Deng Y, Gong Y, Huang J, Bai Y, et al. AGT M235T polymorphism contributes to risk of preeclampsia: evidence from a meta-analysis. Journal of the renin-angiotensin-aldosterone system: JRAAS 2012;13(3):379—86.

[137] Poch E, Gonzalez D, Giner V, Bragulat E, Coca A, de La Sierra A. Molecular basis of salt sensitivity in human hypertension. Evaluation of renin-angiotensin-aldosterone system gene polymorphisms. Hypertension 2001;38(5):1204—9.

[138] Wang Y, Li B, Zhao W, Liu P, Zhao Q, Chen S, et al. Association study of G protein-coupled receptor kinase 4 gene variants with essential hypertension in northern Han Chinese. Ann Human Genet 2006;70(Pt 6):778—83.

[139] Sanada H, Yatabe J, Midorikawa S, Hashimoto S, Watanabe T, Moore JH, et al. Single-nucleotide polymorphisms for diagnosis of salt-sensitive hypertension. Clin Chem 2006;52(3):352—60.

[140] Caprioli J, Mele C, Mossali C, Gallizioli L, Giacchetti G, Noris M, et al. Polymorphisms of EDNRB, ATG, and ACE genes in salt-sensitive hypertension. Can J Physiol Pharmacol 2008;86(8):505—10.

[141] Dos Santos EA, Dahly-Vernon AJ, Hoagland KM, Roman RJ. Inhibition of the formation of EETs and 20-HETE with 1-aminobenzotriazole attenuates pressure natriuresis. Am J Physiol Reg Int Comp Physiol 2004;287(1):R58—68.

[142] Williams JS, Hopkins PN, Jeunemaitre X, Brown NJ. CYP4A11 T8590C polymorphism, salt-sensitive hypertension, and renal blood flow. J Hypertens 2011;29(10):1913—8.

[143] Ledford H. Africa yields two full human genomes. Nature 2010;463(7283):857.

[144] Lee BH, Cho HY, Lee H, Han KH, Kang HG, Ha IS, et al. Genetic basis of Bartter syndrome in Korea. Nephrol Dial Transplant 2012;27(4):1516—21.

[145] Sile S, Velez DR, Gillani NB, Narsia T, Moore JH, George Jr AL, et al. CLCNKB-T481S and essential hypertension in a Ghanaian population. J Hypertens 2009;27(2):298—304.

[146] Kelly TN, Hixson JE, Rao DC, Mei H, Rice TK, Jaquish CE, et al. Genome-wide linkage and positional candidate gene study of blood pressure response to dietary potassium

intervention: the genetic epidemiology network of salt sensitivity study. Circ Cardiovasc Genet 2010;3(6):539—47.

[147] He Y, Han L, Li W, Shu X, Zhao C, Bi M, et al. Effects of the calcium-sensing receptor A986S polymorphism on serum calcium and parathyroid hormone levels in healthy individuals: a meta-analysis. Gene 2012;491(2):110—15.

[148] Shakhssalim N, Kazemi B, Basiri A, Houshmand M, Pakmanesh H, Golestan B, et al. Association between calcium-sensing receptor gene polymorphisms and recurrent calcium kidney stone disease: a comprehensive gene analysis. Scand J Urol Nephrol 2010;44(6):406—12.

[149] Bacsi K, Hitre E, Kosa JP, Horvath H, Lazary A, Lakatos PL, et al. Effects of the lactase 13910 C/T and calcium-sensor receptor A986S G/T gene polymorphisms on the incidence and recurrence of colorectal cancer in Hungarian population. BMC Cancer 2008;8:317.

[150] Dai Q, Shrubsole MJ, Ness RM, Schlundt D, Cai Q, Smalley WE, et al. The relation of magnesium and calcium intakes and a genetic polymorphism in the magnesium transporter to colorectal neoplasia risk. Am J Clin Nutr 2007;86(3):743—51.

[151] Mei Z, Cogswell ME, Looker AC, Pfeiffer CM, Cusick SE, Lacher DA, et al. Assessment of iron status in US pregnant women from the National Health and Nutrition Examination Survey (NHANES), 1999—2006. Am J Clin Nutr 2011;93(6):1312—20.

[152] Gambling L, Kennedy C, McArdle HJ. Iron and copper in fetal development. Semin Cell Dev Biol 2011 Aug;22(6):637—44.

[153] Brissot P, Bardou-Jacquet E, Jouanolle AM, Loreal O. Iron disorders of genetic origin: a changing world. Trends Mol Med 2011 Dec;17(12):707—13.

[154] Lucotte G, Mercier G. Celtic origin of the C282Y mutation of hemochromatosis. Genet Test 2000;4(2):163—9.

[155] Mayr R, Janecke AR, Schranz M, Griffiths WJ, Vogel W, Pietrangelo A, et al. Ferroportin disease: a systematic meta-analysis of clinical and molecular findings. J Hepatol 2010;53(5):941—9.

[156] Frank KM, Schneewind O, Shieh WJ. Investigation of a researcher's death due to septicemic plague. N Engl J Med 2011;364(26):2563—4.

[157] Galan SR, Kann PH, Gress TM, Michl P. Listeria monocytogenes-induced bacterial peritonitis caused by contaminated cheese in a patient with haemochromatosis. Zeitschrift fur Gastroenterologie 2011;49(7):832—5.

[158] Baker MA, Wilson D, Wallengren K, Sandgren A, Iartchouk O, Broodie N, et al. Polymorphisms in the gene that encodes the iron transport protein ferroportin 1 influence susceptibility to tuberculosis. J Inf Dis 2012;205(7):1043—7.

[159] Burke W, Imperatore G, McDonnell SM, Baron RC, Khoury MJ. Contribution of different HFE genotypes to iron overload disease: a pooled analysis. Genet Med 2000;2(5):271—7.

[160] Ellervik C, Tybjaerg-Hansen A, Nordestgaard BG. Total mortality by transferrin saturation levels: two general population studies and a metaanalysis. Clin Chem 2011;57(3):459—66.

[161] Gerhard GS, Chokshi R, Still CD, Benotti P, Wood GC, Freedman-Weiss M, et al. The influence of iron status and genetic polymorphisms in the HFE gene on the risk for postoperative complications after bariatric surgery: a prospective cohort study in 1,064 patients. Patient Saf Surg 2011;5(1):1.

[162] Aranda N, Viteri FE, Montserrat C, Arija V. Effects of C282Y, H63D, and S65C HFE gene mutations, diet, and life-style factors on iron status in a general Mediterranean population from Tarragona, Spain. Ann Hematol 2010;89(8):767—73.

[163] Heath AL, Roe MA, Oyston SL, Gray AR, Williams SM, Fairweather-Tait SJ. Blood loss is a stronger predictor of iron status in men than C282Y heterozygosity or diet. Journal of the American College of Nutrition 2008;27(1):158—67.

[164] McLaren CE, Garner CP, Constantine CC, McLachlan S, Vulpe CD, Snively BM, et al. Genome-wide association study identifies genetic loci associated with iron deficiency. PLoS One 2011;6(3):e17390.

[165] Tanaka T, Roy CN, Yao W, Matteini A, Semba RD, Arking D, et al. A genome-wide association analysis of serum iron concentrations. Blood 2010;115(1):94—6.

[166] Grasberger H, Refetoff S. Genetic causes of congenital hypothyroidism due to dyshormo-nogenesis. Curr Opin Pediatr 2011 Aug;23(4):421−8.

[167] Böttcher Y, Eszlinger M, Tönjes A, Paschke R. The genetics of euthyroid familial goiter. Trends Endocrinol Metab 2005;16(7):314−9.

[168] Singer J, Eszlinger M, Wicht J, Paschke R. Evidence for a more pronounced effect of genetic predisposition than environmental factors on goitrogenesis by a case control study in an area with low normal iodine supply. Horm Metab Res 2011;43(5):349−54.

[169] Scott DA, Wang R, Kreman TM, Sheffield VC, Karniski LP. The Pendred syndrome gene encodes a chloride-iodide transport protein. Nat Genet 1999;21(4):440−3.

[170] King JC. Zinc: an essential but elusive nutrient. Am J Clin Nutr 2011;94(2):679S−84S.

[171] John LB, Ward AC. The Ikaros gene family: transcriptional regulators of hematopoiesis and immunity. Molecular Immunology 2011;48(9−10):1272−8.

[172] Schmitt S, Kury S, Giraud M, Dreno B, Kharfi M, Bezieau S. An update on mutations of the SLC39A4 gene in acrodermatitis enteropathica. Hum Mutat 2009;30(6):926−33.

[173] Kharfi M, El Fekih N, Aounallah-Skhiri H, Schmitt S, Fazaa B, Kury S, et al. Acrodermatitis enteropathica: a review of 29 Tunisian cases. Int J Dermatol 2010;49(9):1038−44.

[174] Klevay LM. Lack of a recommended dietary allowance for copper may be hazardous to your health. Journal of the American College of Nutrition 1998;17(4):322−6.

[175] Chowanadisai W, Lonnerdal B, Kelleher SL. Identification of a mutation in SLC30A2 (ZnT-2) in women with low milk zinc concentration that results in transient neonatal zinc deficiency. J Biol Chem 2006;281(51):39699−707.

[176] Nicolson TJ, Bellomo EA, Wijesekara N, Loder MK, Baldwin JM, Gyulkhandanyan AV, et al. Insulin storage and glucose homeostasis in mice null for the granule zinc transporter ZnT8 and studies of the type 2 diabetes-associated variants. Diabetes 2009;58(9):2070−83.

[177] Kawasaki E. ZnT8 and type 1 diabetes [Review]. Endocrine J 2012;59(7):531−7.

[178] Kanoni S, Nettleton JA, Hivert MF, Ye Z, van Rooij FJ, Shungin D, et al. Total zinc intake may modify the glucose-raising effect of a zinc transporter (SLC30A8) variant: a 14-cohort meta-analysis. Diabetes 2011;60(9):2407−16.

[179] Giacconi R, Cipriano C, Muti E, Costarelli L, Maurizio C, Saba V, et al. Novel -209A/G MT2A polymorphism in old patients with type 2 diabetes and atherosclerosis: relationship with inflammation (IL-6) and zinc. Biogerontology 2005;6(6):407−13.

[180] EASL Clinical Practice Guidelines: Wilson's disease. J Hepatol 2012;56(3):671−85.

[181] Barik A, Mishra B, Shen L, Mohan H, Kadam RM, Dutta S, et al. Evaluation of a new copper(II)-curcumin complex as superoxide dismutase mimic and its free radical reactions. Free Radic Biol Med 2005;39(6):811−22.

[182] Gromadzka G, Chabik G, Mendel T, Wierzchowska A, Rudnicka M, Czlonkowska A. Middle-aged heterozygous carriers of Wilson's disease do not present with significant phenotypic deviations related to copper metabolism. Journal of Genetics 2010;89(4):463−7.

[183] Fuchs SA, Harakalova M, van Haaften G, van Hasselt PM, Cuppen E, Houwen RH. Application of exome sequencing in the search for genetic causes of rare disorders of copper metabolism. Metallomics: Integ Biometal Sci 2012 Jul 28;4(7):606−13.

[184] Combs Jr GF, Watts JC, Jackson MI, Johnson LK, Zeng H, Scheett AJ, et al. Determinants of selenium status in healthy adults. Nutrition J 2011;10:75.

[185] Cominetti C, de Bortoli MC, Purgatto E, Ong TP, Moreno FS, Garrido Jr AB, et al. Associations between glutathione peroxidase-1 Pr0198Leu polymorphism, selenium status, and DNA damage levels in obese women after consumption of Brazil nuts. Nutrition 2011;27(9): 891−6.

[186] Chen J, Cao Q, Qin C, Shao P, Wu Y, Wang M, et al. GPx-1 polymorphism (rs1050450) contributes to tumor susceptibility: evidence from meta-analysis. J Canc Res Clin Oncol 2011;137(10):1553−61.

[187] Xiong YM, Mo XY, Zou XZ, Song RX, Sun WY, Lu W, et al. Association study between polymorphisms in selenoprotein genes and susceptibility to Kashin-Beck disease. Osteoarthritis and cartilage/OARS. Osteoarthritis Research Society 2010;18(6):817−24.

[188] Reiss J, Hahnewald R. Molybdenum cofactor deficiency: Mutations in GPHN, MOCS1, and MOCS2. Hum Mutat 2011;32(1):10−8.

[189] Veldman A, Santamaria-Araujo JA, Sollazzo S, Pitt J, Gianello R, Yaplito-Lee J, et al. Successful treatment of molybdenum cofactor deficiency type A with cPMP. Pediatrics 2010; 125(5):e1249—54.

[190] Johnson JL, Coyne KE, Rajagopalan KV, Van Hove JL, Mackay M, Pitt J, et al. Molybdopterin synthase mutations in a mild case of molybdenum cofactor deficiency. Am J Med Genet 2001;104(2):169—73.

[191] Touati G, Rusthoven E, Depondt E, Dorche C, Duran M, Heron B, et al. Dietary therapy in two patients with a mild form of sulphite oxidase deficiency. Evidence for clinical and biological improvement. J Inherit Metab Dis 2000;23(1):45—53.

[192] Eck P, Erichsen HC, Taylor JG, Yeager M, Hughes AL, Levine M, et al. Comparison of the genomic structure and variation in the two human sodium-dependent vitamin C transporters, SLC23A1 and SLC23A2. Hum Genet 2004;115(4):285—94.

[193] Savini I, Rossi A, Pierro C, Avigliano L, Catani MV. SVCT1 and SVCT2: key proteins for vitamin C uptake. Amino Acids 2008;34(3):347—55.

[194] Corpe CP, Tu H, ECk P, Wang J, Faulhaber-Walter R, Schnermann J, et al. Vitamin C transporter Slc23a1 links renal reabsorption, vitamin C tissue accumulation, and perinatal survival in mice. J Clin Invest 2010;120(4):1069—83.

[195] Erichsen HC, Engel SA, ECk PK, Welch R, Yeager M, Levine M, et al. Genetic variation in the sodium-dependent vitamin C transporters, SLC23A1, and SLC23A2 and risk for preterm delivery. Am J Epidemiol 2006;163(3):245—54.

[196] Block G, Shaikh N, Jensen CD, Volberg V, Holland N. Serum vitamin C and other biomarkers differ by genotype of phase 2 enzyme genes GSTM1 and GSTT1. Am J Clin Nutr 2011;94(3):929—37.

[197] Cahill LE, Fontaine-Bisson B, El-Sohemy A. Functional genetic variants of glutathione S-transferase protect against serum ascorbic acid deficiency. Am J Clin Nutr 2009;90(5): 1411—17.

[198] Linster CL, Van Schaftingen E, Vitamin C. Biosynthesis, recycling and degradation in mammals. The FEBS journal 2007;274(1):1—22.

[199] Delanghe JR, Langlois MR, De Buyzere ML, Na N, Ouyang J, Speeckaert MM, et al. Vitamin C deficiency: more than just a nutritional disorder. Genes Nutr 2011;6(4):341—6.

[200] Savy M, Hennig BJ, Doherty CP, Fulford AJ, Bailey R, Holland MJ, et al. Haptoglobin and sickle cell polymorphisms and risk of active trachoma in Gambian children. PLoS One 2010;5(6):e11075.

[201] Chiossi G, Neri I, Cavazzuti M, Basso G, Facchinetti F. Hyperemesis gravidarum complicated by Wernicke encephalopathy: background, case report, and review of the literature. Obstetrical & Gynecological Survey 2006;61(4):255—68.

[202] Luigetti M, Sabatelli M, Cianfoni A. Wernicke's encephalopathy following chronic diarrhoea. Acta Neurologica Belgica 2011;111(3):257.

[203] Sechi G, Serra A. Wernicke's encephalopathy: new clinical settings and recent advances in diagnosis and management. Lancet Neurol 2007;6(5):442—55.

[204] Blass JP, Gibson GE. Abnormality of a thiamine-requiring enzyme in patients with Wernicke-Korsakoff syndrome. N Engl J Med 1977;297(25):1367—70.

[205] Coy JF, Dressler D, Wilde J, Schubert P. Mutations in the transketolase-like gene TKTL1: clinical implications for neurodegenerative diseases, diabetes and cancer. Clin Lab 2005;51(5—6):257—73.

[206] Kono S, Miyajima H, Yoshida K, Togawa A, Shirakawa K, Suzuki H. Mutations in a thiamine-transporter gene and Wernicke's-like encephalopathy. N Engl J Med 2009;360(17):1792—4.

[207] Heap LC, Pratt OE, Ward RJ, Waller S, Thomson AD, Shaw GK, et al. Individual susceptibility to Wernicke-Korsakoff syndrome and alcoholism-induced cognitive deficit: impaired thiamine utilization found in alcoholics and alcohol abusers. Psychiatric genetics 2002; 12(4):217—24.

[208] Guerrini I, Thomson AD, Gurling HM. Molecular genetics of alcohol-related brain damage. Alcohol Alcohol 2009;44(2):166—70.

[209] Thomson AD, Marshall EJ. The natural history and pathophysiology of Wernicke's Encephalopathy and Korsakoff's Psychosis. Alcohol Alcohol 2006;41(2):151—8.

[210] Folstein MF, Folstein SE, McHugh PR. "Mini-mental state." A practical method for grading the cognitive state of patients for the clinician. J Psychiatr Res 1975;12(3):189—98.

[211] Dalla Barba G. Different patterns of confabulation. Cortex 1993;29(4):567—81.

[212] Fregly AR, Smith MJ, Graybiel A. Revised normative standards of performance of men on a quantitative ataxia test battery. Acta Otolaryngol 1973;75(1):10—6.

[213] Sullivan EV, Rohlfing T, Pfefferbaum A. Pontocerebellar volume deficits and ataxia in alcoholic men and women: no evidence for "telescoping." Psychopharmacology (Berl) 2010;208(2):279—90.

[214] Chiong MA, Sim KG, Carpenter K, Rhead W, Ho G, Olsen RK, et al. Transient multiple acyl-CoA dehydrogenation deficiency in a newborn female caused by maternal riboflavin deficiency. Mol Genet Metab 2007;92(1—2):109—14.

[215] Ho G, Yonezawa A, Masuda S, Inui K, Sim KG, Carpenter K, et al. Maternal riboflavin deficiency, resulting in transient neonatal-onset glutaric aciduria Type 2, is caused by a microdeletion in the riboflavin transporter gene GPR172B. Hum Mutat 2011;32(1):E1976—84.

[216] Bosch AM, Abeling NG, Ijlst L, Knoester H, van der Pol WL, Stroomer AE, et al. Brown-Vialetto-Van Laere and Fazio Londe syndrome is associated with a riboflavin transporter defect mimicking mild MADD: a new inborn error of metabolism with potential treatment. J Inherit Metab Dis 2011;34(1):159—64.

[217] Oduho GW, Han Y, Baker DH. Iron deficiency reduces the efficacy of tryptophan as a niacin precursor. J Nutr 1994;124(3):444—50.

[218] Consolazio CF, Johnson HL, Krzywicki HJ, Witt NF. Tryptophan-niacin interrelationships during acute fasting and caloric restriction in humans. Am J Clin Nutr 1972;25(6):572—5.

[219] Horwitt MK, Harper AE, Henderson LM. Niacin-tryptophan relationships for evaluating niacin equivalents. Am J Clin Nutr 1981;34(3):423—7.

[220] Fukuwatari T, Shibata K. Effect of nicotinamide administration on the tryptophan-nicotinamide pathway in humans. Int J Vitam Nutr Res 2007;77(4):255—62.

[221] Pique-Duran E, Perez-Cejudo JA, Cameselle D, Palacios-Llopis S, Garcia-Vazquez O. Pellagra: a clinical, histopathological, and epidemiological study of 7 cases. Actas Dermo-sifiliograficas 2012;103(1):51—8.

[222] Tanaka T, Scheet P, Giusti B, Bandinelli S, Piras MG, Usala G, et al. Genome-wide association study of vitamin B6, vitamin B12, folate, and homocysteine blood concentrations. Am J Hum Genet 2009;84(4):477—82.

[223] Hazra A, Kraft P, Lazarus R, Chen C, Chanock SJ, Jacques P, et al. Genome-wide significant predictors of metabolites in the one-carbon metabolism pathway. Hum Mol Genet 2009;18(23):4677—87.

[224] Afman LA, Trijbels FJ, Blom HJ. The H475Y polymorphism in the glutamate carboxypeptidase II gene increases plasma folate without affecting the risk for neural tube defects in humans. J Nutr 2003;133(1):75—7.

[225] Ashfield-Watt PA, Pullin CH, Whiting JM, Clark ZE, Moat SJ, Newcombe RG, et al. Methylenetetrahydrofolate reductase 677C—>T genotype modulates homocysteine responses to a folate-rich diet or a low-dose folic acid supplement: a randomized controlled trial. Am J Clin Nutr 2002;76(1):180—6.

[226] Yazdanpanah N, Uitterlinden AG, Zillikens MC, Jhamai M, Rivadeneira F, Hofman A, et al. Low dietary riboflavin but not folate predicts increased fracture risk in postmenopausal women homozygous for the MTHFR 677 T allele. Journal of Bone and Mineral Research: the official journal of the American Society for Bone and Mineral Research 2008;23(1):86—94.

[227] Wilson CP, Ward M, McNulty H, Strain JJ, Trouton TG, Horigan G, et al. Riboflavin offers a targeted strategy for managing hypertension in patients with the MTHFR 677TT genotype: a 4-y follow-up. Am J Clin Nutr 2012;95(3):766—72.

[228] McNulty H, Dowey le RC, Strain JJ, Dunne A, Ward M, Molloy AM, et al. Riboflavin lowers homocysteine in individuals homozygous for the MTHFR 677C->T polymorphism. Circulation 2006;113(1):74—80.

[229] Martin YN, Salavaggione OE, Eckloff BW, Wieben ED, Schaid DJ, Weinshilboum RM. Human methylenetetrahydrofolate reductase pharmacogenomics: gene resequencing and functional genomics. Pharmacogenet Genomics 2006;16(4):265–77.

[230] Selzer RR, Rosenblatt DS, Laxova R, Hogan K. Adverse effect of nitrous oxide in a child with 5,10-methylenetetrahydrofolate reductase deficiency. N Engl J Med 2003;349(1):45–50.

[231] Hustad S, Ueland PM, Vollset SE, Zhang Y, Bjorke-Monsen AL, Schneede J. Riboflavin as a determinant of plasma total homocysteine: effect modification by the methylenetetrahydrofolate reductase C677T polymorphism. Clin Chem 2000;46(8 Pt 1):1065–71.

[232] Suormala T, Gamse G, Fowler B. 5,10-Methylenetetrahydrofolate reductase (MTHFR) assay in the forward direction: residual activity in MTHFR deficiency. Clin Chem 2002;48(6 Pt 1): 835–43.

[233] Royston BD, Nunn JF, Weinbren HK, Royston D, Cormack RS. Rate of inactivation of human and rodent hepatic methionine synthase by nitrous oxide. Anesthesiology 1988;68(2):213–16.

[234] Nagele P, Zeugswetter B, Wiener C, Burger H, Hupfl M, Mittlbock M, et al. Influence of methylenetetrahydrofolate reductase gene polymorphisms on homocysteine concentrations after nitrous oxide anesthesia. Anesthesiology 2008;109(1):36–43.

[235] Matsuzawa J, Matsui M, Konishi T, Noguchi K, Gur RC, Bilker W, et al. Age-related volumetric changes of brain gray and white matter in healthy infants and children. Cereb Cortex 2001;11(4):335–42.

[236] Christensen B, Guttormsen AB, Schneede J, Riedel B, Refsum H, Svardal A, et al. Preoperative methionine loading enhances restoration of the cobalamin-dependent enzyme methionine synthase after nitrous oxide anesthesia. Anesthesiology 1994;80(5):1046–56.

[237] Lin X, Lu D, Gao Y, Tao S, Yang X, Feng J, et al. Genome-wide association study identifies novel loci associated with serum level of vitamin B12 in Chinese men. Hum Mol Genet 2012 Jun 1;21(11):2610–17.

[238] Azevedo M, Eriksson S, Mendes N, Serpa J, Figueiredo C, Resende LP, et al. Infection by Helicobacter pylori expressing the BabA adhesin is influenced by the secretor phenotype. J Pathol 2008;215(3):308–16.

[239] Oussalah A, Besseau C, Chery C, Jeannesson E, Gueant-Rodriguez RM, Anello G, et al. Helicobacter pylori serologic status has no influence on the association between fucosyltransferase 2 polymorphism (FUT2 461 G->A) and vitamin B-12 in Europe and West Africa. Am J Clin Nutr 2012;95(2):514–21.

[240] Mollin DL, Baker SJ, Donlach I. Addisonian pernicious anaemia without gastric atrophy in a young man. British Journal of Haematology 1955;1(3):278–90.

[241] Masnou H, Domenech E, Navarro-Llavat M, Zabana Y, Manosa M, Garcia-Planella E, et al. Pernicious anaemia in triplets. A case report and literature review. Gastroenterologia y Hepatologia 2007;30(10):580–2.

[242] Cowan TM, Blitzer MG, Wolf B. Technical standards and guidelines for the diagnosis of biotinidase deficiency. Genet Med 2010;12(7):464–70.

[243] Weber P, Scholl S, Baumgartner ER. Outcome in patients with profound biotinidase deficiency: relevance of newborn screening. Develop Med Child Neurol 2004;46(7):481–4.

[244] Durance T. Residual Avid in Activity in Cooked Egg White Assayed with Improved Sensitivity. Journal of Food Science 1991;56(3):707–9.

[245] Mock DM, Stratton SL, Horvath TD, Bogusiewicz A, Matthews NI, Henrich CL, et al. Urinary excretion of 3-hydroxyisovaleric acid and 3-hydroxyisovaleryl carnitine increases in response to a leucine challenge in marginally biotin-deficient humans. J Nutr 2011;141(11):1925–30.

[246] Mock DM, Mock NI, Nelson RP, Lombard KA. Disturbances in biotin metabolism in children undergoing long-term anticonvulsant therapy. J Pediatr Gastroenterol Nutr 1998;26(3):245–50.

[247] Fierce Y, de Morais Vieira M, Piantedosi R, Wyss A, Blaner WS, Paik J. In vitro and in vivo characterization of retinoid synthesis from beta-carotene. Arch Biochem Biophys 2008;472(2):126–38.

[248] Ferrucci L, Perry JR, Matteini A, Perola M, Tanaka T, Silander K, et al. Common variation in the beta-carotene 15,15'-monooxygenase 1 gene affects circulating levels of carotenoids: a genome-wide association study. Am J Hum Genet 2009;84(2):123–33.

[249] Lietz G, Oxley A, Leung W, Hesketh J. Single nucleotide polymorphisms upstream from the beta-carotene 15,15′-monoxygenase gene influence provitamin A conversion efficiency in female volunteers. J Nutr 2012;142(1):161S—5S.

[250] Amengual J, Lobo GP, Golczak M, Li HN, Klimova T, Hoppel CL, et al. A mitochondrial enzyme degrades carotenoids and protects against oxidative stress. FASEB J 2011;25(3):948—59.

[251] Vage DI, Boman IA. A nonsense mutation in the beta-carotene oxygenase 2 (BCO2) gene is tightly associated with accumulation of carotenoids in adipose tissue in sheep (Ovis aries). BMC Genet 2010;11:10.

[252] El Kares R, Manolescu DC, Lakhal-Chaieb L, Montpetit A, Zhang Z, Bhat PV, et al. A human ALDH1A2 gene variant is associated with increased newborn kidney size and serum retinoic acid. Kidney Int 2010;78(1):96—102.

[253] Gilbert T, Merlet-Benichou C. Retinoids and nephron mass control. Pediatr Nephrol 2000;14(12):1137—44.

[254] Zhang Z, Quinlan J, Hoy W, Hughson MD, Lemire M, Hudson T, et al. A common RET variant is associated with reduced newborn kidney size and function. J Am Soc Nephrol 2008;19(10):2027—34.

[255] Clark AT, Bertram JF. Molecular regulation of nephron endowment. Am J Physiol 1999;276(4 Pt 2):F485—97.

[256] Brenner BM, Garcia DL, Anderson S. Glomeruli and blood pressure. Less of one, more the other? Am J Hypertension 1988;1(4 Pt 1):335—47.

[257] Lindqvist A, Sharvill J, Sharvill DE, Andersson S. Loss-of-function mutation in carotenoid 15,15′-monooxygenase identified in a patient with hypercarotenemia and hypovitaminosis A. J Nutr 2007;137(11):2346—50.

[258] Cohen. Observations on carotenemia. Ann Intern Med 1958;48(2):219—27.

[259] Sharvill DE. Familial hypercarotinaemia and hypovitaminosis A. Proc R Soc Med 1970;63(6):605—6.

[260] Zhu J, Deluca HF, Vitamin D. 25-hydroxylase—Four decades of searching, are we there yet? Arch Biochem Biophys 2012 Jul 1;523(1):30—6.

[261] Cheng JB, Levine MA, Bell NH, Mangelsdorf DJ, Russell DW. Genetic evidence that the human CYP2R1 enzyme is a key vitamin D 25-hydroxylase. Proc Natl Acad Sci U S A 2004;101(20):7711—15.

[262] Ramagopalan SV, Dyment DA, Cader MZ, Morrison KM, Disanto G, Morahan JM, et al. Rare variants in the CYP27B1 gene are associated with multiple sclerosis. Ann Neurol 2011;70(6):881—6.

[263] Huang J, Xie ZF. Polymorphisms in the vitamin D receptor gene and multiple sclerosis risk: A meta-analysis of case-control studies. J Neurol Sci 2012;313(1-2):79—85.

[264] Jablonski NG, Chaplin G. Colloquium paper: human skin pigmentation as an adaptation to UV radiation. Proc Natl Acad Sci U S A 2010;107(Suppl. 2):8962—8.

[265] Lalueza-Fox C, Rompler H, Caramelli D, Staubert C, Catalano G, Hughes D, et al. A melanocortin 1 receptor allele suggests varying pigmentation among Neanderthals. Science 2007;318(5855):1453—5.

[266] Dessinioti C, Antoniou C, Katsambas A, Stratigos AJ. Melanocortin 1 receptor variants: functional role and pigmentary associations. Photochem Photobiol 2011;87(5):978—87.

[267] Liu F, Struchalin MV, Duijn K, Hofman A, Uitterlinden AG, Duijn C, et al. Detecting low frequent loss-of-function alleles in genome wide association studies with red hair color as example. PLoS One 2011;6(11):e28145.

[268] Wang TJ, Zhang F, Richards JB, Kestenbaum B, van Meurs JB, Berry D, et al. Common genetic determinants of vitamin D insufficiency: a genome-wide association study. Lancet 2010;376(9736):180—8.

[269] McGrath JJ, Saha S, Burne TH, Eyles DW. A systematic review of the association between common single nucleotide polymorphisms and 25-hydroxyvitamin D concentrations. J Steroid Biochem Mol Biol 2010;121(1—2):471—7.

[270] Fu L, Yun F, Oczak M, Wong BY, Vieth R, Cole DE. Common genetic variants of the vitamin D binding protein (DBP) predict differences in response of serum 25-hydroxyvitamin D [25(OH)D] to vitamin D supplementation. Clin Biochem 2009;42(10—11):1174—7.

[271] Ji GR, Yao M, Sun CY, Li ZH, Han Z. BsmI, TaqI, ApaI and FokI polymorphisms in the vitamin D receptor (VDR) gene and risk of fracture in Caucasians: a meta-analysis. Bone 2010;47(3):681–6.

[272] Lee JE. Circulating levels of vitamin D, vitamin D receptor polymorphisms, and colorectal adenoma: a meta-analysis. Nutr Res Pract 2011;5(5):464–70.

[273] Rollison DE, Cole AL, Tung KH, Slattery ML, Baumgartner KB, Byers T, et al. Vitamin D intake, vitamin D receptor polymorphisms, and breast cancer risk among women living in the southwestern. US Breast Cancer Res Treat 2012 Apr;132(2):683–91.

[274] Kang TJ, Jin SH, Yeum CE, Lee SB, Kim CH, Lee SH, et al. Vitamin D Receptor Gene TaqI, BsmI and FokI polymorphisms in Korean patients with tuberculosis. Immune Netw 2011;11(5):253–7.

[275] Schlingmann KP, Kaufmann M, Weber S, Irwin A, Goos C, John U, et al. Mutations in CYP24A1 and idiopathic infantile hypercalcemia. N Engl J Med 2011;365(5):410–21.

[276] Pillai DK, Iqbal SF, Benton AS, Lerner J, Wiles A, Foerster M, et al. Associations between genetic variants in vitamin D metabolism and asthma characteristics in young African Americans: a pilot study. J Invest Med 2011;59(6):938–46.

[277] Chardon L, Sassolas A, Dingeon B, Michel-Calemard L, Bovier-Lapierre M, Moulin P, et al. Identification of two novel mutations and long-term follow-up in abetalipoproteinemia: a report of four cases. Eur J Pediatr 2009;168(8):983–9.

[278] Zamel R, Khan R, Pollex RL, Hegele RA. Abetalipoproteinemia: two case reports and literature review. Orphanet J Rare Dis 2008;3:19.

[279] Peretti N, Sassolas A, Roy CC, Deslandres C, Charcosset M, Castagnetti J, et al. Guidelines for the diagnosis and management of chylomicron retention disease based on a review of the literature and the experience of two centers. Orphanet J Rare Dis 2010;5:24.

[280] Gotoda T, Arita M, Arai H, Inoue K, Yokota T, Fukuo Y, et al. Adult-onset spinocerebellar dysfunction caused by a mutation in the gene for the alpha-tocopherol-transfer protein. N Engl J Med 1995;333(20):1313–18.

[281] Bardowell SA, Stec DE, Parker RS. Common variants of cytochrome P450 4F2 exhibit altered vitamin E-{omega}-hydroxylase specific activity. J Nutr 2010;140(11):1901–6.

[282] Jain V, Wood SJ, Feiveson AH, Black FO, Paloski WH. Diagnostic accuracy of dynamic pos-turography testing after short-duration spaceflight. Aviat Space Environ Med 2010;81(7):625–31.

[283] Kohlmeier M, Salomon A, Saupe J, Shearer MJ. Transport of vitamin K to bone in humans. J Nutr 1996;126(Suppl. 4):1192S–6S.

[284] Vermeer C. Vitamin K: the effect on health beyond coagulation—an overview. Food Nutr Res 2012:56.

[285] Kohlmeier M, Saupe J, Schaefer K, Asmus G. Bone fracture history and prospective bone fracture risk of hemodialysis patients are related to apolipoprotein E genotype. Calcif Tissue Int 1998;62(3):278–81.

[286] Peter I, Crosier MD, Yoshida M, Booth SL, Cupples LA, Dawson-Hughes B, et al. Associations of APOE gene polymorphisms with bone mineral density and fracture risk: a meta-analysis. Osteoporosis International 2011;22(4):1199–209.

[287] Kohlmeier M, da Costa KA, Fischer LM, Zeisel SH. Genetic variation of folate-mediated one-carbon transfer pathway predicts susceptibility to choline deficiency in humans. Proc Natl Acad Sci U S A 2005;102(44):16025–30.

[288] Zeisel SH. Nutritional genomics: defining the dietary requirement and effects of choline. J Nutr 2011;141(3):531–4.

[289] Song J, da Costa KA, Fischer LM, Kohlmeier M, Kwock L, Wang S, et al. Polymorphism of the PEMT gene and susceptibility to nonalcoholic fatty liver disease (NAFLD). FASEB J 2005;19(10):1266–71.

[290] Johnson AR, Lao S, Wang T, Galanko JA, Zeisel SH. Choline Dehydrogenase Polymorphism rs12676 Is a Functional Variation and Is Associated with Changes in Human Sperm Cell Function. PLoS One 2012;7(4):e36047.

CHAPTER 5

How does Nutrigenetics Influence Long-Term Health?

Terms of the Trade

- Epistasis: Modification of a nutrition—genetic interaction by variation at another locus.
- Hemizygosity: State where only one of the usual two copies of a gene is present in cells.
- Neural tube defects: Anencephaly, spina bifida, and related birth defects.

ABSTRACT

Nutritional factors play a major role in the development of many common conditions and diseases and equally often in their prevention and treatment. Often, a nutritional factor makes a difference only in some individuals and not in others. A proper understanding of the genes involved in nutritional metabolism and regulation of nutritional states often helps to understand how common genetic variants change the response to important nutrition factors. This section explores nutrigenetic variation in the context of complex diseases and conditions to illustrate how many of these interactions we already know about.

5.1 GOOD NUTRITION MEANS DIFFERENT THINGS FOR DIFFERENT PEOPLE

Now we will get seriously into the health consequences of nutrition choices. When two people eat exactly the same way, their health outcomes can be quite different. This is because each individual's metabolism is tuned differently by the genome, as the previous chapters should have made clear. For an increasing number of health conditions we can now predict the likely individual response to nutritional interventions and start using genetic information to chart individual options. Please note that many of the gene—nutrient interactions mentioned are initial observations that need to be corroborated by additional

Nutrigenetics. http://dx.doi.org/10.1016/B978-0-12-385900-6.00005-8

investigations. The purpose of mentioning them is to illustrate the mechanisms through which such interactions work.

5.2 IT STARTS RIGHT AT THE BEGINNING— PRECONCEPTUAL AND PRENATAL NUTRITION

5.2.1 Out to lunch

Imagine a couple on their lunch break. They are happily chatting away, making plans for settling down and starting a family. The last thing on their mind will probably be what their lunch might do for their first child, still just a glint in their eyes. So let's see how much influence the parents' nutrition actually has on the future of the child to be.

First, there is the issue of fertility. Could they even have children if they wanted to? You read earlier that women with untreated celiac disease (CD) are less fertile than unaffected women [1]. The good news is that the avoidance of gluten appears to restore fertility for women with CD [1]. Gluten is the binder protein in barley, rye, wheat, and other closely related grains. It is the stuff that holds pizza dough together. It is also in beer (when it is made from barley or wheat) and in many commercial products, including lip balm, sunscreen lotion, and the glue of self-adhesive envelopes. Amaranth, corn (maize), millet, potato, sorghum, rice, sorghum, tapioca, and wild rice are examples of gluten-free starches. Gluten-free flour is made from a mixture of such gluten-free starches. Oats are also gluten-free, but may be unintentionally mixed with a small percentage of gluten-containing grains. Gluten does not seem to affect the fertility of men [2]. This is just one well-documented example showing the impact of nutrition on fertility. Other foods and nutrients are also important and may similarly involve individual genetic predisposition.

Next, let's take a brief look at the gender of the child. The amount of fat in food strongly affects the male–female sex ratio in older mice [3]. Mice fed a high-fat diet give birth to twice as many males as females. When the mother is placed on a high-carbohydrate diet, female offspring outnumber males by more than two to one. This does not mean that fat has similarly dramatic effects on human sex ratios, but illustrates the principle. In humans, recent weight gain has been demonstrated to shift the sex ratio by a few percent in favor of more males [4].

Third, nutrition of a newly pregnant mother strongly influences the risk of birth defects in her child. The greatest risk is during the first few weeks of pregnancy because this is when the organs are formed. Choline, folate, riboflavin, vitamin B6, and vitamin B12 are important because they support the synthesis of phospholipids and DNA bases. The same nutrients are also needed for proper DNA methylation. The mother's optimal folate intake reduces the risk of neural tube defects (anencephaly and spina bifida), heart defects, and orofacial clefts (cleft lip and cleft palate). The legally mandated addition of folic acid to flour and baked goods in the USA and Canada has reduced the rate of neural tube defects by more than a quarter and may be lowered even further by added folate intake [5].

5.2.2 Nutrigenetics of birth defect risks

Not every woman has the same risk of having a child with a birth defect, even under the same nutritional circumstances. Several common genetic variants influence the nutrient amounts that prevent birth defects most effectively (Figure 5.1; Table 5.1). So far, we know most about the increased risk of women with two copies of the *5,10-methylene tetrahydrofolate reductase* (*MTHFR*; EC1.5.1.20; OMIM 607093) 677T rs1801133 variant [6]. The mRNA of the 677T variant is less stable than the more common 677C variant, producing smaller amounts of the MTHFR enzyme. Low MTHFR activity is associated with a higher concentration of the potentially toxic metabolite homocysteine and with a diminished capacity to provide methyl groups for DNA methylation. The lower amount of enzyme produced with the 677T mRNA can be offset by an increased availability of folate, the cofactor FAD, and

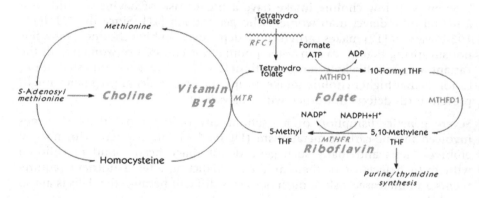

FIGURE 5.1
Variants of several genes with a role in one-carbon metabolism predispose to birth defects in women with low intake of folate or other critical nutrients. ADP, adenosine diphosphate; ATP, adenosine triphosphate; THF, tetrahydrofolate.

Table 5.1	Several Common Genetic Variants Influence how much of a Particular Nutritional Factor is Needed to Minimize the Risk of Birth Defects		
Genotype	**Prevalence (%)**	**Nutritional Factor**	**Reference**
MTHFR 677T/T	10–30	(+) folate (+) riboflavin (+) vitamin B12	[8]
RFC 80G/G	20–30	(+) folate	[9]
MTHFD1 1958A/G and A/A	15–65	(+) choline	[13]

The plus sign before a listed nutritional factor indicates that people with the condition need more of the factor.

vitamin B12 [7,8]. This means that women with the 677T/T genotype need about 50% more folate than women with the other genotypes and also need to guard more carefully against low intakes of riboflavin and vitamin B12 than others.

The *reduced folate carrier* gene (*RFC1*; OMIM 102579) controls the availability of folate to cells. The carrier protein encoded by the *RFC1* variant 80G (rs1051266) is less effective than the protein derived from the more common 80A variant. It is easy to understand then that in the absence of fortification or supplement use women with two copies of the 80G variant have more than twice the risk of having a child with a birth defect than women with the 80A variant [9].

A partially overlapping group of birth defect-related gene variants potentially leads to unmet choline needs in the embryo. Humans have all the genes to produce choline, but most people cannot make enough to meet their requirements and depend on additional intakes with food [10]. This dependency on choline from food is a potential risk for pregnant women, despite the fact that their high estrogen levels during pregnancy boost choline synthesis. Women with low choline intake have a higher risk of having a child with a neural tube defect than women who get enough [11]. Since the *MTHFD1* 1958A variant [12] makes carriers more dependent on high choline intake, it is not surprising that at least in some populations children of women with the variant are more likely to have a birth defect [13]. The as yet untested implication is that higher choline intake during the first weeks of pregnancy might prevent birth defects in women with this variant.

Severe genetic disruption of just about any of the more than 30 genes involved in one-carbon metabolism (Figure 5.1) can interfere with normal embryogenesis and many such gene defects have been found in children with birth defects or in their mothers. Predicting altered nutrient requirements and increased risk of birth defects is difficult because the defects are so rare [9].

5.2.3 Prenatal nutrition

Of course, nutrition continues to be important throughout the pregnancy. As the fetus grows bigger, much larger amounts of many critical nutrients are needed. Inadequate supplies make an early termination of pregnancy more likely. As in other critical situations, genetic variants increase the vulnerability to a nutrition mismatch. One example is the common 19 bp insertion/deletion variant in intron 1 of *dihydrofolate reductase* (*DHFR*; EC1.5.1.3; OMIM 126060). DHFR is particularly critical for rapidly multiplying cells because the enzyme regenerates dihydrofolate, the metabolic product of thymidylate synthesis for mRNA (Figure 5.2).

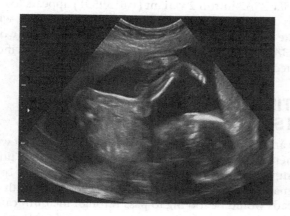

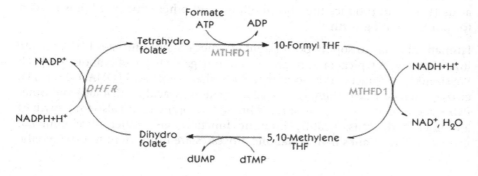

FIGURE 5.2
Dihydrofolate reductase (DHFR) regenerates the dihydrofolate metabolite left over from thymidine synthesis. ADP, adenosine diphosphate; ATP, adenosine triphosphate; NAD, nicotinamide adenine dinucleotide; THF, tetrahydrofolate.

Pregnant women with two copies of the 19 bp *DHFR* variant are more likely to have a preterm delivery than women without it [14]. This variant is less active than the more common version and reactivation of both dihydrofolate and folic acid is more sluggish. There is some indication that using dietary supplements with folic acid may not be helpful because greatly increased concentrations of unmetabolized free folic acid may interfere with normal metabolism. The solution in this case is to rely instead on folate from foods like oranges and broccoli.

Vitamin C is another important nutrient for a healthy pregnancy. Women with low vitamin C intake have an increased risk of premature membrane rupture and preterm delivery [15]. Two sodium-dependent ascorbate transporters,

SLC23A1 and SLC23A2, manage the distribution of vitamin C to and between tissues. Both have variants with potential impact on pregnancy outcome. The presence of an *SLC23A2* intron 2 variant (rs6139591) appears to double the risk of spontaneous preterm delivery [16]. We don't know yet whether higher vitamin C intake improves pregnancy outcome specifically in women with the high-risk *SLC23A2* variant. The efficacy of the genotype-specific intervention must be validated to make this information clinically useful.

5.3 BOOSTING COGNITIVE DEVELOPMENT OF INFANTS

Birth is already an exceptionally difficult process as it is, fraught with lethal risks for both mother and child. There is no question that the human brain is outsized in comparison to the brains of our primate cousins or the brains of more distant mammalian relatives. If infants were born with more mature brains, their heads would be too big to pass through the birth canal. Because of this size limitation the infant's brain has to more than double in volume during the first year of life and continue to grow significantly for another year (Figure 5.3). Complex lipids constitute a large percentage of the added mass. A healthy infant needs more than 600 mg of essential polyunsaturated fatty acids (PUFA) to produce the phospholipids and other structural lipids needed for normal brain growth.

Human milk has to supply both essential omega-3 and omega-6 PUFA for the infant's needs. A typical ratio of omega-3 to omega-6 fatty acid is about 1:8 [17]. We should not forget at this point that docosahexaenoic acid (DHA, C22:6 ω3), eicosapentaenoic acid (EPA, C20:5 ω3), arachidonic acid (C20:4 ω6) and other PUFA are needed not only for optimal brain development, but also for growth of all other body cells, healthy immune function, and other developmental processes. The infant's own metabolic processes are insufficient to generate the

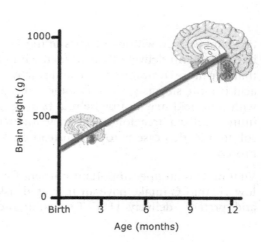

FIGURE 5.3
A newborn infant's brain approximately doubles during the first year of life and then continues to grow during the second year.

necessary amounts of DHA for building the white matter of the human brain from other omega-3 fatty acids. This means that milk (or infant formula) has to provide enough of all of these fatty acids. If the infant does not get enough, the growth of its brain will be slowed and cognitive development delayed [18]. Such delayed development may not be fully compensated for by later growth. While ensuring adequate supplies is critically important, we have to guard against falling into the other extreme of overfeeding. There is the distinct possibility that inappropriately high intakes can reduce some aspects of cognitive development. Long-term follow-up of the effects of fish oil supplementation during the first few months of breastfeeding in Danish women suggested a slightly unfavorable effect on some higher executive functions [19].

Ultimately, the question is whether a breastfeeding mother provides the optimal amounts of all the fatty acids her infant needs. The mother's food intake pattern is the first determinant. However, even if she does not consume much of a needed fatty acid, she is usually still able to make it from a precursor. Let's focus again on the omega-3 fatty acid DHA, of which infants need a lot. The most direct way would be the consumption of DHA-rich food sources such as cold-water fish or fish oil, but most women do not get much DHA from food. This leaves alpha-linolenic acid (ALA; C18:3 ω3) as the main omega-3 fatty acid in seeds (particularly flax seed), vegetable oils, and vegetables. The mother's metabolism then has to convert ALA to DHA, a fairly involved multistep process (Figure 5.4). Note the two delta6-desaturase steps at the beginning and toward the end of the sequence. Both are catalyzed by an enzyme produced by the FADS2 gene (OMIM 606149). This gene is noteworthy because several common genetic variants appear to modify the activity of the expressed enzyme. The minor allele of one variant locus in particular (rs174575 located in intron 1) may be associated with slowed conversion of ALA to DHA [20]. Lack of confirmation in other studies means that this interaction is not ready for practical use but should be considered in further research.

5.4 CANCER

5.4.1 Cancer, nutrition, and genetic predisposition

Cancer is the insufficiently controlled, disruptive, and destructive proliferation of clonal cells that now kills about as many Americans as cardiovascular disease. Cancer is not one disease but many different ones, each with its own behaviors and outcomes. Men and women share only some types of cancers. Others are obviously specific to a gender. Even for those cancers that occur in both women and men, risk factors often play out very differently, even in opposite directions. The majority of cancers arise at the intersection of nutrition and genetic predisposition. This should not be surprising, since we get with our food many compounds that cause DNA damage, promote cancer cell proliferation, or interfere with mechanisms of cancer control. Another reason is that the accelerated cellular growth rate of cancer ties diseased tissues more closely than healthy tissues to a steady nutrient supply.

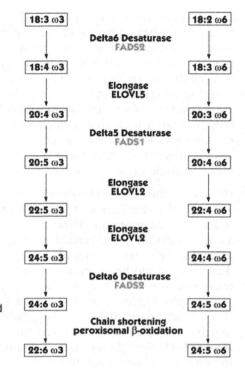

FIGURE 5.4
Mother and child produce from the available precursors various very long-chain polyunsaturated fatty acids they need. The desaturase enzymes FADS1 and FADS2 play key roles in these conversions.

The following outlines can only give a taste of the many specific mechanisms that contribute to cancer incidence and disease outcomes (Table 5.2). The pace of research is explosive in this field and new insights are reported almost daily.

5.4.2 DNA synthesis and epigenetic control of cell proliferation

Folate-mediated one-carbon metabolism plays multiple important roles in normal and abnormal cell division. The synthesis of the purine bases, adenine and guanine, uses 10-formyl tetrahydrofolate. Methylation of thymidine uses 5,10-methylene tetrahydrofolate, and methylation of DNA and histones depends on 5-methyl-tetrahydrofolate for the production of the universal methylation agent S-adenosylmethionine (SAM). The demethylation that drives the unwinding of chromatin and opening up of gene promoter regions in the nucleus releases significant amounts of formaldehyde, which is highly toxic and has to be sequestered immediately as 5,10-tetrahydrofolate [21].

Each of these functions is known to be affected jointly by genetic predisposition and nutrient status. For instance, low activity of DHFR or low folate availability is associated with increased incorporation of uracil instead of thymidine into the

Table 5.2		Examples of Genotype-specific Interventions for Long-term Prevention of Cancer			
Gene Variant	rs Number	Frequency	Intervention	Genotype-specific Benefit	Reference
DHFR 19 del	none	0.19	Folic acid < 200 µg/day	−33% breast cancer	[23]
MPO −463A	rs2333227	0.22	Fruits/vegs > 5 servings	−70% breast cancer	[26]
XRCC1 26304T	rs1799782	0.06	Vitamin C > 131 mg/day	−37% breast cancer	[36]
MGMT 417G	–	0.24	Fruits/vegs > 5 servings	−50% breast cancer	[38]
Fast *NAT1/NAT2*	several	0.50	Meat < 2 servings/week	−60% colorectal cancer	[32]
ESR2 −13950C	rs2987983	0.32	Isoflavones > 2 mg/day	−37% prostate cancer	[40]
CUBN 233T	rs1907362	0.07	Vitamin D 1000 IU	−26% breast cancer	[42]
PTGS2 +6365C	rs5275	0.60	Fatty ocean fish once/week	−71% prostate cancer	[46]

The benefits shown are always the maximal effect expected when switching from the worst possible intake category to the presumably beneficial diet pattern long before potential cancer development. Actual benefits will be only a fraction of this maximum. *rs, reference single nucleotide polymorphism (SNP); vegs, vegetables.*

DNA of dividing cells. Higher uracil content destabilizes DNA, accelerates the rate of mutations, and increases cancer risk. On the other hand, low production of the essential methylation cofactor SAM skews normal DNA and histone methylation patterns, which can increase cancer risk due to DNA breaks and inappropriate gene expression. Low intake of choline and other methyl donor nutrients greatly increases rates of liver and other cancers in animal models [22].

The complexity of the multiple levels of gene-nutrient interactions becomes apparent when examining how folate nutrition influences breast cancer risk. First of all, low folate intake slightly increases breast cancer risk but excessive intake may also increase it [23]. Next, the *MTHFR* genotype 677T/T (rs1801131) appears to slightly increase the folate intake level that is needed to avoid increased risk [24]. Use of a folate-containing supplement, on the other hand, increases breast cancer risk by more than 50% in women with two copies of the *DHFR* 19 del variant [23]. One copy appears to increase risk by a smaller percentage. No such risk increase is seen in women without the variant. There is also no reason to believe that high intake of food folate with oranges or broccoli increases breast cancer risk in women with a *DHFR* 19 del variant. On the contrary, such foods are likely to reduce breast cancer risk in those as in other women.

5.4.3 DNA damage by free radicals

Free radicals are highly reactive molecules with one or two unpaired electrons. Large amounts of free radicals are generated during oxidative phosphorylation, by xanthine oxidase, in interactions of iron and copper with oxygen, and in other perfectly normal reactions. Another important source of free radicals is leukocytes, some of which literally blast a highly corrosive mix of free radicals at bacteria and other invasive organisms. Granulocytes and monocytes use the enzyme myeloperoxidase (*MPO*; EC 1.11.1.7; OMIM 606989) to generate hypochlorous acid for the antibacterial blast. Sustained release of these corrosive free radicals is one of the reasons why chronic inflammation is harmful.

More than 20% of Americans carry a variant (−463G>A; rs2333227) in the promoter region of *MPO* that is associated with lower expression of the gene and reduced MPO enzyme activity [25]. In young women with the −463A allele, high fruit and vegetable intake is associated with a 70% lower breast cancer risk than those with a low intake of these foods [26]. Women without this *MPO* variant do not seem to have such a clear benefit with high fruit and vegetable intake. It is likely that this genotype-specific response also applies to cancers at some other sites. The benefit of genotype-specific advice will be much smaller, of course, if the woman already has a sizeable fruit and vegetable intake. One should also not lose sight of the fact that high fruit and vegetable consumption provides many benefits unrelated to cancer prevention, regardless of the individual genotype.

Along the same lines is the role of the ubiquitous enzyme catalase (EC 1.11.1.6) encoded by *CAT* (OMIM 115500). This enzyme is an important part of our defense against free radicals, in this case by splitting the powerful oxidizer hydrogen peroxide into harmless oxygen and water molecules. It is important to understand that hydrogen peroxide, like other free radicals, sustains tissue health by functioning in cell signaling and as an enzyme cofactor. Hydrogen peroxide only becomes a problem when its concentration is not properly controlled and it starts damaging vital structures. There is even increasing evidence that insufficiently checked hydrogen peroxide concentration due to a local lack of catalase causes the graying of hair [27]. Catalase is not absolutely essential because other enzymes and antioxidants still provide protection but the coverage gets thinner and depends on optimal nutritional status. The common variant *CAT* −262C>T (rs1001179) encodes an enzyme variant (allozyme) with reduced activity [28]. Contrary to what one might expect, high antioxidant consumption with fruits and vegetables reduces the already relatively low breast cancer risk of women with the *CAT* −262C/C genotype but not in those with one or two T alleles [29]. This finding is an important reminder that we still lack a full understanding of the effects of antioxidants in foods and how they create synergies with free radical-quenching enzymes.

5.4.4 Phase 1 and phase 2 enzymes

Foods contain a huge number of chemicals, many of which are potentially harmful. We are not just talking about industrial chemicals but also about a plethora of natural poisons that plants use as a defense against being eaten. This may sound worse than it is because our body can usually handle these plant chemicals (they are often called *phytochemicals*). The removal of cancer-causing chemicals works the same way, but the need for quick elimination is more obvious. Most of the action takes place in the small intestine and the liver. Detoxification happens in two phases (Figure 5.5). The first phase uses enzymes

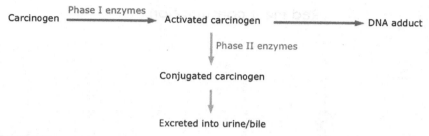

FIGURE 5.5
Inactivation and removal of an ingested carcinogen often proceeds in two phases: first the compound is activated by phase 1 enzymes; and then it is conjugated to a metabolite that aids its excretion with bile or urine. If the activated carcinogen is not conjugated immediately, it may reach nuclear DNA and form a DNA adduct. Such adducts can lead to incorporation of the wrong base and thus generate a mutation during the next cell division.

that prepare the chemicals by hydroxylating or otherwise activating them. Most of these phase 1 enzymes are microsomal cytochromes. The second phase links (conjugates) the activated chemicals to glycine, glucuronic acid, glutathione, or another metabolite. This conjugation usually makes the chemicals more suitable for excretion into bile or urine [30]. These mechanisms also work for the elimination of various normal metabolic waste products. Bilirubin, for instance, needs to be glucuronidated for excretion.

5.4.4.1 ACETYLATION

Among the phase 1 enzymes that activate carcinogens are N-acetyl transferase 1 (*NAT1*; EC 2.3.1.5; OMIM 108345) and 2 (*NAT2*; OMIM 612182). These enzymes activate various carcinogens that arise from roasting meats. People with particularly high activity are classified as fast acetylators; those with low activity are slow acetylators. The distinction is of interest, because high phase 1 NAT activity gives the phase 2 conjugation activity less chance to make these carcinogens safe or to promote their excretion before they can do harm. With the phase 1 and phase 2 enzymes it is all about balance. The practical relevance was seen in a number of studies. With the same intake level of red meat (beef, lamb, pork, or venison), women with the NAT1 fast acetylator type were 43% more likely to get breast cancer than those with slow acetylators [31]. The effect of NAT2 acetylator status was similar, but smaller. A similar interaction of red meat consumption level with acetylator status was observed for colorectal cancer (Figure 5.6) [32]. Of course, the solution for fast acetylators is obvious: lower meat intake will reduce their cancer risk. Knowing their predictable response to red meat consumption may help them to adjust dietary habits to their personal vulnerability.

5.4.4.2 GLUCURONIDES

As with so many other processes involved in carcinogenesis, the efficiency of carcinogen removal differs greatly between individuals, not least due to

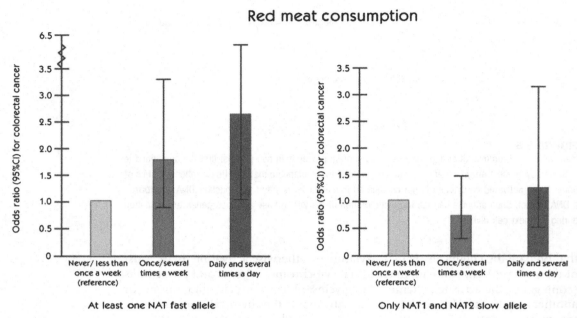

Red meat consumption

FIGURE 5.6
Individuals with high N-acetyl transferase (NAT) activities are more susceptible to the cancer-causing effects of red meat (beef, lamb, pork, and venison). Here, carriers of at least one fast acetylator variant of either *NAT1* or *NAT2* are grouped together and compared with carriers of low-activity variants for both enzymes.

common genetic variants. There are many different genes that encode UDP-glycosyltransferases (EC 2.4.1.17). We focus here on a particular one (encoded by the *UGT1A1* gene; OMIM 191740), which is of importance for the elimination of compounds that cause cancer of the breast, colon, and other sites. A common variant, *UGT1A1 *28*, has reduced activity and appears to increase the risk of colorectal cancer [33]. So far, that is important, but the nutritional aspect is still missing, unless we count exposure to the carcinogens. It is very convenient that *UGT1A1* also is responsible for bilirubin glucuronidation and clearance, which gives us a readily available indicator for any changes in UGT1A1 activity. This measure shows that cruciferous vegetables, such as broccoli, boost UGT1A1 activity in women with the *UGT1A1 *28* variant, but not in those without it [34]. No effect was seen in men. The increase does not seem to bring the activity to the level of people without the variant, but these observations point to opportunities for genotype-specific menu adjustments to reduce cancer risk. Eventually, the goal will be to identify the specific foods that are most effective for people with such low-function variants.

5.4.4.3 GLUTATHIONE CONJUGATES

Glutathione is another important metabolite that gets chemicals ready for excretion. Several genes encode glutathione transferases (EC 2.5.1.18) that link glutathione to carcinogens. The GST mu 1 (*GSTM1*; OMIM 138350) and GST theta 1 (*GSTT1*; OMIM 600436) genes encode two of these glutathione transferases. For a reason that is not entirely clear, around half of the people in diverse populations carry so-called *null* alleles for these genes. These null alleles cannot encode active enzymes because large segments of the gene are deleted. It has therefore been expected that null allele carriers have a higher cancer risk. This has been found in some studies, but not in many others. One of the reasons for such inconsistent results seems to be that some food constituents promote the activity of these transferases unless the gene is deleted. In a population of Spanish agricultural workers with a particularly high incidence of lung cancer due to smoking and other exposures, high intake of carotenoid-rich foods reduced the risk for individuals with functional glutathione transferase genes but not for people with null alleles [35]. These findings demonstrate once again that genetic and nutritional risk factors need to be considered together if we ever want to understand their proper role. They can also help to identify which nutritional interventions are most likely to be effective in a particular individual and which will probably be less beneficial.

5.4.5 Polynucleotide repair

The human genome has a whole battery of mechanisms that can correct a chemical change in the DNA after exposure to mutagens. Several genes excise faulty bases and give the affected DNA strand a second chance.

The *X-ray repair cross complementing group 1* (*XRCC1*; OMIM 194360) gene encodes a protein that stimulates excision repair. The gene got its tedious name as the result of a long series of similarly tedious experiments that tried to find out which genes might make hamsters more resistant to X-ray irradiation. This gene provides very important protection but does not work with equal efficiency in all of us. The variant 26304C>T (Arg194Trp; rs1799782) in exon 6 of *XRCC1* is a common low-efficiency variant. The frequency of the 26304T allele is relatively low in people of African (4%) and Caucasian (6–7%) background but is more common in people with Latin-American (12%) or Asian (34%) heritage. It is not clear why such a variant should be common at all, since there is a potential for increased mutation and ultimately increased cancer risk. The balancing upside (and there are always an upside and a downside with such common variants) may be related to reproductive success or infant survival but its nature is not known. The Long Island Breast Cancer Study, a case-control study [36], suggests risk reduction (odds ratio 0.58) with at least five servings of fruits and vegetables in carriers of the 194Trp allele but not in those without it (odds ratio 0.94). Similar, but slightly weaker, differences were seen in those with a vitamin C intake of > 131 mg/day (0.63 vs. 1.01) and a vitamin E intake of > 29 mg/day (0.65 vs. 1.10). Another low-function *XRCC1* variant, rs25487, which encodes arginine instead of glutamine at position 399 of the mature enzyme, similarly predicted whether the antioxidant lycopene (which provides the red color in tomatoes and water melons, particularly rich sources of this chemical) was associated with a lower risk of prostate cancer [37]. Men without an allele encoding the fully functional enzyme appear to have a 79% lower prostate cancer risk when they eat at least a couple of tablespoons of tomato sauce or a small wedge of water melon each day.

When larger chemicals, such as food-derived carcinogens and industrial pollutants, bind to DNA and create DNA adducts, our genome is in real trouble. Among the most common DNA adducts are the O(6)-alkyl-guanines. These DNA adducts are usually misread during the next cell division and cause the insertion of a wrong base into the DNA sequence. At that point, the damage becomes irreversible and may be the starting point for cancer. The key enzyme for fixing DNA adducts is O(6)-methylguanine DNA methyltransferase (EC 2.1.1.63), encoded by *MGMT* (OMIM 156569). The common *MGMT* variant 427A>G (rs2308321) replaces isoleucine in position 143 with valine (Ile143Val) with as yet unclear consequences for MGMT function. The bad news for women with one or two copies of the *MGMT* 427G allele is that they have a slightly higher breast cancer risk than women without this allele. The good news is that they can reduce their risk by about half if they eat more than five servings of fruits and vegetables a day [38]. Women without the 427G allele do not seem to have this kind of benefit from high fruit and vegetable intake. There is a caveat, though, as with all nutrition—cancer interactions. We know that most breast cancers and other cancers form many years before they are detected. This means that nutrition prevention will not necessarily help with an existing cancer. Prevention is best done early on and should be maintained for a lifetime. On the other hand, one should never assume that it is too late for preventing a new cancer. What better alternative is there for cancer prevention?

5.4.6 Tumor promotion and hormones

Most common cancers grow very slowly over many years before they become a significant health threat by growing into adjacent tissue or spreading to distant sites. The majority of old men have prostate cancer, but most of them will not be killed by this cancer because it often grows very slowly [39]. The question with this and many other cancers is whether nutrition can slow their growth enough to minimize their health impact. We know now about an increasing number of specific nutritional factors that reduce the observed cancer risk by preventing symptomatic disease during the normal life span of the affected population.

5.4.6.1 ESTROGEN

Medical treatment of prostate cancer often aims to reduce the ratio of testosterone to estrogen concentration. This is one of the reasons for using phytoestrogens in soy for the suppression of the growth of as yet undiagnosed prostate tumors as a primary prevention strategy. The binding of phytoestrogens to estrogen receptor beta (*ESR2*; OMIM 601663) appears to inhibit 5-alpha reductase, aromatase, 17beta-hydroxysteroid dehydrogenase, and other enzymes affecting estrogen metabolism. A number of studies have not found a worthwhile benefit from soy for prostate cancer prevention. But look at what this non-effect finding turns into when we take into account the common *ESR2* variant −13950T>C (rs2987983), which occurs in about a third of Americans. A case-control study of prostate cancer in Sweden found opposite effects of increasing soy intakes depending on genotype (Figure 5.7). Prostate cancer risk in men with the −13950T/T genotype increased with higher soy consumption levels. But higher soy intake decreased risk in men with one or two copies of the −13950C allele [40]. Assuming that this finding holds true for most men, not just these Swedish study participants, genetic testing would be really important. The same recommendation for all men simply would not be helpful in such a circumstance because it would be harmful for many either way. If the recommendation is to avoid higher soy intake, men with the C allele will miss out on the opportunity to reduce their prostate cancer risk. If higher soy consumption is recommended for everybody, on the other hand, the men with the T/T genotype would have a higher risk. What course of action would you recommend?

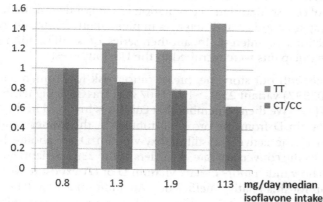

FIGURE 5.7

Men with different genotypes respond to increasing isoflavone intake in opposing directions. These phytochemicals in soy appear to progressively increase prostate cancer risk in men with the *ESR2* −13950T/T genotype, but decrease the risk of men with one or two C alleles [40]. About one in three American adults carry one or two ESR2 −13950C alleles.

Long-term diet modification (increased soy consumption or use of a low-dosed isoflavone supplement) of the approximately one in three adults with the −13950C allele might reduce their risk of prostate cancer by as much as 37%, while there may be a substantial risk increase for the others. Given that about 1% of all adult male deaths are due to prostate cancer, reduction of premature mortality risk with test-directed diet change may accrue to as much as 0.3% (37% of 1%).

5.4.6.2 *VITAMIN D*

Of course, we know that vitamin D is neither an essential nutrient nor an 'amin' (it is a sterol). For most people, the vast majority of vitamin D is made in the skin during exposure to ultraviolet (UV) light. A nice day at the beach produces more than 100,000 IU of vitamin D in a young person, whereas a glass of vitamin D-fortified milk contains about 100 IU. The problem arises in people with little sun exposure, either because they stay mostly indoors or live at high latitude where the sun is too weak for many months of the year to produce any vitamin D in the skin. The few good food sources (eggs, fatty ocean fish, and fortified milk) are usually not enough to make up for the lack of UV-rich sunlight.

There has been a heated debate about the ability of optimal vitamin D status to prevent cancer, particularly breast cancer and prostate cancer. The concept was initially suggested to explain why women in the northern half of the USA had almost twice the cancer risk of women living further south [41]. Since skin exposure in the northern half of the USA can generate significant amounts of vitamin D only during 7—8 months of the year, vitamin D concentrations in blood tend to be lower, particularly during the winter months.

Being in the habit, we immediately start thinking about the various gene variants that influence vitamin D status. A large case-control study of breast cancer in Caucasian women in Canada looked into the role of a panel of vitamin D-related polymorphisms.

Why would it be desirable to restrict this investigation to Caucasian women? The answer is, of course, that skin pigmentation is a major determinant of vitamin D synthesis capacity and great differences in pigmentation intensity could obscure the interactions of interest. In another series of studies, for instance, only Hispanic participants were recruited in the US Southwest.

To continue with our storyline, breast cancer risk for women with the *CUBN* (OMIM 602997) variant 233G>T (rs1907362) was 36% higher than the risk of the others [42]. We then remember that cubilin is involved in the uptake of 25-hydroxyvitamin D from the proximal tubules in the kidneys and delivery for conversion to the active 1,25-dihydroxy-vitamin D metabolite [43]. Reduced uptake limits the conversion rate in carriers of the 233T variant. For now, we can only guess how much more dietary vitamin D or UV exposure would be needed to make up for this relative inefficiency. An additional 1000 IU of vitamin D is known to be a safe dose for healthy people and seems a reasonable choice.

Impact of the common *VDR Fok1* variant F>f (rs2228570) is quite variable, which may be due to its extensive interaction with other variants in the *VDR* gene and with multiple nutrients, including vitamin D and calcium. The *Fok1* f variant affects the start codon of *VDR* and its protein product appears to mediate 1,25-dihydroxy-vitamin D action on vitamin D-responsive elements in the DNA less effectively than does F allele-derived protein. This basically means that more vitamin D is needed to achieve its affect [44]. Individuals with one or two f alleles were almost five times as likely to have a disabling mutation in the tumor suppressor gene *P53* (OMIM 191170) in one case-control study of colorectal cancer [45], showing the potential impact of such variation.

5.4.6.3 EICOSANOIDS AND INFLAMMATION

For cancer of the breast and prostate, the proinflammatory eicosanoid products of omega-6 fatty acids, particularly of arachidonic acid, help existing cancer cells to grow more aggressively. The analogous eicosanoids of the omega-3 fatty acid lineage do the opposite and tend to slow growth. These effects are not very powerful, but matter enough to influence long-term outcome. Both types of eicosanoids are produced by prostaglandin-endoperoxide synthases (EC 1.14.99.1). One form, *PTGS1* (also called COX-1, OMIM 176805), is always active. A second one, *PTGS2* (also called COX-2, OMIM 600262), is inducible by inflammation and various growth factors and hormones. This is the one that makes all the difference because higher activity means more rapid cancer cell proliferation. Eating fatty ocean fish (herring, salmon, and the like) once a week may help men with the *PTGS2* variant 6365T>C (rs5275) to reduce their prostate cancer risk by about 70%, but there does not seem to be any benefit to men without the variant [46]. Because about six out of ten men carry the responsive allele, it seems like a good idea for men to have salmon or one of its cousins every now and then.

5.5 DIABETES

5.5.1 Nutrition-related risk of developing type 2 diabetes mellitus

Finding genetic risk factors for the diet-induced (and therefore preventable) development of type 2 diabetes mellitus (DM2) is the Holy Grail (or at least one of several) of nutrigenetic research. As in the Percival legend, the search continues, but some important insights have been gained.

The most consistently replicated genetic variants with a link to DM2 risk are located in the peroxisome proliferator-associated receptor-gamma (*PPARG*; OMIM 601487). The variant encoding an alanine instead of proline in position 12 (Pro12Ala, 49C>G; rs1801282) has drawn particular interest because it is associated with increased insulin resistance and a tendency to deposit fat in adipose tissue (adipogenesis). However, much of the data come from case-control and other cross-sectional studies. Taken together, individuals with the relatively rare (7–8% of Americans have one copy) allele 49G might have about 16% lower DM2 risk than people without it [47]. Individuals with the 49G allele seem to be slightly less sensitive to the effect of overweight and overeating on diabetes risk but the difference is much too small to give them license to indulge.

A meta-analysis with a large assembly of nondiabetic cohorts found, as expected, that high whole-grain consumption was associated with lower fasting glucose and insulin concentrations [48]. A genome-wide search was conducted for genetic variants that were least sensitive to the insulin-sparing effects of whole-grain foods. The one that stood out was a variant (rs780094) in a noncoding region (near a splice site in intron 18) of the gene for the glucokinase regulatory protein (*GCKR*; OMIM 600842). The protein produced by this gene is a regulator of glucose metabolism, which inhibits the key enzyme glucokinase (EC 2.7.1.1; OMIM 138079). The variant appears to be less responsive to fructose-6-phospate, a key intermediate of glycolysis. The extent of the effect is too small to be useful for genotype-specific nutritional management, but shows how nutrigenetic research can pinpoint important interactions. Variants in the *transcription factor 7-like 2* gene (*TCF7L2*; OMIM 602228) also modify the ability of whole-grain consumption to prevent D2M. TCF7L2 is another known regulator of glucose balance, which acts by controlling glucagon gene expression. What is interesting about the *TCF7L2* variants is that they also show that fat intake influences insulin sensitivity and diabetes risk [49].

Of course, obesity is a major risk factor for the development of D2M and numerous studies bear this out. As should be expected, many obesity susceptibility genes are linked to a later D2M risk.

5.5.2 Nutrition-related mortality with type 2 diabetes mellitus

Sirtuin 1 (silent mating type information regulation 2 homolog 1; EC 3.5.1.-), encoded by the gene *SIRT1* (OMIM 604479), belongs to a family of nicotinamide adenine dinucleotide (NAD)-dependent histone deacetylases. What is unusual about this deacetylase is that it splits NAD and attaches the acetate from the acetylated histone to the adenosine diphosphate-ribose part of the NAD. This mechanism links DNA expression of protective regulators directly to the availability of NAD. An important thing to know about the *SIRT1* gene is that its homolog *SIR2* makes round worms and fruit flies live longer and protects mice against cancer, dementia, and heart disease [50]. Now, it may be of interest to biologists or exterminators that pests with better *SIR2* live longer and stay on top of their game until a ripe old age, but what does it mean for humans? It turns out that some people with diabetes might be able to benefit from this research (Figure 5.8).

In one large prospective study [51], D2M patients with the more common forms of three *SIRT1* variants (rs7895833, rs1467568, and rs497849) had a shortened life

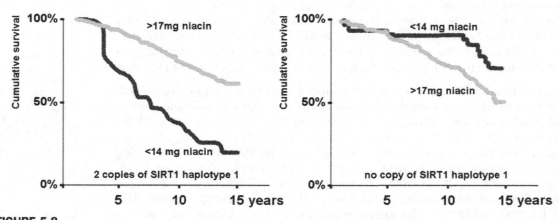

FIGURE 5.8

Patients with type 2 diabetes mellitus with two copies of the common *SIRT1* haplotype 1 (on the left) have a greatly increased long-term risk of death if they have low niacin intake (less than 14 mg/day). Patients without this relatively common genetic predisposition (on the right) do not have this remarkable sensitivity to low niacin intake [51].

span when their niacin intake was in the lowest tertile (less than about 14 mg/day). For participants with two copies of the most common haplotype 1 across the three variants (one out of six participants), the avoidable risk of death with low niacin (less than about 14 mg/day; lowest tertile) intake was more than five times higher than with niacin in the highest tertile of more than 17 mg/day (Figure 5.8). Those with just one copy of this haplotype still had more than twofold excess risk at low intake levels. For carriers of this haplotype, getting enough niacin may be a matter of life and death. It should be pointed out that searching for a niacin effect on mortality would be in vain, just as searching for a genotype or haplotype effect on its own would be. As so often, the remarkably strong benefit of adequate niacin will only become apparent when both factors are taken into account.

5.6 HEART DISEASE

5.6.1 Lipoprotein response to lipid intake

The number of genetic variants that influence the response to dietary fat and sterol intake is extensive and only a few can be addressed here. Several have been described in the previous section. We will consider only variants where knowledge of the genotype will lead to different actions. After all, genetic testing makes little sense if the outcome of an intervention is the same for everyone with the different alleles of a gene variant. Even if somebody with the allele benefits slightly more from an intervention than somebody with the other allele, we don't really need to know about it because we will still give both individuals the same intervention.

5.6.1.1 CHOLESTEROL

Cholesterol intake raises serum cholesterol noticeably in a few people but not in most others (Table 5.3). A large egg contains about 240 mg cholesterol and an

Gene Variant	rs Number	Frequency	Intervention (amount/day)	Estimated Change in Serum Cholesterol	Reference
APOE 4/4	rs429358 + rs7412	0.02	Remove 300 mg cholesterol	−0.46 mmol/L	[52]
APOE 3/4 + 4/4	rs429358 + rs7412	0.12	Replace 9% saturated fat	−0.75 mmol/L	[52]
APOE 3/4 + 4/4	rs429358 + rs7412	0.12	Limit alcohol consumption	−0.45 mmol/L	[55]
MTP −493T/T	rs1800591	0.13	Remove 300 mg cholesterol	−0.56 mmol/L	[53]
FADS1C/T and T/T	rs174546	0.53	Add 1000 mg ALA	−0.2 mmol/L	[56]

Table 5.3 Estimated Potential Changes in Serum Cholesterol Concentration in Response to the Indicated Nutritional Changes

The variant frequencies do not add up to 100%, since some people may carry more than one of the responder variants. ALA, alpha-linolenic acid; rs, reference single nucleotide polymorphism (SNP).

average-sized hot dog (100 g/3.5 oz.) contains 45 mg. Increases in total cholesterol are largely due to increases in low-density lipoprotein (LDL) cholesterol and to a lesser extent in high-density lipoprotein (HDL) cholesterol. The cholesterol-responsive minority appears to consist mainly of individuals with the apolipoprotein E (*APOE*; OMIM 107741) 4/4 genotype [52] and those with the −493T/T genotype for the microsomal triglyceride transfer protein (*MTTP*; OMIM 157147) [53]. Apolipoprotein A-IV (*APOA4*; OMIM 107690) is another important player in lipid metabolism. This component of HDL activates lipoprotein modifiers, including the lipoprotein lipase activator apolipoprotein C-II and the enzyme lecithin-cholesterol acyltransferase (*LCAT*; OMIM 606967). Many people carry an *APOA4* variant (Glu360His; rs5110) that encodes an apolipoprotein A-IV version with histidine instead of glutamine in position 360. Carriers of the apo A-IV 360His variant have a significantly diminished LDL cholesterol response to cholesterol intake [54].

5.6.1.2 POLYUNSATURATED FATS

People with one or two T alleles of the rs174546 FADS1 variant have lower cholesterol concentration with higher consumption of the omega-3 fatty acid ALA. Flaxseed oil is the richest ALA source, with about 1375 mg in half a teaspoon (2.5 mL). Canola oil contains about 1000 mg per teaspoon.

Apolipoprotein A-I, the protein produced by the *APOA1* gene (OMIM 107680), is the structural protein of HDL. Each HDL particle contains exactly one apolipoprotein A-I molecule. Higher PUFA intake might reduce risk by increasing HDL cholesterol levels in people with the *APOA1* variant −75A (rs670). The approximately 30% carriers of the −75A allele can expect an increase in HDL cholesterol of about 15% when they switch from a diet with less than 4% of PUFA from total energy intake to more than 8%. That HDL cholesterol-raising effect will be about twice as large in the one in 30 people with two copies of the *APOA1* −75A allele [57].

Apolipoprotein B (OMIM 107730) is the structural protein of VLDL and its downstream product LDL. Each of these lipoproteins again contains just one apolipoprotein B molecule. Several common variants influence the extent to which PUFA lower LDL cholesterol concentration [58]. However, none of them identifies nonresponders or otherwise adds to current clinical practice. We would give the same nutrition prescription to all, regardless of the presence or absence of a variant.

5.6.1.3 ALCOHOL

The APOE*4 allele predisposes to a significant LDL cholesterol-raising effect of alcohol, which is not observed in people without it [55].

5.6.2 Nutrients that affect nonlipid risk factors

One of the most important contributors to cardiovascular risk is high blood pressure, which is elevated by high sodium intake in salt-sensitive individuals.

The genetic factors involved in response to salt consumption have been discussed in detailed in Chapter 4.

Contrary to long-held assumptions, slightly elevated homocysteine concentrations have only a very small effect on coronary heart disease [59]. This explains why folate supplementation has shown little or no benefit in most studies, not even for people who are homozygous for the *MTHFR* 677T variant. An adequate supply of riboflavin, the rarely mentioned other cofactor of MTHFR, may be more important for preventing hypertension and increased cardiovascular risk of people with the 677T/T genotype [60]. It is always important to remember that health outcomes related to riboflavin and *MTHFR* genotypes tend to differ between populations on each side of the Atlantic because foods are fortified with riboflavin in the USA and Canada but not in European countries.

5.6.3 Deadly Java

Caffeine powers daily life for many people. Many have a few cups of coffee a day, others prefer their favorite caffeinated soda, and some take caffeine shots directly to cope with exhausting work, driving, or study sessions. But what does this do to our health? Many have looked into this question and seemed to get as many answers as there are studies. The concern on the mind of many is that high caffeine intake may increase the risk of heart attacks, possibly through a temporary increase in blood pressure or other acute mechanism. A large long-term prospective cohort study suggested that this should not be a major concern [61]. But there is more happening under the surface than these findings suggest.

People metabolize and eliminate caffeine from their system at very different rates. About half are fast metabolizers with highly active cytochrome P450 1A2 (encoded by *CYP1A2*; OMIM 124060), the enzyme that inactivates caffeine in the liver. The other half has low-activity variants that leave caffeine circulating in blood for many hours. Heritability of the caffeine-metabolizing activity has been estimated to over 0.7 [62].

A recent prospective case-control study took a more probing, nutrigenetic approach than the previous investigations. Nearly all new (incident) heart attacks were captured in several districts in Costa Rica and matched with comparable healthy controls. All participants were asked about their coffee consumption. The study found only a modest link of caffeine consumption with heart attacks, despite the inclusion of more than 2000 cases [63]. Basically, less than three cups of coffee (about 300 mg of caffeine) had no significant impact on heart attack risk, and only with four or more cups was there a modest increase in risk.

And that could have been the end of it, but the investigators wanted to know whether the *CYP1A2* genotype might play a role. What they found was that fast caffeine metabolizers (those with two alleles of the *1A allele) had no hint of a risk even at the highest levels of caffeine consumption (Figure 5.9). However, heart attack risk in the slow metabolizers (those with one or two of the

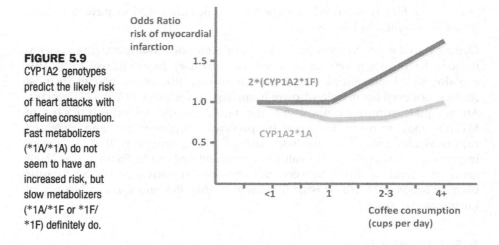

FIGURE 5.9 CYP1A2 genotypes predict the likely risk of heart attacks with caffeine consumption. Fast metabolizers (*1A/*1A) do not seem to have an increased risk, but slow metabolizers (*1A/*1F or *1F/ *1F) definitely do.

low-activity *1F alleles) was increased by 36% with 2—3 cups of coffee per day, and by 64% with more than three cups.

Here we have a striking example where simple genotype-directed guidance is likely to make a big difference. Of course, we could now tell everybody to avoid most caffeine. But acceptance of such a suggestion is often low. Instead, this is what you can tell a middle-aged man with the slow metabolizer genotype: His caffeine addiction increases his risk of a heart attack far more than his cholesterol-lowering statin medication can reduce it! That might get his attention and convince him to switch to decaffeinated coffee after his first cup.

5.6.4 Nitric oxide

Smooth muscle cells in the walls of arteries can change the flow of blood in a particular body region. Cold temperature causes the constriction of small arteries in the extremities, which conserves heat in the body core. Blood flow to other organs and tissues is controlled in a similar fashion to adjust to local needs. A persistently high degree of vascular constriction increases blood pressure, makes the heart work harder, and promotes cardiovascular disease in the long term. This is why it is important to have an agent that reminds vascular smooth muscle cells to relax again after a while. This agent is nitric oxide (NO) produced locally from L-arginine by endothelial nitric oxide synthase (eNOS; EC 1.14.13.39) encoded by the gene *NOS3* (OMIM 163729). The common *NOS3* variant 894G>T (Glu298Asp; rs1799983) changes the responsiveness of the gene to stimulation. Individuals with two copies of the T allele have a substantially (31%) increased heart attack risk [141]. The allele also appears to increase vulnerability to pregnancy-induced hypertension and premature separation of the placenta (abruptio placentae) due to the excessive blood pressure [142].

Flavonoid-rich fruits and vegetables can promote eNOS activity, relax constricted blood vessels, and ease the flow of blood through them. Unfortunately, the beneficial effect of fruits and vegetables seems to be absent in most people with an 894T allele [143]. High flavonoid intake in one group of healthy adults increases the widening of small arteries in response to acetylcholine stimulation in individuals without the 894T allele but not at all in those with the allele. The 894T allele is present in a majority of individuals with Caucasian ancestry, but is less common in people with African or Asian roots.

This example demonstrates once again the importance of adequate fruit and vegetable consumption for human health, but also that not everybody will benefit equally. What is more, fruit and vegetable benefits through other mechanisms will involve different genes and therefore different individual susceptibilities.

5.7 THROMBOEMBOLISM

Picture this. You are finally on your way to your dream vacation to Paris! Your travel starts with an overnight flight from the USA to France. As you are getting off the plane, you feel a stabbing pain between your shoulder blades and in the chest. Then you get increasingly short of breath and your heart starts racing [64]. You are getting weaker by the minute and eventually ask to be taken to the nearest hospital. There they tell you that a blood clot is blocking a major branch of your right lung artery—a life-threatening condition. How did this happen?

Sometimes blood clots form in peripheral veins, most often in the deep veins of the calves. Sitting motionless for a long time during travel makes it easier for clots to form because blood just sits there and does not get moved along by muscle action. The risk is that eventually the blood flow moves the clot toward the heart (right chamber) and from there into a pulmonary artery (Figure 5.10). Because the pulmonary arteries divide into much narrower branches, the clot gets stuck there and blocks a branch. This is called *pulmonary embolism* [65].

The Centers for Disease Control and Prevention (CDC) tell us that every year about one in 20,000 people in their fifties dies because of pulmonary embolism and that this rate rises with increasing age [66]. Mortality rates in people with African ancestry, in obese people, and in women using hormonal contraception are somewhat higher.

Of course, we are interested here in genetic risk factors that interact with nutrition and we will not be disappointed. Prothrombin (*F2*; OMIM 176930) is a precursor of the enzyme that initiates blood clotting by cleaving fibrinogen. A 20210G>A polymorphism (rs1799963) in the 3′ untranslated region (3′ UTR) appears to increase prothrombin expression and promote blood clotting [67]. About 3% of Americans have the homozygous 20210A/A genotype [68]. The variant appears to be less common in some other parts of the world. Results from one prospective cohort (The Women's Genome Health Study) of 22,413 white women followed over a 10-year period confirm that the *F2* 20210A allele increases venous thromboembolism risk more than threefold over carriers of the 20210G/G genotype [69].

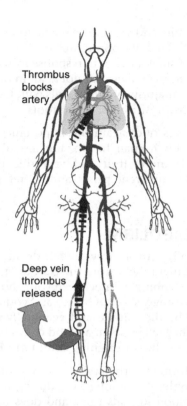

Thrombus
blocks
artery

Deep vein
thrombus
released

FIGURE 5.10
Blood clots can form in deep veins of the legs and then travel through the heart until they encounter the progressively narrower blood vessels in the lungs and block a major segment of the lung's blood supply. Such pulmonary embolism can strike unexpectedly and be fatal within a very short time.

Blood coagulation factor V (*F5*; OMIM 612309) activates prothrombin. About 5% of Americans carry one or two *F5* 1691A alleles (factor V Leiden variant; rs6025), which predisposes them to a more than threefold increased risk of deep vein thrombosis and pulmonary embolism [68] at a rate of one or two per 1000 person years. The previously mentioned prospective cohort study also found that one copy of the F5 Leiden variant (rs6025) increases venous thromboembolism risk more than twofold [69]. The T allele of another nearby variant (rs4524) consistently appears to confer an additional 33% risk for deep vein thrombosis [70].

The good news is that among female carriers of two 20210A alleles those using a moderately dosed vitamin E supplement (about 300 IU/day) had a much lower (−67%) long-term risk of deep vein thrombosis and pulmonary embolism than the ones relying only on the vitamin E from food (around 10 IU/day). Added vitamin E provided little protection for women with the other genotypes [68]. Of course, they already had a much lower risk even without the extra vitamin E. A very similar risk reduction with supplemental vitamin E was seen in women with the *F5* Leiden variant [68].

What does this mean for the prevention of deep vein thrombosis and thromboembolism? As will be discussed in a later chapter, everybody could now decide to start taking a dietary supplement with 200 IU of vitamin E (since no good food sources are available with the requisite amounts). Alternatively, we could rely on genetic testing to decide whether to use a supplement or not. This assumes, of course, that we are sufficiently confident in the strength of the evidence for a genotype-specific benefit.

5.8 ASTHMA AND ALLERGIES

Asthma is an inflammatory disease that constricts the small bronchi and makes exhaling hard. The inflammation is mainly caused by allergic reactions to any of numerous environmental exposures, such as aspirin, dust mites, and toxins from infectious disease [71]. Oxygen free radicals and antioxidants are significant players in the battlefield of asthma-related inflammation. Both exercise and ozone exposure increase the level of free radicals [71]. MPO is responsible for the release of copious amounts of free radicals that activated macrophages direct at bacteria, irritants, and viruses.

The body uses enzymes and defensive antioxidant compounds to protect against the corrosive effect of free radicals. Manganese superoxide dismutase (MnSOD; EC 1.15.1.1) and catalase convert free radicals into less harmful metabolites. Glutathione, which the body constantly has to produce in large quantities, is a major defense against many forms of free radicals and also recharges other antioxidant molecules. The large family of glutathione S-transferases is particularly interesting because some of these enzymes remove toxic lipids that have been peroxidized by free radicals. Last, but not least, the antioxidant defenses are topped off by diverse natural antioxidants in foods. These helpful compounds include carotenoids, flavonoids, vitamin C, and vitamin E. Colorful fruits and vegetables are most likely to be very good sources of such antioxidants. Very high-dosed vitamin C and vitamin E in dietary supplements may actually be counterproductive, because a vitamin C or E molecule becomes a dangerous free radical itself after an encounter with free radicals. When the vitamin concentration is high, there is then more opportunity to infect its neighbor and this can quickly overwhelm the glutathione system [72].

Given that the protective enzymes work together with antioxidants from foods, one should wonder whether genetic variants dictate optimal antioxidant intake for reducing asthma symptoms. Few studies have taken into account such interactions and much more work needs to be done.

Let's explore in more detail just one of these genes, *glutathione S-transferase M1* (*GSTM1*; EC 2.5.1.8; OMIM 138350). GSTM1 stands out in this group of GST enzymes because it stops and cleans up damage inflicted by free radicals on membrane lipids. The enzyme attaches glutathione to peroxidized fatty acids and thereby initiates their metabolism (Figure 5.11). Rapid containment of the damage is critical, because peroxidized lipids are free radicals that keep a chain reaction going. This is a bit like a forest fire in the membranes that just keeps

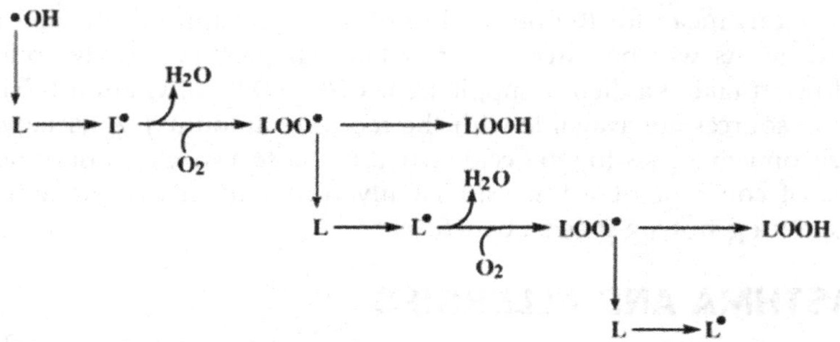

FIGURE 5.11
The unpaired electron of a hydroxy free radical (•) hits a highly unsaturated membrane lipid (L) and starts a peroxidation chain reaction. Follow the course of the unpaired electron as it goes from one lipid to the next, always leaving behind an oxidized lipid (LOO•) that is a free radical itself. GSTM1 can stop the reaction by attaching a glutathione molecule to the damaged lipid. The fat-soluble vitamin E can also stop the chain reaction. Both fat-soluble and water-soluble antioxidants from foods quench free radicals before they can reach the membrane lipids and cause damage.

burning until it is put out. GSTM1 enzyme molecules are the firefighters that stop the fire. Now consider that nearly half of all Americans carry two *GSTM1* null (nonfunctional) alleles. This is like having no firefighters in your district. In this situation, prevention becomes paramount and antioxidants from foods are like regular rain that keeps the ground and the trees moist and less combustible.

Clinical data indicate that this simplified analogy holds some truth. Supplementation with vitamins C (250 mg/day) and E (75 IU/day) reduced asthma symptoms in children with two *GSTM1* null alleles but not significantly in children with fully functioning *GSTM1* alleles [73]. About 40% of the Mexican children in this study did not have a functional *GSTM1* allele. Three lessons can be learned from this study. First, a benefit from generous (but not excessive) antioxidant intake would be overlooked without taking genotype into account. In the National Health and Nutrition Examination Surveys III, vitamin C intake was not a good predictor of asthma symptoms [74], similar to the findings of a few other smaller studies [75]. When it comes to vitamin E, only one out of six studies suggested that a slightly increased intake for all would reduce asthma risk in children; the others found no difference [75]. Second, not everybody may benefit, but it is still a very large group whose condition may be improved with a simple nutritional adjustment. If 40% of all children with asthma got some relief with a little bit of extra vitamin C and vitamin E, their improved quality of life and the cost savings would definitely be worth it. Such nutritional intake adjustments need not rely on dietary supplements because fruits, vegetables, and whole grains can provide the necessary amounts just as well and confer additional benefits for general health. Third, there is still a lack of attention to gene–nutrition interactions. We have to wait for rigorous follow-up tests of this important and practically relevant interaction.

5.9 GOUT

Charles Dickens described gout as a 'patrician demon' that is among the dignities of illustrious bloodlines [76]. This noble bloodline certainly must go back a very long time. Actually it started many millions of years ago for all of us when the ability to convert uric acid to the less problematic metabolite allantoin with the help of uricase (EC 1.7.3.3) increasingly declined in primates [77,78]. Humans do not have a functional *uricase* (*UOX*; OMIM 191540) gene anymore, which means that uric acid is the final metabolite of purine metabolism (Figure 5.12). Unmetabolized uric acid is a potent antioxidant. Some have speculated that our higher uric acid concentration provides better protection against free radicals and that this helped to continuously increase primate life spans to our own extraordinary life expectancy of 60—80 years and more.

Charles Dickens wrote about the debatably noble heritage of gout in his 1853 epic novel, *Bleak House*.

> Sir Leicester receives the gout as a troublesome demon, but still a demon of the patrician order. All the Dedlocks, in the direct male line, through a course of time during and beyond which the memory of man goeth not to the contrary, have had the gout. It can be proved, sir. Other men's fathers may have died of the rheumatism, or may have taken base contagion from the tainted blood of the sick vulgar; but, the Dedlock family have communicated something exclusive, even to the levelling process of dying, by dying of their own family gout. It has come down, through the illustrious line, like the plate, or the pictures, or the place in Lincolnshire. It is among their dignities. Sir Leicester is, perhaps, not wholly without an impression, though he has never resolved it into words, that the angel of death in the discharge of his necessary duties may observe to the shades of the aristocracy, "My lords and gentlemen, I have the honor to present to you another Dedlock certified to have arrived per the family gout."

Reabsorption of uric acid from the proximal convoluted tubule depends on the urate/anion exchanger URAT1 (SLC22A12, OMIM 607096) and the glucose transporter SLC2A9 (OMIM 606142) [79,80]. The adenosine triphosphate (ATP)-dependent transporter *ABCG2* (OMIM 603756) [81] and the sodium-driven transporter *NPT4* (OMIM 611034) [82] actively secrete uric acid into the tubule. However, this system may not be sufficient for high uric acid production due to high purine intake or accelerated purine breakdown (as occurs with high fructose consumption or obesity). If we produce more than we can clear via the kidneys, uric acid concentration in blood increases.

Uric acid concentration in blood and tissues can easily rise to levels that exceed solubility limits and then uric acid crystals form in soft tissues. Uric acid is also

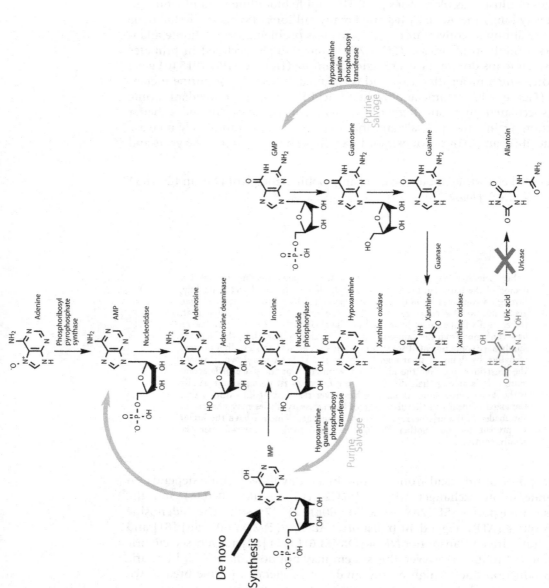

FIGURE 5.12

Uric acid is the main product of the breakdown of purine bases in humans, because their gene for uricase is defective. The intermediate breakdown products hypoxanthine and guanine can be salvaged (arrows) and refurbished as purine bases as long as the hypoxanthine-guanine phosphoribosyl transferase is working efficiently. AMP, adenosine monophosphate.

oxidized by macrophage MPO to 5-hydroxyisourate and other metabolites that act as free radicals [83]. This means that as soon as the uric acid crystals around the joints and tendons trigger an inflammatory response, macrophages come in to create a blast of damaging free radicals. Gout also increases the risk of hypertension, kidney stones and kidney disease. To this day gout can be a terrible burden for 1−2% of men in industrialized countries. It is less common that women are affected because they tend to excrete uric acid more efficiently [84].

Strong genetic predisposition together with diet is responsible for the high uric acid levels in blood (hyperuricemia) that cause the attacks. As Dickens and many writers before him knew quite well, not everybody is equally at risk. The main genetic risk factor is apparently male gender. There are probably several reasons for this (including the ability of women to excrete uric acid more efficiently, as indicated above) and they are not yet sufficiently understood. Hypoxanthine-guanine phosphoribosyltransferase (HPRT; EC 2.4.2.8), a key enzyme for the salvage (recycling) of purines, is located on the X chromosome. Complete disruption of the gene causes Lesch-Nyhan syndrome (OMIM 308000), associated with both gout and mental dysfunction. Variants with 10−20% residual activity still cause severe gout and kidney stone disease [85−88] because with less efficient recycling more purines end up on the uric acid waste pile. Take a moment to study that part of the pathway in Figure 5.12. A large number of *HPRT* gene variants are known but have not been studied in detail. It is quite conceivable that many relatively rare variants are common enough in combination to explain the higher prevalence of gout in men. After all, there are literally thousands of potential mutations that could reduce *HPRT* gene function. Because men are hemizygous for the gene (meaning that they have no backup copy), a loss-of-function variant would always impact their uric acid metabolism and risk of gout.

Common URAT1 and SLC17A3 variants modify serum uric acid concentrations in European and Asian populations. The *ABCG2* variant Q141 (rs2231142) just about doubles gout risk in Caucasians [81,89,90]. The *SLC2A9* variant rs13129697 has nearly as much effect [90]. Asians have another *SLC2A9* variant, Arg265His (rs3733591), which raises uric acid concentrations and increases risk of gout [91]. These are just a few specific examples of the numerous variants in uric acid transporters [92,93] that together determine uric acid excretion.

The existence of such a genetic predisposition is, of course, only one part of the picture. On the other side are the nutritional factors that help individuals with increased risk to avoid the manifestation of this risk. Obesity [94] and excessive consumption of fructose [95], alcohol [94] and animal-derived foods, particularly meat and seafood [96] but not low-fat dairy foods [96], are important factors. Low vitamin C intake is a risk factor [97] and optimizing vitamin C intake significantly above normally recommended amounts reduces risk [98]. It seems likely that responsiveness to high vitamin C intake differs considerably between individuals.

Genetic predisposition and nutritional factors intersect because in most cases neither of them alone would give rise to symptomatic disease on its own.

Regularly having large meat portions, getting a lot of fructose from sweetened beverages, and being obese is a recipe for developing gout if your uric acid excretion capacity is limited. At the same time, all your friends may make exactly the same unhealthy choices and have perfectly normal serum uric acid concentrations and never get a touch of gout. As we noted earlier, life is not fair.

Gout in Maori New Zealanders

High uric acid concentration in blood and symptomatic gout appear to be more common in Polynesians than in other populations [99] and continues to be on the rise in Maoris, the indigenous population of New Zealand [100]. The prevalence in Polynesian and Maori men now exceeds 10%. Many young and physically fit men are affected and can suffer excruciating pain from arthritis during a gout flare up. Gout does not just cause pain for the sufferers but is also associated with cardiovascular

constructed on the basis of the variants rs16890979 (1 = C, 2 = T), rs5028843 (1 = G, 2 = A), rs11942223 (1 = T, 2 = C), and rs12510549 (1 = T, 2 = C). The table below groups haplotypes by their rs11942223 allele component (the third part of the haplotype designation). Note that the SLC2A9 haplotype 1/1/1/1 increases gout risk and the haplotypes 2/2/2/2, 2/2/2/1, 1/2/1/1 and 1/1/2/1 protect against gout risk in both populations.

Haplotype	Odds Ratio Maori (Confidence Interval)	Odds Ratio Caucasian (Confidence Interval)
1/1/1/1	5.05 (2.10–11.5)	2.10 (1.40–3.03)
1/1/1/2	8.75 (1.22–63.0)	0.51 (0.20–1.30)
1/2/1/1	< 0.1 (not determined)	0.61 (0.24–1.59)
2/2/2/2	< 0.1 (not determined)	0.67 (0.44–1.03)
2/2/2/1	0.22 (0.05–0.96)	0.14 (0.03–0.57)
1/1/2/1	< 0.1 (not determined)	< 0.1 (not determined)

Less common haplotypes are not listed.

disease, diabetes, dyslipidemia, hypertension, or renal disease in most of them [101].

A major reason for the high prevalence appears to be less efficient clearance of uric acid in their kidneys [102], probably partly due to shared variation in the SLC2A9 gene. More than half of all people with Maori ancestry carry two copies of the rs11942223 variant, which is most closely related to elevated serum uric acid concentrations and gout. However, this variant is not more common in people with Maori ancestry than in people with Caucasian ancestry. A recent set of case-control studies (healthy people from the same population matched by age and sex to patients with gout) compared the effect of SLC2A9 variants on gout risk in Northeast America (Massachusetts) and New Zealand. Haplotypes were

This does not exclude variants in the numerous other genes involved in uric acid production and clearance from contributing to the high gout risk of Maoris. The impact of multiple genes and haplotypes on presentation and severity of symptomatic gout is population-specific and certainly needs further clarification [103].

In the end, it is most likely that the high purine intake and high obesity-related uric acid production in the population interact with genetic predispositions that are very similar to the ones found in Caucasians. Some in either population are protected from the gout-inducing effects of their excess body fat and overconsumption of hyperuremic foods. Nonetheless, all of them should probably work on their lifestyle and nutrition because gout will not be their only problem.

5.10 THOSE CLINGY LOVE HANDLES

5.10.1 The obesity epidemic

About a third of all Americans are obese today and a larger percentage will be obese tomorrow. We have known for a while that our tendency to gain weight depends to a large degree on inherited factors. The heritability of our body mass index (BMI) appears to be around 65%. However, genetic variants that were identified by scanning the whole genome account for less than 2% of the population-wide weight differences. An important reason is that many genetic variants influence the way we respond to our environment. Just consider how the wave of obesity has swept across the USA in the last two decades. Compare the CDC maps with the obesity rates from 1990 and 2010 (Figure 5.13). None of the US states for which data are available had obesity rates of 15% or higher in the year 1990. Fast forward 20 years, and you see that obesity rates have climbed to over 20% in every single state and are much higher in most states.

> **A REMINDER**
> Heritability is the extent in percent to which a trait is inherited.

The genetic makeup of the American people certainly did not change, but their environment did. This means that we have to look for genetic variants that make it more likely that the carrier responds to an environmental change with weight gain.

Epistasis is another important concept that helps to explain why it is so hard to put our hands on these elusive genetic causes of obesity (and other nutritional conditions). It means quite simply that the effect of a variant at one locus becomes apparent only if the individual also carries a certain variant at another locus. Without that variant in the second locus, the variant at the first locus has

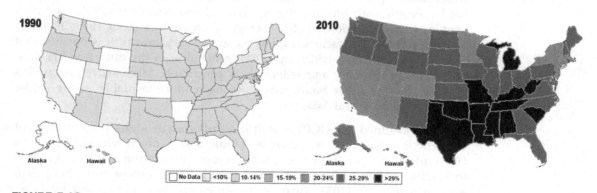

FIGURE 5.13

The percentage of obese Americans rose sharply between 1990 and 2010 in every single state, according to the Centers of Disease Control (http://www.cdc.gov/obesity/data/trends.html). This rapid change conclusively disproves the idea that genetic variants on their own are responsible for today's obesity rates, because they did not change.

no effect. This kind of tangled web is what we are really talking about when we complain about the multifactorial genetics of obesity.

5.10.2 Resting energy expenditure

"I have a slow metabolism" is often the first thing we hear (or say) when it comes to excess body fat. We all know about people who seem to be eating like there is no tomorrow but never get fat. The idea is that they don't have the *thrifty* genes that have helped the ancestors of some of us to make it through lean times. Now we carriers of the thrifty genes have to pay the price for our ancestors' survival and every extra hamburger sticks right to our ribs (and elsewhere). Could this be true?

Before the excitement gets out of hand, let's just say that some people are indeed using calories a little more efficiently than others. But the details are very complex and many of the most intriguing observations have not been adequately replicated. This means that the ground is still shifting and the following outlines may eventually be contradicted by newer data.

First, we should look at a group of mitochondrial anion carrier proteins that make ATP production less efficient and increase heat production. The UCPs basically work like a side channel in a water mill that guides excess water around the mill's blades and prevents it from turning the wheel. In the mitochondria, it is the protons from oxidative phosphorylation that are leaked through UCPs to the cytosol without generating ATP.

Uncoupling protein 1 (*UCP1*; OMIM 113730) in brown adipose tissue helps babies to keep warm without shivering. The uncoupling proteins 2 (*UCP2*; OMIM 601693) and 3 (*UCP3*; OMIM 602044) in skeletal muscle and other tissues are mostly of importance in adults.

A significant portion of resting energy expenditure is spent on unproductive heat generation, just to maintain a healthy body temperature. Greater UCP activity could easily explain why someone expends more energy without exercising any harder. People with African ancestry have a lower resting metabolic rate than people of European descent. The average difference was about 160 kcal/day in one study [104]. The factors responsible are not yet known. It is obvious that close coupling of fuel oxidation to ATP generation in the mitochondria optimizes energy efficiency and reduces heat generation. More active uncoupling proteins, on the other hand, may have been more favorable in the colder climates of Europe and Asia.

It is often assumed that UCP1 is of little significance in adults, but the effects of the genetic variants tell us otherwise (Table 5.4). Women with the common *UCP1* allele −3826G (rs1800592; reduces expression of the gene and is related to insulin resistance) have 12% lower resting energy expenditure (REE) than women without it [105]. Most of the difference is accounted for by lower utilization of fat, which seems unfortunate. Not only does the G allele seem to be a thrifty variant but it may also do its level best to protect accumulated fat in its carrier. This might have helped previous generations to survive starvation

Table 5.4	Genetic Variants in Three Uncoupling Proteins are Often Found to be Related to Energy Expenditure, Obesity, and Insulin Resistance		
Gene Variant	**rs Number**	**Effect**	**Reference**
UCP1 −3826A/G	rs1800592	Lower REE in women with G	[105]
UCP1 −1766A/G	−	More abdominal fat with G	[107]
UCP1 −112A/C	−	Greater insulin resistance with C	[108]
UCP1 Ala64Thr	−	Greater waist–hip ratio with T	[108]
UCP2 −866G/A	rs659366	Greater waist circumference with A/A	[109]
UCP2 −866G/A	rs659366	Greater insulin resistance with A/A	[110]
UCP2 164C/T	rs660339	Greater waist circumference with T/T	[109]
UCP3 −55C/T	rs1800849	Greater fat mass with T	[111]

REE, resting energy expenditure; rs, reference single nucleotide polymorphism (SNP).

times but does not play well in today's environment of persistent abundance. Additional variants in these uncoupling protein genes tell similar stories.

At least two common variants of the *UCP2* gene are associated with higher waist circumference in homozygous carriers [106]. Similarly, the *UCP3* −3826A/G polymorphism has a strong effect on resting metabolism in young Japanese women. Resting energy expenditure is about 14% higher in women with the A/A genotype than in those with the G/G genotype; those with the A/G genotype have intermediate REE [105]. These associations have to be considered with caution, remembering that the results of attempts to link *UCP2* or *UCP3* variants to energy expenditure or obesity measure are not always consistent. Even strong effects on key parameters such as REE do not always translate into weight or body composition differences because multiple regulatory systems can counteract alterations of such metabolic set points. These findings show, however, individual vulnerabilities that may lead to excessive fat accumulation when several nutritional stressors come together.

Variation in the gene encoding the proinflammatory signal protein interleukin 6 (*IL6*; OMIM 147620) also contributes to the differences in energy expenditure [112]. The common variant −174G>C (rs1800795) reduces expression of *IL-6*. Energy expenditure is 7.5% lower in people with two copies of this *IL6* −174C allele than in people with one G allele. We now see more and more that such differences have practical importance. Gastric banding for reducing intake in obese patients was much more successful in one cohort for the approximately 50% of individuals with at least one copy of the low-activity *IL-6* allele [113]. They lost almost 40% more fat mass than patients without the C allele. This difference shows again that it is not just willpower that makes people lose the most weight (or not gain it in the first place). Genetic predisposition is a very powerful force, particularly when it encounters an unsuitable nutritional environment (nutritope) with a plentiful selection of energy-rich foods. Health care providers and other professionals involved in weight management need to understand the specific risks and predispositions of each individual and just work harder with those at greatest risk for failure. Measurement of resting energy expenditure as part of the initial work-up for new patients and clients can help to tailor interventions most effectively and manage expectations. Comprehensive evaluation of variants known to be related to resting energy expenditure may be helpful in situations where direct measurement is not available or is too costly.

5.10.3 Active energy expenditure

All you have to do to lose excess weight is exercise more and eat less, right? As it turns out, genetic variants have a lot to do with the inclination to get off the couch and start shedding some fat. According to a very large European twin study, more than half of the variance in voluntary exercise appears to be inherited [114]. The scope of genetic traits with influence on exercise levels is broad, including personality traits and mood as much as physical factors. A genome-wide association study (GWAS) of leisure exercise in Dutch and Caucasian-American adults pointed to variants in dozens of genes [115]. All effects were relatively modest and only some of these relationships will probably hold up to further scrutiny. The important take-home message is that there are genetic reasons why it is particularly harder for some people to get moving. It will be helpful for them and their healthcare providers to understand these barriers and what it takes to overcome them.

PAPSS2 (OMIM 603005) is one of the genes likely to impact spontaneous leisure activities. The enzyme encoded by this gene produces activated sulfate, which is needed for the synthesis of chondroitin 6'-sulfate and other sulfated molecules in cartilage and other soft tissue. People may be more prone to aches and pains after exercise because they cannot efficiently repair their worn joint cartilage. A *PAPSS2* variant may make them more susceptible to low sulfate intake (mainly in the form of cysteine and methionine) in food. In particular, people with a vegetarian or vegan lifestyle tend to have lower intakes of sulfur-containing amino acids. This shows the kind of conceptual connections that a nutrigenetic perspective can point to.

Another likely candidate is the leptin receptor (*LEPR*; OMIM 601007) in the brain (the hypothalamus to be precise), which responds to the hormone-like peptide leptin (*LEP*; OMIM 164160) from fat cells. It is noteworthy that the leptin system also appears to be involved in regulating voluntary exercise levels.

5.10.4 Appetite regulation

With a nearly unlimited food supply for most people in the Western world, it really comes down to how much we eat. Some of us are more vulnerable to the temptations of overeating than others. We have very complex mechanisms that regulate intake. The appetite control system favors overeating because from an evolutionary perspective there are few downsides to overeating but many to undereating. Most importantly, starvation is a great danger for fertility, successful completion of the pregnancy, and raising a robust and disease-resistant child to reproductive maturity. Should it surprise us then that some people have an innate difficulty in losing weight? This may be an imagined difficulty, but that imagination is at least partially driven by genetic variants in our brains, such as the −759C variant (rs3813929) in the promoter region of the serotonin receptor 5-HT2c (*HTR2C*; OMIM 312861) gene [116].

The health consequences of overeating and obesity that we see today are a very recent phenomenon and have had limited, if any, impact on the sequence of appetite-regulating genes (yet). There is little doubt that our appetite grows whenever we need to meet the challenge of plenty. Yogi Berra must have recognized that deeper truth when he reportedly said to a waitress, "You better cut the pizza in four pieces because I'm not hungry enough to eat six." That said, some resist the excesses better than others. We should assume that genetic variation occurs in many, if not most, of the genes involved in appetite regulation. Only some of this variation has been investigated; the biological relevance of less has been confirmed and almost none of it has informed clinical practice, yet.

5.10.4.1 A GENE NAMED FATSO

The obesity susceptibility gene most consistently linked to fat mass and obesity [117] is called *FTO* (OMIM 610966), which was originally short for fatso because the gene is so large (417,979 bases long). It now uses the more respectable name of *fat and obesity-related* gene. The *FTO* gene encodes an iron-containing dioxygenase that repairs alkylated DNA and ribonucleic acid (RNA). How this activity relates to weight control is unknown. People with the *FTO* variant rs1421085 eat more often per day than people without it [118]. The 10% of patients with an *FTO* G allele (rs16945088) lost 13% less weight (32 vs. 28 kg) than others 2 years after gastric band surgery [119]. This suggests that carriers of this G allele are born with an appetite that is particularly hard to control. The one thing that seems to make a difference is lower saturated fat intake [120].

5.10.4.2 ADIPOSE TISSUE-DERIVED HORMONES

The first appetite-signaling hormone to be discovered, leptin (LEP, OMIM 164160) conveyed a good deal of wishful thing with its name. The name can be loosely translated as *thin-maker*, based on the Classic Greek word for thin (λεπτοσ). The hope was to inject this hormone and then see the pounds melt away. It did not quite work out that way because leptin has so many different functions, which means that a lot of undesirable side effects come with such use. Another likely obesity susceptibility candidate is the leptin receptor (LEPR, OMIM 601007) in the brain (the hypothalamus to be precise), which responds to the hormone-like peptide leptin from fat cells. The amount of leptin released into blood is proportional to the amount of stored fat. Lower blood concentrations signal to the hypothalamus that we need to eat more because we are in danger of losing weight.

A very small number of cases of extreme obesity have been found to be caused by defective *LEP* or *LEPR* genes. However, the contribution of such variants to obesity in the general population is minimal. The leptin system is obviously too important for many vital regulatory functions to permit the rise of common defective variants.

5.10.4.3 HYPOTHALAMUS

A central player in the satiety response in the brain is melanocortin, which binds to the melanocortin 4 receptor (*MC4R*; OMIM 155541). Feeding leads through a series of events to the activation of specialized pro-opiomelanocortin neurons, which release melanocortin. This neurotransmitter binds to MC4R on hypothalamic neurons that depress appetite. This is the satiety axis on which nicotine acts and depresses appetite. Genetic disruption of the *MC4R* gene causes extreme hunger, binge eating, and overeating behavior. Among the common *MC4R* variants, the C allele of rs17782313 (present in nearly half of most populations) stands out because it is consistently related to slightly increased BMI. The difference is small in most studies, about 0.3 BMI points [121]. Nonetheless, the allele-specific effect is apparently important enough to be associated with a 14% excess risk of diabetes.

The second main axis for intake control in the brain relates to signaling that induces appetite. Endocannabinoids are a key component of this appetite-inducing axis. These hormone-like compounds, which are natural products of our fat metabolism such as anandamide, 2-arachidonoylglycerol, and oleamide, act on the cannabinoid receptor 1 (*CNR1*; OMIM 114610) in the hypothalamus and thereby increase appetite. Of course, the names come from the fact that delta-9-tetrahydrocannabinol (THC) and other psychoactive constituents of *Cannabis sativa* mimic the effect of the endocannabinoids. Like the endocannabinoids, THC and related compounds increase appetite and have been shown to stimulate appetite in patients with anorexia due to cancer or other chronic disease [122]. Endocannabinoids are broken down by fatty acid amide hydrolase (*FAAH*; EC 3.5.1.99; OMIM 602935) to ensure that we don't stay hungry forever.

Common variants of *CNR1* have been found to influence how much the carrier eats and drinks [123]. Men with the *CNR1* genotype 1422A/A (rs1049353) tend to have a higher waist-to-hip ratio and waist circumference [124]. Unfortunately, this finding of higher risk for obesity does not readily offer a nutritional solution. Nonetheless, it may be important for health care providers to know whether their patient is prone to weight gain because of a genetic predisposition.

An analogous example is that of the *FAAH* 385C/A polymorphism (Pro129Thr). The 385A allele was previously found to be overrepresented in drug addicts, indicating that this variant actually affects craving pathways [125]. People with a 385A allele have consistently higher BMI, although the difference is only around 0.3 [126]. Nonetheless, such modest differences are likely to add up across the different variants that people carry.

A final actor in this hypothalamic dance of neurons signaling appetite and satiety is the SH2B adapter protein 1 (*SH2B1*; OMIM 608937). The name is as arcane as its precise function is elusive. Deletion of the mouse gene causes insulin and leptin resistance, hyperphagia (overeating), and obesity [127]. People with chromosome deletions that include the *SH2B1* gene also suffer from severe obesity and hyperphagia. We mainly know about the role of *SH2B1* because variants in this gene are consistently related to obesity in GWAS. More studies are needed to clarify how this gene works and what nutritional measures might help to counteract weight gain in people with high-risk variants.

5.10.5 Saturated fat makes some of us fat

The government of Denmark does not want its residents to eat so much saturated fat and has therefore started taxing this tempting commodity at a rate of about US$3 (16 Kroner) per kg. Considering the LDL cholesterol-raising properties of saturated fat, this measure could improve the health of money-conscious consumers, but the effect on body weight is not so simple. Several polymorphic variants predispose a minority to weight gain with high saturated fat intake. Most people without these variants do not seem to gain more weight with saturated fat than with unsaturated fat.

The *APOA2* gene (OMIM 107670) encodes the apolipoprotein A-II lipoprotein (ApoAII), a major constituent of HDL that is mainly produced in the liver and small intestine. About 10—15% of American adults [128] carry the variant −265T>C (rs5082), which is located in the promoter region of the *APOA2* gene and reduces ApoAII protein production. ApoAII appears to promote insulin secretion and play a role in protection against the development of metabolic syndrome [129].

Adults with the ApoA2 −265C/C genotype consume 100—200 kcal/ day more than carriers of the T allele. Low saturated fat intake is associated with about two points lower BMI than with a typical intake (> 22 g/day) in carriers with the ApoA2 −265C/C genotype, but not in those with the other genotypes [127]. These strong relationships were replicated in several studies comprising different ethnicities and geographic regions using different study designs [130].

The *circadian locomotion output cycles kaput* gene (*CLOCK*; OMIM 601851) encodes a protein with surmised transcription factor characteristics. It is the master regulator of the circadian rhythm and alters absorption of fat from the intestine by turning off the microsomal triglyceride transfer protein (*MTP*; OMIM 157147). Both mice and humans with a homozygous defect in the *CLOCK* gene are hyperphagic (overeating) and obese. About 6% of Americans carry two copies of the 3111T>C variant (rs1801260) in the 3′−UTR. The haplotype CGA (rs3749474:rs4580704:rs1801260) is associated with lower than average blood pressure, decreased likelihood of metabolic syndrome, and higher insulin sensitivity [131]. Waist circumference increases with high saturated fat intake (> 12% of total energy) in carriers of the *CLOCK* 3111C allele, but not in noncarriers [131]. The difference in weight loss related to genotype is relatively small: 8 kg for C/C + T/C vs. 10.4 kg for T/T [132].

Finally, variants in the previously discussed *FTO* gene respond to lower consumption of saturated fat [133]. Children with a copy of the A allele of the *FTO* variant rs9939609 are more than twice as likely to be obese as children without this allele, but only if they eat a lot of saturated fat.

It is still unclear how combinations of these and other variants influence individual responses to saturated fat consumption. What we know is that an extra serving of saturated fat is particularly unhelpful for specific and identifiable groups of people.

5.10.6 Carbohydrates

The persistent debate about the right amount of carbohydrates for a healthy lifestyle continues to be a boon for book authors and a matter of confusion for health care providers. Inconsistent findings are part of the reason why it is difficult to come up with a clear recommendation. As is so often found for other nutrients, individual responses prevent a simple answer that fits everybody.

Perilipin (*PLIN*; OMIM 170290) is a protein in fat tissue that coats lipid droplets and prevents their hydrolysis by shielding them from hormone-sensitive lipase. There is a common *PLIN* variant 11482G>A (rs894160) that appears to modify its ability to slow fat mobilization from adipose tissue. Almost one-third of Americans have at least one copy, which is enough to make a difference. Women (but not men) with the variant are more resistant to weight loss than others [134]. This may be at least in part due to the fact that insulin resistance increases when these women restrict their carbohydrate intake. Such low-carbohydrate diets are particularly unfortunate when the intake of saturated fat is high. This is what can happen when there is too much of an emphasis on meat and cheese without the bun.

These findings suggest that women with a *PLIN* 11482A allele should not use a low-carbohydrate diet and should take pains to lower their saturated fat consumption to less than 10% of their total energy budget.

5.11 FATTY LIVER DISEASE

The accumulation of fat in liver cells is about more than just that little bit of extra fat. It is the result of an imbalance between the amount of fatty acids coming into the liver and the amount exported with very-low-density lipoproteins (VLDL). The input side is easy to understand: these are the fatty acids from food and from mobilized adipose tissue fat that are not used by muscle as an energy source. The significance of a high fat intake as a source of excess fatty acids in the blood is obvious.

Let's look at Mr. H. Dumpty, an obese 50-year-old man. His daily consumption of 2700 kcal, with a third of these calories coming from fat, will push about 100 g of fat into his blood. Obesity increases the concentration of fatty acids in his blood because more fat is mobilized from adipose tissue in direct proportion to the amount of fat stored there. We find that Mr. Dumpty weighs 80 kg with a BMI of 35.6. We will assume that his fat mass as a percentage of total body weight is about 25%. This means that his 20 kg of adipose tissue will release at least 1% of its stored fat every day, which adds up to about 200 g fat/day. Eventually, 200 g fat will get stored again in adipose tissue and the remaining 100 g fat will be used as energy fuel, but in the meantime a lot of it has to be taken up by the liver. Exercise helps significantly because it removes fatty acids from the circulation and uses them to fuel muscle activity. Brisk walking and hiking for 4 hours will use about 100 g of fat and lighten the burden of circulating fatty acids.

The output side is a little more complicated. Because the liver has a limited capacity for burning fatty acids, all of the remainder has to be pushed out again as triglycerides and phospholipids packaged into VLDL. Because so much goes through the liver each day, even a small percentage left behind will quickly develop into a large problem. Any genetic variation that changes the capacity for burning fatty acids or that affects packaging into VLDL particles will change the output rate.

The variants most consistently linked to a risk of fatty liver affect the activity of adiponutrin (*PNPLA3*; EC 3.1.1.3; OMIM 609597). PNPLA3 is a transmembrane protein in the liver and in adipose tissue that becomes most active after eating, presumably to help with moving any excess fat around safely. One *PNPLA3* variant, I148M (C>G rs738409), is particularly interesting because it is associated with an increased risk of developing fatty liver, hepatitis, and liver cancer [135]. Of course, we are interested because the *PNPLA3* rs738409 variant determines how well someone with fatty liver responds to nutritional intervention [136]. Patients with two copies of the rs738409 G allele will lose nearly half of their excess liver fat within a week when they follow a diet of 1000 kcal/day with low carbohydrate content. The change in people with two copies of the rs738409 C allele is much smaller. Individuals with the rs738409 G/G genotype may also benefit from maintaining a low ratio of dietary omega-6 to omega-3 fatty acids. A 50% difference in this intake ratio appears to be associated with half as much fat in the liver. No such intake-related effect is found for the other genotypes. Such distinctions provide an important roadmap for the genotype-specific treatment of the growing number of patients with fatty liver.

Another important risk factor for the development of fatty liver is insufficient choline intake. Men with the *MTHFD1* 1958A allele need about twice as much choline from food as men without the allele [12]. This makes them much more vulnerable to developing fatty liver because it is hard to get the extra amount of choline. Would you be able to tell them that a modest extra dose of choline might help them? As mentioned before, eggs are a particularly choline-rich food, with about 120 mg choline in one large egg.

5.12 BONE HEALTH

5.12.1 Strong bones

We all have heard that people from some families are *strong-boned* and from others not so much. Genetic studies actually tell us that there is considerable truth to this notion. About half or more of the variability in bone strength (judging by bone mineral density) is inherited. A GWAS in a New England cohort of men (Framingham Heart Study) found links to several genes related to nutrient metabolism, including *MTHFR* and *VDR* [137], which straightaway suggests a nutrigenetic interaction. But, as we have seen before, nutrigenetic relationships tend to remain hidden until the nutritional component is taken into account.

Gender and age are important determinants of bone structure and mineral content. This means that one should never assume that gene variants have the same impact in men and women. This should not surprise us because estrogens strongly influence many bone-related events.

5.12.2 Calcium

The first nutrient that comes to mind when thinking of bone health is calcium, not least because our bones contain more than 1 kg of calcium and have to maintain that amount to stay strong. Adequate intake is particularly important during puberty and adolescence, when bone is adding large amounts of the mineral. The benefits later in life are more subtle and take long to become noticeable. Very high calcium intakes will usually not prevent osteoporosis or bone fracture better than adequate amounts.

The most commonly consumed calcium-rich foods in America and Europe are milk and dairy products. The problem is that many people avoid dairy foods because they believe that they cannot digest the lactose in such foods and will develop severe abdominal symptoms like bloating, cramps, and diarrhea. People with self-reported lactose intolerance tend to get less calcium than others [138]. Eventually, this lower calcium intake translates into lower bone mineral mass in bones. Men with the normal, nonpersistence lactase genotype −13910C/C have 3% less bone density than men with the lactase-promoting genotype T/T [139]. Another consequence of milk avoidance due to perceived lactose intolerance is a lower intake of vitamin D, since fortified milk is the main source in the USA and Canada (but not in Europe, where vitamin D is not routinely added to milk).

5.12.3 Vitamin D

Vitamin D helps with bone growth, and mineralization, and constant rebuilding in a number of ways. For one, calcium absorption from the small intestines is greatly improved by vitamin D. This means that more than twice as much calcium becomes available with good vitamin status than with poor status. As discussed in Chapter 4, variants of genes involved in vitamin D metabolism and action determine whether a given amount of sunshine or level of dietary intake is sufficient to establish good vitamin D status. Let us just look at the *Fok1* variants (F/f, C>T, Met1Thr; rs2228570) of the *VDR* gene. The f variant encodes a threonine instead of a methionine as the first amino acid of the receptor and makes the mature VDR protein more resistant to the harmful effects of low vitamin D supply. Children with the f/f genotype appear to absorb calcium less well and build bone less actively than children with the F/F genotype [44]. Can it surprise us then that young adults with the f/f genotype have an increased risk of stress fractures [140]? This suggests that it is particularly important for children with the f/f genotype to get enough vitamin D and calcium to achieve optimal bone growth and mineralization.

Practice questions

How can a variant of the *fatty acid amide hydrolase* gene (*FAAH*) increase obesity risk?

Why are genetic factors not responsible for the recent epidemic of obesity in children?

What kind of genetic test can tell a 50-year-old man whether eating soy will protect him against prostate cancer?

What kind of genotype-specific intervention may protect some people from pulmonary embolism?

Who should be made aware that replacing saturated dietary fat with unsaturated fat might help them to avoid some weight gain?

Why is supplementation with modest amounts of vitamin C and E helpful for some people with asthma, but not for others?

How can flax seed oil lower LDL cholesterol better in people with one *FADS1* allele than with the other one?

Why do women with a *UCP1* −3826G allele have lower resting energy expenditure than women without it?

Provide some plausible explanations as to why in the Victorian era members of the British nobility were more likely to have gout than their lowborn neighbors.

In which women is occasional beer consumption able to interfere with fertility?

Why does high fruit and vegetable intake provide some protection against breast cancer in women with the low-activity *MPO* allele −463A but not in those without it?

SUMMARY AND SEGUE TO THE NEXT CHAPTER

Many people may suffer and eventually even die from chronic diseases because their food choices do not match their body's genetic profile. If they had known, they might have done something about it. But nutrition information seems to change all the time, particularly when it comes to nutrition and genes. The next chapter will discuss why we so often get this impression and what you should know before ordering a nutrigenetic test.

References

[1] Zugna D, Richiardi L, Akre O, Stephansson O, Ludvigsson JF. A nationwide population-based study to determine whether coeliac disease is associated with infertility. Gut 2010;59(11): 1471–5.

[2] Zugna D, Richiardi L, Akre O, Stephansson O, Ludvigsson JF. Celiac disease is not a risk factor for infertility in men. Fertil Steril 2011;95(5):1709–13. e1–3.

[3] Rosenfeld CS, Roberts RM. Maternal diet and other factors affecting offspring sex ratio: a review. Biol Reprod 2004;71(4):1063–70.

[4] Villamor E, Sparen P, Cnattingius S. Interpregnancy weight gain and the male-to-female sex ratio of the second pregnancy: a population-based cohort study. Fertil Steril 2008;89(5): 1240–4.

[5] Ahrens K, Yazdy MM, Mitchell AA, Werler MM. Folic acid intake and spina bifida in the era of dietary folic acid fortification. Epidemiology 2011;22(5):731–7.

[6] Shaw GM, Lu W, Zhu H, Yang W, Briggs FB, Carmichael SL, et al. 118 SNPs of folate-related genes and risks of spina bifida and conotruncal heart defects. BMC Med Genet 2009;10:49.

[7] Hustad S, Ueland PM, Vollset SE, Zhang Y, Bjorke-Monsen AL, Schneede J. Riboflavin as a determinant of plasma total homocysteine: effect modification by the methylenetetrahydrofolate reductase C677T polymorphism. Clin Chem 2000;46(8 Pt 1):1065–71.

[8] Moat SJ, Ashfield-Watt PA, Powers HJ, Newcombe RG, McDowell IF. Effect of riboflavin status on the homocysteine-lowering effect of folate in relation to the MTHFR (C677T) genotype. Clin Chem 2003;49(2):295–302.

[9] Finnell RH, Shaw GM, Lammer EJ, Rosenquist TH. Gene-nutrient interactions: importance of folic acid and vitamin B12 during early embryogenesis. Food Nutr Bull 2008; 29(Suppl. 2):S86–98. discussion S99–100.

[10] Zeisel SH. Nutritional genomics: defining the dietary requirement and effects of choline. J Nutr 2011;141(3):531–4.

[11] Shaw GM, Carmichael SL, Yang W, Selvin S, Schaffer DM. Periconceptional dietary intake of choline and betaine and neural tube defects in offspring. Am J Epidemiol 2004; 160(2):102–9.

[12] Kohlmeier M, da Costa KA, Fischer LM, Zeisel SH. Genetic variation of folate-mediated one-carbon transfer pathway predicts susceptibility to choline deficiency in humans. Proc Natl Acad Sci U S A 2005;102(44):16025–30.

[13] Parle-McDermott A, Kirke PN, Mills JL, Molloy AM, Cox C, O'Leary VB, et al. Confirmation of the R653Q polymorphism of the trifunctional C1-synthase enzyme as a maternal risk for neural tube defects in the Irish population. Eur J Hum Genet 2006;14(6):768–72.

[14] Johnson WG, Scholl TO, Spychala JR, Buyske S, Stenroos ES, Chen X. Common dihydrofolate reductase 19-base pair deletion allele: a novel risk factor for preterm delivery. Am J Clin Nutr 2005;81(3):664–8.

[15] Siega-Riz AM, Promislow JH, Savitz DA, Thorp Jr JM, McDonald T. Vitamin C intake and the risk of preterm delivery. Am J Obstet Gynecol 2003;189(2):519–25.

[16] Erichsen HC, Engel SA, ECk PK, Welch R, Yeager M, Levine M, et al. Genetic variation in the sodium-dependent vitamin C transporters, SLC23A1, and SLC23A2 and risk for preterm delivery. Am J Epidemiol 2006;163(3):245–54.

[17] Innis SM, King DJ. trans Fatty acids in human milk are inversely associated with concentrations of essential all-cis n-6 and n-3 fatty acids and determine trans, but not n-6 and n-3, fatty acids in plasma lipids of breast-fed infants. Am J Clin Nutr 1999;70(3):383–90.

[18] Guesnet P, Alessandri JM. Docosahexaenoic acid (DHA) and the developing central nervous system (CNS)—Implications for dietary recommendations. Biochimie 2011;93(1):7–12.

[19] Cheatham CL, Nerhammer AS, Asserhoj M, Michaelsen KF, Lauritzen L. Fish oil supplementation during lactation: effects on cognition and behavior at 7 years of age. Lipids 2011;46(7):637–45.

[20] Xie L, Innis SM. Genetic variants of the FADS1 FADS2 gene cluster are associated with altered (n-6) and (n-3) essential fatty acids in plasma and erythrocyte phospholipids in women during pregnancy and in breast milk during lactation. J Nutr 2008;138(11):2222–8.

[21] Luka Z, Moss F, Loukachevitch LV, Bornhop DJ, Wagner C. Histone demethylase LSD1 is a folate-binding protein. Biochemistry 2011;50(21):4750–6.

[22] Zeisel SH. Dietary choline deficiency causes DNA strand breaks and alters epigenetic marks on DNA and histones. Mutat Res 2012;733(1–2):34–8.

[23] Xu X, Gammon MD, Wetmur JG, Rao M, Gaudet MM, Teitelbaum SL, et al. A functional 19-base pair deletion polymorphism of dihydrofolate reductase (DHFR) and risk of breast cancer in multivitamin users. Am J Clin Nutr 2007;85(4):1098–102.

[24] Maruti SS, Ulrich CM, Jupe ER, White E. MTHFR C677T and postmenopausal breast cancer risk by intakes of one-carbon metabolism nutrients: a nested case-control study. Breast Cancer Research: BCR 2009;11(6):R91.

[25] Haslacher H, Perkmann T, Gruenewald J, Exner M, Endler G, Scheichenberger V, et al. Plasma myeloperoxidase level and peripheral arterial disease. Eur J Clin Invest 2012;42(5):463–9.

[26] Ahn J, Gammon MD, Santella RM, Gaudet MM, Britton JA, Teitelbaum SL, et al. Myeloperoxidase genotype, fruit and vegetable consumption, and breast cancer risk. Cancer Res 2004;64(20):7634–9.

[27] Schallreuter KU, Salem MM, Hasse S, Rokos H. The redox—biochemistry of human hair pigmentation. Pigment Cell & Melanoma Research 2011;24(1):51–62.

[28] Nadif R, Mintz M, Jedlicka A, Bertrand JP, Kleeberger SR, Kauffmann F. Association of CAT polymorphisms with catalase activity and exposure to environmental oxidative stimuli. Free Radical Research 2005;39(12):1345–50.

[29] Ahn J, Gammon MD, Santella RM, Gaudet MM, Britton JA, Teitelbaum SL, et al. Associations between breast cancer risk and the catalase genotype, fruit and vegetable consumption, and supplement use. Am J Epidemiol 2005;162(10):943–52.

[30] Saracino MR, Lampe JW. Phytochemical regulation of UDP-glucuronosyltransferases: implications for cancer prevention. Nutrition and Cancer 2007;59(2):121–41.

[31] Egeberg R, Olsen A, Autrup H, Christensen J, Stripp C, Tetens I, et al. Meat consumption, N-acetyl transferase 1 and 2 polymorphism and risk of breast cancer in Danish postmenopausal women. Eur J Cancer Prev 2008;17(1):39–47.

[32] Lilla C, Verla-Tebit E, Risch A, Jager B, Hoffmeister M, Brenner H, et al. Effect of NAT1 and NAT2 genetic polymorphisms on colorectal cancer risk associated with exposure to tobacco smoke and meat consumption. Cancer Epidemiol Biomarkers Prev 2006;15(1):99–107.

[33] Hiljadnikova Bajro M, Josifovski T, Panovski M, Jankulovski N, Kapedanovska Nestorovska A, Matevska N, et al. Promoter length polymorphism in UGT1A1 and the risk of sporadic colorectal cancer. Cancer Genetics 2012;205(4):163–7.

[34] Chang JL, Bigler J, Schwarz Y, Li SS, Li L, King IB, et al. UGT1A1 polymorphism is associated with serum bilirubin concentrations in a randomized, controlled, fruit and vegetable feeding trial. J Nutr 2007;137(4):890–7.

[35] Gervasini G, San Jose C, Carrillo JA, Benitez J, Cabanillas A. GST polymorphisms interact with dietary factors to modulate lung cancer risk: study in a high-incidence area. Nutrition and Cancer 2010;62(6):750–8.

[36] Shen J, Gammon MD, Terry MB, Wang L, Wang Q, Zhang F, et al. Polymorphisms in XRCC1 modify the association between polycyclic aromatic hydrocarbon-DNA adducts, cigarette smoking, dietary antioxidants, and breast cancer risk. Cancer Epidemiol Biomarkers Prev 2005;14(2):336–42.

[37] Goodman M, Bostick RM, Ward KC, Terry PD, van Gils CH, Taylor JA, et al. Lycopene intake and prostate cancer risk: effect modification by plasma antioxidants and the XRCC1 genotype. Nutrition Cancer 2006;55(1):13—20.

[38] Shen J, Terry MB, Gammon MD, Gaudet MM, Teitelbaum SL, Eng SM, et al. MGMT genotype modulates the associations between cigarette smoking, dietary antioxidants and breast cancer risk. Carcinogenesis 2005;26(12):2131—7.

[39] Eylert MF, Persad R. Management of prostate cancer. Br J Hosp Med (Lond) 2012;73(2): 95—9.

[40] Hedelin M, Balter KA, Chang ET, Bellocco R, Klint A, Johansson JE, et al. Dietary intake of phytoestrogens, estrogen receptor-beta polymorphisms and the risk of prostate cancer. Prostate 2006;66(14):1512—20.

[41] Garland FC, Garland CF, Gorham ED, Young JF. Geographic variation in breast cancer mortality in the United States: a hypothesis involving exposure to solar radiation. Preventive Medicine 1990;19(6):614—22.

[42] Anderson LN, Cotterchio M, Cole DE, Knight JA. Vitamin D-related genetic variants, interactions with vitamin D exposure, and breast cancer risk among Caucasian women in Ontario. Cancer Epidemiol Biomarkers Prev 2011;20(8):1708—17.

[43] Kaseda R, Hosojima M, Sato H, Saito A. Role of megalin and cubilin in the metabolism of vitamin D(3). Therapeutic Apheresis Dialysis 2011;15(Suppl. 1):14—17.

[44] Abrams SA, Griffin IJ, Hawthorne KM, Chen Z, Gunn SK, Wilde M, et al. Vitamin D receptor Fok1 polymorphisms affect calcium absorption, kinetics, and bone mineralization rates during puberty. Journal Bone Mineral Research 2005;20(6):945—53.

[45] Slattery ML, Wolff RK, Herrick JS, Caan BJ, Samowitz W. Calcium, vitamin D, VDR genotypes, and epigenetic and genetic changes in rectal tumors. Nutrition Cancer 2010;62(4): 436—42.

[46] Hedelin M, Chang ET, Wiklund F, Bellocco R, Klint A, Adolfsson J, et al. Association of frequent consumption of fatty fish with prostate cancer risk is modified by COX-2 polymorphism. Int J Cancer 2007;120(2):398—405.

[47] Gouda HN, Sagoo GS, Harding AH, Yates J, Sandhu MS, Higgins JP. The association between the peroxisome proliferator-activated receptor-gamma2 (PPARG2) Pro12Ala gene variant and type 2 diabetes mellitus: a HuGE review and meta-analysis. Am J Epidemiol 2010;171(6): 645—55.

[48] Nettleton JA, McKeown NM, Kanoni S, Lemaitre RN, Hivert MF, Ngwa J, et al. Interactions of dietary whole-grain intake with fasting glucose- and insulin-related genetic loci in individuals of European descent: a meta-analysis of 14 cohort studies. Diabetes Care 2010;33(12): 2684—91.

[49] Ruchat SM, Elks CE, Loos RJ, Vohl MC, Weisnagel SJ, Rankinen T, et al. Evidence of interaction between type 2 diabetes susceptibility genes and dietary fat intake for adiposity and glucose homeostasis-related phenotypes. Journal of Nutrigenetics Nutrigenomics 2009; 2(4—5):225—34.

[50] Herranz D, Serrano M. SIRT1: recent lessons from mouse models. Nat Rev Cancer 2010; 10(12):819—23.

[51] Zillikens MC, van Meurs JB, Sijbrands EJ, Rivadeneira F, Dehghan A, van Leeuwen JP, et al. SIRT1 genetic variation and mortality in type 2 diabetes: interaction with smoking and dietary niacin. Free Radic Biol Med 2009;46(6):836—41.

[52] Sarkkinen E, Korhonen M, Erkkila A, Ebeling T, Uusitupa M. Effect of apolipoprotein E polymorphism on serum lipid response to the separate modification of dietary fat and dietary cholesterol. Am J Clin Nutr 1998;68(6):1215—22.

[53] Wolff E, Vergnes MF, Defoort C, Planells R, Portugal H, Nicolay A, et al. Cholesterol absorption status and fasting plasma cholesterol are modulated by the microsomal triacylglycerol transfer protein -493 G/T polymorphism and the usual diet in women. Genes Nutr 2011;6(1):71—9.

[54] Ordovas JM, Lopez-Miranda J, Mata P, Perez-Jimenez F, Lichtenstein AH, Schaefer EJ. Gene-diet interaction in determining plasma lipid response to dietary intervention. Atherosclerosis 1995;118(Suppl.):S11—27.

[55] Ordovas JM. Gene-diet interaction and plasma lipid responses to dietary intervention. Biochemical Society Transactions 2002;30(2):68–73.

[56] Dumont J, Huybrechts I, Spinneker A, Gottrand F, Grammatikaki E, Bevilacqua N, et al. FADS1 genetic variability interacts with dietary alpha-linolenic acid intake to affect serum non-HDL-cholesterol concentrations in European adolescents. J Nutr 2011;141(7):1247–53.

[57] Ordovas JM, Corella D, Cupples LA, Demissie S, Kelleher A, Coltell O, et al. Polyunsaturated fatty acids modulate the effects of the APOA1 G-A polymorphism on HDL-cholesterol concentrations in a sex-specific manner: the Framingham Study. Am J Clin Nutr 2002;75(1):38–46.

[58] Rantala M, Rantala TT, Savolainen MJ, Friedlander Y, Kesaniemi YA. Apolipoprotein B gene polymorphisms and serum lipids: meta-analysis of the role of genetic variation in responsiveness to diet. Am J Clin Nutr 2000;71(3):713–24.

[59] Clarke R, Bennett DA, Parish S, Verhoef P, Dotsch-Klerk M, Lathrop M, et al. Homocysteine and coronary heart disease: meta-analysis of MTHFR case-control studies, avoiding publication bias. PLoS Med 2012;9(2):e1001177.

[60] Wilson CP, Ward M, McNulty H, Strain JJ, Trouton TG, Horigan G, et al. Riboflavin offers a targeted strategy for managing hypertension in patients with the MTHFR 677TT genotype: a 4-y follow-up. Am J Clin Nutr 2012;95(3):766–72.

[61] Sofi F, Conti AA, Gori AM, Eliana Luisi ML, Casini A, Abbate R, et al. Coffee consumption and risk of coronary heart disease: a meta-analysis. Nutrition, metabolism, and cardiovascular diseases: NMCD 2007;17(3):209–23.

[62] Rasmussen BB, Brix TH, Kyvik KO, Brosen K. The interindividual differences in the 3-demthylation of caffeine alias CYP1A2 is determined by both genetic and environmental factors. Pharmacogenetics 2002;12(6):473–8.

[63] Cornelis MC, El-Sohemy A, Kabagambe EK, Campos H. Coffee, CYP1A2 genotype, and risk of myocardial infarction. JAMA 2006;295(10):1135–41.

[64] Lapostolle F, Surget V, Borron SW, Desmaizieres M, Sordelet D, Lapandry C, et al. Severe pulmonary embolism associated with air travel. N Engl J Med 2001;345(11):779–83.

[65] Bartholomew JR, Schaffer JL, McCormick GF. Air travel and venous thromboembolism: minimizing the risk. Cleve Clin J Med 2011;78(2):111–20.

[66] Lilienfeld DE. Decreasing mortality from pulmonary embolism in the United States, 1979–1996. Int J Epidemiol 2000;29(3):465–9.

[67] Ceelie H, Spaargaren-van Riel CC, Bertina RM, Vos HL. G20210A is a functional mutation in the prothrombin gene; effect on protein levels and 3′-end formation. J Thromb Haemost 2004;2(1):119–27.

[68] Glynn RJ, Ridker PM, Goldhaber SZ, Zee RY, Buring JE. Effects of random allocation to vitamin E supplementation on the occurrence of venous thromboembolism: report from the Women's Health Study. Circulation 2007;116(13):1497–503.

[69] Zee RY, Glynn RJ, Cheng S, Steiner L, Rose L, Ridker PM. An evaluation of candidate genes of inflammation and thrombosis in relation to the risk of venous thromboembolism: The Women's Genome Health Study. Circ Cardiovasc Genet 2009;2(1):57–62.

[70] Bezemer ID, Bare LA, Arellano AR, Reitsma PH, Rosendaal FR. Updated analysis of gene variants associated with deep vein thrombosis. JAMA 2010;303(5):421–2.

[71] Peden DB. The role of oxidative stress and innate immunity in O(3) and endotoxin-induced human allergic airway disease. Immunol Rev 2011;242(1):91–105.

[72] Versari D, Daghini E, Rodriguez-Porcel M, Sattler K, Galili O, Pilarczyk K, et al. Chronic antioxidant supplementation impairs coronary endothelial function and myocardial perfusion in normal pigs. Hypertension 2006;47(3):475–81.

[73] Romieu I, Sienra-Monge JJ, Ramirez-Aguilar M, Moreno-Macias H, Reyes-Ruiz NI, Estela del Rio-Navarro B, et al. Genetic polymorphism of GSTM1 and antioxidant supplementation influence lung function in relation to ozone exposure in asthmatic children in Mexico City. Thorax 2004;59(1):8–10.

[74] Romieu I, Mannino DM, Redd SC, McGeehin MA. Dietary intake, physical activity, body mass index, and childhood asthma in the Third National Health And Nutrition Survey (NHANES III). Pediatr Pulmonol 2004;38(1):31–42.

[75] Nurmatov U, Devereux G, Sheikh A. Nutrients and foods for the primary prevention of asthma and allergy: systematic review and meta-analysis. J Allergy Clin Immunol 2011;127(3):724—33.

[76] Dickens C. Bleak House, Chapter XVI, Tom-All-Alone's. London: Bradbury & Evans. Transcript of the UNC copy can be found at, http://www.ibiblio.org/dickens/html/42049.html; 1852. p.155.

[77] Oda M, Satta Y, Takenaka O, Takahata N. Loss of urate oxidase activity in hominoids and its evolutionary implications. Mol Biol Evol 2002;19(5):640—53.

[78] Johnson RJ, Andrews P, Benner SA, Oliver W, Theodore E. Woodward award. The evolution of obesity: insights from the mid-Miocene. Trans Am Clin Climatol Assoc 2010; 121:295—305. discussion 308.

[79] Le MT, Shafiu M, Mu W, Johnson RJ. SLC2A9—a fructose transporter identified as a novel uric acid transporter. Nephrol Dial Transplant 2008;23(9):2746—9.

[80] Cheeseman C. Solute carrier family 2, member 9 and uric acid homeostasis. Curr Opin Nephrol Hypertens 2009;18(5):428—32.

[81] Woodward OM, Kottgen A, Coresh J, Boerwinkle E, Guggino WB, Kottgen M. Identification of a urate transporter, ABCG2, with a common functional polymorphism causing gout. Proc Natl Acad Sci U S A 2009;106(25):10338—42.

[82] Jutabha P, Anzai N, Kitamura K, Taniguchi A, Kaneko S, Yan K, et al. Human sodium phosphate transporter 4 (hNPT4/SLC17A3) as a common renal secretory pathway for drugs and urate. J Biol Chem 2010;285(45):35123—32.

[83] Meotti FC, Jameson GN, Turner R, Harwood DT, Stockwell S, Rees MD, et al. Urate as a physiological substrate for myeloperoxidase: implications for hyperuricemia and inflammation. J Biol Chem 2011;286(15):12901—11.

[84] Puig JG, Michan AD, Jimenez ML, Perez de Ayala C, Mateos FA, Capitan CF, et al. Female gout. Clinical spectrum and uric acid metabolism. Arch Intern Med 1991;151(4):726—32.

[85] Fujimori S, Hidaka Y, Davidson BL, Palella TD, Kelley WN. Identification of a single nucleotide change in a mutant gene for hypoxanthine-guanine phosphoribosyltransferase (HPRT Ann Arbor). Hum Genet 1988;79(1):39—43.

[86] Davidson BL, Pashmforoush M, Kelley WN, Palella TD. Human hypoxanthine-guanine phosphoribosyltransferase deficiency. The molecular defect in a patient with gout (HPRTAshville). J Biol Chem 1989;264(1):520—5.

[87] Nguyen KV, Naviaux RK, Paik KK, Nyhan WL. Novel Mutations in the Human HPRT Gene. Nucleosides Nucleotides Nucleic Acids 2011;30(6):440—5.

[88] Ea HK, Bardin T, Jinnah HA, Aral B, Liote F, Ceballos-Picot I. Severe gouty arthritis and mild neurologic symptoms due to F199C, a newly identified variant of the hypoxanthine guanine phosphoribosyltransferase. Arthritis Rheum 2009;60(7):2201—4.

[89] Phipps-Green AJ, Hollis-Moffatt JE, Dalbeth N, Merriman ME, Topless R, Gow PJ, et al. A strong role for the ABCG2 gene in susceptibility to gout in New Zealand Pacific Island and Caucasian, but not Maori, case and control sample sets. Hum Mol Genet 2010;19(24):4813—9.

[90] Yang Q, Kottgen A, Dehghan A, Smith AV, Glazer NL, Chen MH, et al. Multiple genetic loci influence serum urate levels and their relationship with gout and cardiovascular disease risk factors. Circ Cardiovasc Genet 2010;3(6):523—30.

[91] Tu HP, Chen CJ, Tovosia S, Ko AM, Lee CH, Ou TT, et al. Associations of a non-synonymous variant in SLC2A9 with gouty arthritis and uric acid levels in Han Chinese subjects and Solomon Islanders. Ann Rheum Dis 2010;69(5):887—90.

[92] Stark K, Reinhard W, Grassl M, Erdmann J, Schunkert H, Illig T, et al. Common polymorphisms influencing serum uric acid levels contribute to susceptibility to gout, but not to coronary artery disease. PLoS One 2009;4(11):e7729.

[93] Kolz M, Johnson T, Sanna S, Teumer A, Vitart V, Perola M, et al. Meta-analysis of 28,141 individuals identifies common variants within five new loci that influence uric acid concentrations. PLoS Genet 2009;5(6):e1000504.

[94] Cea Soriano L, Rothenbacher D, Choi HK, Garcia Rodriguez LA. Contemporary epidemiology of gout in the UK general population. Arthritis Res Ther 2011;13(2):R39.

[95] Choi HK, Willett W, Curhan G. Fructose-rich beverages and risk of gout in women. JAMA 2010;304(20):2270—8.

[96] Choi HK, Atkinson K, Karlson EW, Willett W, Curhan G. Purine-rich foods, dairy and protein intake, and the risk of gout in men. N Engl J Med 2004;350(11):1093—103.

[97] Choi HK, Gao X, Curhan G. Vitamin C intake and the risk of gout in men: a prospective study. Arch Intern Med 2009;169(5):502—7.

[98] Juraschek SP, Miller 3rd ER, Gelber AC. Effect of oral vitamin C supplementation on serum uric acid: A meta-analysis of randomized controlled trials. Arthritis Care Res (Hoboken) 2011 Sep;63(9):1295—306.

[99] Rose BS. Gout in Maoris. Semin Arthritis Rheum 1975;5(2):121—45.

[100] Klemp P, Stansfield SA, Castle B, Robertson MC. Gout is on the increase in New Zealand. Ann Rheum Dis 1997;56(1):22—6.

[101] Hollis-Moffatt JE, Xu X, Dalbeth N, Merriman ME, Topless R, Waddell C, et al. Role of the urate transporter SLC2A9 gene in susceptibility to gout in New Zealand Maori, Pacific Island, and Caucasian case-control sample sets. Arthritis Rheum 2009; 60(11):3485—92.

[102] Simmonds HA, McBride MB, Hatfield PJ, Graham R, McCaskey J, Jackson M. Polynesian women are also at risk for hyperuricaemia and gout because of a genetic defect in renal urate handling. Br J Rheumatol 1994;33(10):932—7.

[103] Hollis-Moffatt JE, Gow PJ, Harrison AA, Highton J, Jones PB, Stamp LK, et al. The SLC2A9 nonsynonymous Arg265His variant and gout: evidence for a population-specific effect on severity. Arthritis Res Ther 2011;13(3):R85.

[104] Manini TM, Patel KV, Bauer DC, Ziv E, Schoeller DA, Mackey DC, et al. European ancestry and resting metabolic rate in older African Americans. Eur J Clin Nutr 2011;65(6):663—7.

[105] Nagai N, Sakane N, Tsuzaki K, Moritani T. UCP1 genetic polymorphism (-3826 A/G) diminishes resting energy expenditure and thermoregulatory sympathetic nervous system activity in young females. Int J Obes (Lond) 2011;35(8):1050—5.

[106] Martinez-Hervas S, Mansego ML, de Marco G, Martinez F, Alonso MP, Morcillo S, et al. Polymorphisms of the UCP2 gene are associated with body fat distribution and risk of abdominal obesity in Spanish population. Eur J Clin Invest 2012;42(2):171—8.

[107] Jia JJ, Tian YB, Cao ZH, Tao LL, Zhang X, Gao SZ, et al. The polymorphisms of UCP1 genes associated with fat metabolism, obesity and diabetes. Molecular Biology Reports 2010;37(3):1513—22.

[108] Herrmann SM, Wang JG, Staessen JA, Kertmen E, Schmidt-Petersen K, Zidek W, et al. Uncoupling protein 1 and 3 polymorphisms are associated with waist-to-hip ratio. J Mol Med (Berl) 2003;81(5):327—32.

[109] Martinez-Hervas S, Mansego ML, de Marco G, Martinez F, Alonso MP, Morcillo S, et al. Polymorphisms of the UCP2 gene are associated with body fat distribution and risk of abdominal obesity in Spanish population. Eur J Clin Invest 2012;42(2):171—8.

[110] Dalgaard LT. Genetic Variance in Uncoupling Protein 2 in Relation to Obesity, Type 2 Diabetes, and Related Metabolic Traits: Focus on the Functional -866G>A Promoter Variant (rs659366). Journal of Obesity 2011;2011:340241.

[111] de Luis DA, Aller R, Izaola O. Gonzalez Sagrado M, Conde R. Association of -55ct Polymorphism of Ucp3 Gene with Fat Distribution, Cardiovascular Risk Factors and Adipocytokines in Patients with Type 2 Diabetes Mellitus. Journal Endocrinological Investigation 2012;35(7):625—8.

[112] Kubaszek A, Pihlajamaki J, Punnonen K, Karhapaa P, Vauhkonen I, Laakso M. The C-174G promoter polymorphism of the IL-6 gene affects energy expenditure and insulin sensitivity. Diabetes 2003;52(2):558—61.

[113] Di Renzo L, Carbonelli MG, Bianchi A, Iacopino L, Fiorito R, Di Daniele N, et al. Body composition changes after laparoscopic adjustable gastric banding: what is the role of -174G>C interleukin-6 promoter gene polymorphism in the therapeutic strategy? Int J Obes (Lond) 2012;36(3):369—78.

[114] Vink JM, Boomsma DI, Medland SE, de Moor MH, Stubbe JH, Cornes BK, et al. Variance components models for physical activity with age as modifier: a comparative twin study in seven countries. Twin Res Hum Genet 2011;14(1):25—34.

[115] De Moor MH, Liu YJ, Boomsma DI, Li J, Hamilton JJ, Hottenga JJ, et al. Genome-wide association study of exercise behavior in Dutch and American adults. Med Sci Sports Exerc 2009;41(10):1887—95.

[116] Pooley EC, Fairburn CG, Cooper Z, Sodhi MS, Cowen PJ, Harrison PJ. A 5-HT2C receptor promoter polymorphism (HTR2C—759C/T) is associated with obesity in women, and with resistance to weight loss in heterozygotes. American journal of medical genetics Part B, Neuropsychiatric Genetics: the official publication of the International Society of Psychiatric Genetics 2004;126B(1):124—7.

[117] Phillips CM, Kesse-Guyot E, McManus R, Hercberg S, Lairon D, Planells R, et al. High dietary saturated fat intake accentuates obesity risk associated with the fat mass and obesity-associated gene in adults. J Nutr 2012;142(5):824—31.

[118] McCaffery JM, Papandonatos GD, Peter I, Huggins GS, Raynor HA, Delahanty LM, et al. Obesity susceptibility loci and dietary intake in the Look AHEAD Trial. Am J Clin Nutr 2012 Jun;95(6):1477—86.

[119] Sarzynski MA, Jacobson P, Rankinen T, Carlsson B, Sjostrom L, Bouchard C, et al. Associations of markers in 11 obesity candidate genes with maximal weight loss and weight regain in the SOS bariatric surgery cases. Int J Obes (Lond) 2011;35(5):676—83.

[120] Moleres A, Ochoa MC, Rendo-Urteaga T, Martinez-Gonzalez MA, Azcona San Julian MC, Martinez JA, et al. Dietary fatty acid distribution modifies obesity risk linked to the rs9939609 polymorphism of the fat mass and obesity-associated gene in a Spanish case-control study of children. Br J Nutr 2012;107(4):533—8.

[121] Qi L, Kraft P, Hunter DJ, Hu FB. The common obesity variant near MC4R gene is associated with higher intakes of total energy and dietary fat, weight change and diabetes risk in women. Hum Mol Genet 2008;17(22):3502—8.

[122] Bedi G, Foltin RW, Gunderson EW, Rabkin J, Hart CL, Comer SD, et al. Efficacy and tolerability of high-dose dronabinol maintenance in HIV-positive marijuana smokers: a controlled laboratory study. Psychopharmacology (Berl) 2010;212(4):675—86.

[123] Bienertova-Vasku J, Bienert P, Slovackova L, Sabilkova L, Piskackova Z, Forejt M, Splichal Z, Zlamal F, Vasku A. Variability in CNR1 locus influences protein intake and smoking status in the Central-European population. Nutritional Neuroscience 2012;15(4): 163—70.

[124] Peeters A, Beckers S, Mertens I, Van Hul W, Van Gaal L. The G1422A variant of the cannabinoid receptor gene (CNR1) is associated with abdominal adiposity in obese men. Endocrine 2007;31(2):138—41.

[125] Flanagan JM, Gerber AL, Cadet JL, Beutler E, Sipe JC. The fatty acid amide hydrolase 385 A/A (P129T) variant: haplotype analysis of an ancient missense mutation and validation of risk for drug addiction. Hum Genet 2006;120(4):581—8.

[126] de Luis DA, Sagrado MG, Aller R, Izaola O, Conde R, Romero E. C358A missense polymorphism of the endocannabinoid degrading enzyme fatty acid amide hydrolase (FAAH) and insulin resistance in patients with diabetes mellitus type 2. Diabetes Research Clinical Practice 2010;88(1):76—80.

[127] Ren D, Zhou Y, Morris D, Li M, Li Z, Rui L. Neuronal SH2B1 is essential for controlling energy and glucose homeostasis. J Clin Invest 2007;117(2):397—406.

[128] Corella D, Peloso G, Arnett DK, Demissie S, Cupples LA, Tucker K, et al. APOA2, dietary fat, and body mass index: replication of a gene-diet interaction in 3 independent populations. Arch Intern Med 2009;169(20):1897—906.

[129] Fryirs MA, Barter PJ, Appavoo M, Tuch BE, Tabet F, Heather AK, et al. Effects of high-density lipoproteins on pancreatic beta-cell insulin secretion. Arteriosclerosis Thrombosis Vascular Biology 2010;30(8):1642—8.

[130] Smith CE, Ordovas JM, Sanchez-Moreno C, Lee YC, Garaulet M. Apolipoprotein A-II polymorphism: relationships to behavioural and hormonal mediators of obesity. Int J Obes (Lond) 2012;36(1):130—6.

[131] Garaulet M, Lee YC, Shen J, Parnell LD, Arnett DK, Tsai MY, et al. CLOCK genetic variation and metabolic syndrome risk: modulation by monounsaturated fatty acids. Am J Clin Nutr 2009;90(6):1466—75.

[132] Garaulet M, Corbalan MD, Madrid JA, Morales E, Baraza JC, Lee YC, et al. CLOCK gene is implicated in weight reduction in obese patients participating in a dietary programme based on the Mediterranean diet. Int J Obes (Lond) 2010;34(3):516—23.

[133] Moleres A, Ochoa MC, Rendo-Urteaga T, Martinez-Gonzalez MA, Azcona San Julian MC, Martinez JA, et al. Dietary fatty acid distribution modifies obesity risk linked to the rs9939609 polymorphism of the fat mass and obesity-associated gene in a Spanish case-control study of children. Br J Nutr 2012;107(4):533—8.

[134] Corella D, Qi L, Sorli JV, Godoy D, Portoles O, Coltell O, et al. Obese subjects carrying the 11482G>A polymorphism at the perilipin locus are resistant to weight loss after dietary energy restriction. J Clin Endocrinol Metab 2005;90(9):5121—6.

[135] Nischalke HD, Berger C, Luda C, Berg T, Muller T, Grunhage F, et al. The PNPLA3 rs738409 148M/M genotype is a risk factor for liver cancer in alcoholic cirrhosis but shows no or weak association in hepatitis C cirrhosis. PLoS One 2011;6(11):e27087.

[136] Sevastianova K, Kotronen A, Gastaldelli A, Perttila J, Hakkarainen A, Lundbom J, et al. Genetic variation in PNPLA3 (adiponutrin) confers sensitivity to weight loss-induced decrease in liver fat in humans. Am J Clin Nutr 2011;94(1):104—11.

[137] Kiel DP, Demissie S, Dupuis J, Lunetta KL, Murabito JM, Karasik D. Genome-wide association with bone mass and geometry in the Framingham Heart Study. BMC Med Genet 2007;8(Suppl. 1):S14.

[138] Obermayer-Pietsch BM, Bonelli CM, Walter DE, Kuhn RJ, Fahrleitner-Pammer A, Berghold A, et al. Genetic predisposition for adult lactose intolerance and relation to diet, bone density, and bone fractures. J Bone Min Res 2004;19(1):42—7.

[139] Tolonen S, Laaksonen M, Mikkila V, Sievanen H, Mononen N, Rasanen L, et al. Lactase gene c/t(-13910) polymorphism, calcium intake, and pQCT bone traits in Finnish adults. Calcif Tissue Int 2011;88(2):153—61.

[140] Chatzipapas C, Boikos S, Drosos GI, Kazakos K, Tripsianis G, Serbis A, et al. Polymorphisms of the vitamin D receptor gene and stress fractures. Horm Metab Res 2009;41(8):635—40.

[141] Casas JP, Bautista LE, Humphries SE, Hingorani AD. Endothelial nitric oxide synthase genotype and ischemic heart disease: meta-analysis of 26 studies involving 23028 subjects. Circulation 2004;109(11):1359—65.

[142] Hillermann R, Carelse K, Gebhardt GS. The Glu298Asp variant of the endothelial nitric oxide synthase gene is associated with an increased risk for abruptio placentae in pre-eclampsia. J Human Genetics 2005;50(8):415—9.

[143] George TW, Waroonphan S, Niwat C, Gordon MH, Lovegrove JA. The Glu298Asp single nucleotide polymorphism in the endothelial nitric oxide synthase gene differentially affects the vascular response to acute consumption of fruit and vegetable puree based drinks. Molecular Nutrition Food Res 2012;56(7):1014—24.

CHAPTER 6

How Can we Know What the Latest Findings Mean?

Terms of the Trade

- **Alleles:** Alternative DNA sequences at a locus.
- **Autosomal:** Related to or inherited through one of the numbered chromosomes.
- **Codominant:** Mode of inheritance where two traits can be apparent at the same time.
- **Concordance:** Extent of the agreement of a trait or phenotype among siblings.
- **Cosegregation:** The tendency of traits or loci to be inherited together.
- **Dominant:** Mode of inheritance where the trait always overrides the other (recessive) trait in the heterozygous state.
- **Epistasis:** Where one locus affects the appearance of the traits at a distant locus.
- **Genotype:** Combination of inherited variants at a gene locus.
- **Knock-in model:** Strain of animal with targeted replacement of a genome sequence.
- **Knockout model:** Strain of animal with targeted deletion of a gene.
- **Linkage disequilibrium:** Statistical association of two alleles or loci.
- **LoD score:** Logarithm of odds; measure of the distance between two traits or loci.
- **Mendelian randomization:** A study design where outcomes are compared by genotype.
- **Monozygotic twins:** Siblings grown from the same fertilized egg.
- **Phenotype:** External or biological appearance.
- **QTL:** Quantitative trait locus (plural loci).
- **Recessive:** Mode of inheritance where the trait is always overridden by the other (dominant) trait in the heterozygous state.
- **Segregation analysis:** Investigation of the ratios of traits across generations.

ABSTRACT

The first prerequisite for assessing the validity and relevance of nutrigenetic findings, conclusions, and hypotheses is to understand commonly used research methods. You have to wend your way through a maze of reports on in vitro studies, animal experiments, population studies, and clinical trials. Genetic

Nutrigenetics. http://dx.doi.org/10.1016/B978-0-12-385900-6.00006-X

researchers have developed a powerful array of new tools from genetically altered cells and animals to high-throughput genotyping and sequencing techniques. Of course, investigations in nonhuman systems can only clarify mechanisms and help to build plausible hypotheses for well-designed human studies. It is the back and forth between human studies and work with model systems that has empowered the explosive growth of nutrition genetic science. Learning about laboratory methods has to be followed by building an appreciation of data analysis procedures. It is particularly important to understand common sources of errors and misinterpretations. Many initial reports of nutrigenetic interactions are almost unavoidably wrong. Independent replication of study results in different populations is needed before we can draw firm conclusions and use the findings in practice. And, finally, a firm foundation in the philosophy of science is needed to extract meaning from all of that complexity.

6.1 TOOLS FOR NUTRIGENETIC RESEARCH AND PRACTICE

With the growth of genetic tools, an age-old path has returned for unraveling the role of nutrition in human health and disease. Modern nutrition science has come to mean the use of established mechanistic models to predict the effects of foods and nutrients. This has meant a strong research focus on cell culture and animal models. Research hypotheses were, and still are, largely incremental, by systematically expanding these mechanistic models. Bringing human genetics into the mix helps us to jump out of that neat box of planned experiments and generates many unexpected and sometimes unimagined findings. Numerous pathways and functions have been discovered without ever looking for them. There is so much that modern human biology and medicine owes to classical genetics, largely due to the careful exploration of individuals and families with rare inherited diseases and inborn errors of metabolism. While this disease-based research path has taken us very far toward an understanding of human metabolism, the rapid accumulation of vast amounts of precise and specific genetic data is accelerating our journey to previously unimagined levels. Serendipitous findings have become commonplace events. Tracking the twists and turns of nutrition research is now a dizzying occupation.

The following section highlights important approaches for developing novel and productive hypotheses and then testing them with appropriate rigor. Many new and as yet unanticipated methods and approaches are evolving as the science of nutrigenetics continues to grow explosively. Previously valued procedures are rapidly eclipsed, even as you read this. A good example is the sudden availability of affordable genome sequence information, which has become an industrial commodity. DNA sequencing costs have fallen far more rapidly than anybody could have predicted even a few years ago (Figure 6.1) and DNA sequencing has become the domain of factory-like production facilities. Tedious analysis of individual variants is rapidly becoming a thing of the past and nutrition research is moving on to more interesting challenges.

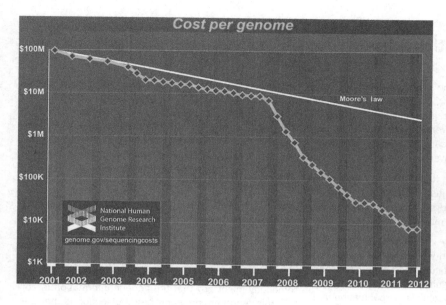

FIGURE 6.1
The cost of sequencing the entire human genome has declined steadily, following a reduction by half every 18 months or so, paralleling developments in computer performance (Moore's law). But, in 2008, the costs started falling precipitously at the rate of an order of magnitude per year.

6.2 IN VITRO AND ANIMAL MODELS

6.2.1 What is the best model system?

This is not the best place to teach general laboratory or research methods, because of their complexity and rapid change. We will just touch on a few examples of approaches that are commonly used to test nutrigenetic hypotheses.

There are two different approaches to using nonhuman model systems for the investigation of gene–nutrient interactions.

6.2.1.1 FORWARD GENETICS

Forward genetics is the more traditional approach, which searches for the genetic basis of specific traits or diseases. Think about alkaptonuria (OMIM 203500), the inborn error of metabolism that attracted Archibald Garrod's interest and started biochemical genetics in 1902. The condition was easy enough to detect because the urine of affected individuals turns dark brown or black after exposure to air for a while. The slow road to identifying the DNA variants responsible for this disease started with painstaking family studies that narrowed the search to microsatellite markers on the large arm of chromosome 3. The dragnet further narrowed to a region of about 16 million bases that contains the *HGD* gene (OMIM 203500) encoding the enzyme homogentisate 1,2-dioxygenase (EC 1.13.11.5). That made a lot of sense because the tyrosine metabolite homogentisate accumulates without this enzyme and excessive excretion of homogentisate is the hallmark of alkaptonuria. The next step was the cloning and sequencing of the *HGD* gene in a fungus, *Aspergillus nidans*, which has convenient features that makes it another member of the geneticists' model zoo. Once

A Lame Statistics Joke

You probably heard this one too many times already. Ten average guys hang out in a bar, feeling quite poor. Then Bill Gates walks in and the average wealth of the people in the bar jumps to billions. Everybody understands the problem with that scenario because Bill Gates' presence does not make any of those average guys richer.

But let's turn this scenario on its head. Does Bill Gates get any poorer (on average) just because he is in the presence of some average guys? Not really, unless he decides to share some of his money.

This fallacy of inappropriately lumping together individuals fails the same way when we analyze a population group and conclude that a particular food or nutrient does not have a (statistically significant) specified health effect. The fact that most people in a population are lactose tolerant means nothing to the individuals who cannot digest lactose. The flaw of a generalizing approach is that it assumes that all members of the group respond the same. If this is not the case, conclusions or recommendations break down and stop being helpful.

Consider the impact of (caffeinated) coffee consumption on the risk of a heart attack in middle-aged men. Case-control studies indicated that consuming large amounts of caffeinated coffee increases the risk of heart attacks (nonfatal myocardial infarction), but a much larger long-term prospective cohort study did not bear out that link [1]. The mystery about these contradictory findings became clearer when it was shown that carriers of the cytochrome P450 1A2 *1F variant (*CYP1A2*1F*; rs762551) have an increased risk, while men without it do not [2]. Cytochrome P450 1A2 is the liver enzyme that metabolizes and inactivates caffeine. The enzyme encoded by the *CYP1A2*1F* allele, which was found in nearly half of the men, has low activity and allows caffeine to remain active for an extended time.

Once we know that a genetic factor strongly affects a particular nutrition-related outcome, we have to take it into account in all future evaluations. This is not different from taking into account the gender of participants in a study of heart attack risk because we already know that women and men respond differently to risk factors. Now somebody might say that the genetic information is not available. Another commonly heard objection is that the study would not have enough power when the participants are split into smaller subgroups with different genotypes. The response in both cases should be that we simply cannot draw reliable conclusions about outcomes that are known to be influenced by this genetic factor without taking it into account.

It is important to understand that this concerns only genetic factors with demonstrated relevance, not factors without a significant factual basis for an assumed impact. Unsupported speculation is not enough to invalidate research findings.

the DNA sequence of the HGD gene in *A. nidans* was deciphered, it was much easier to do the same with the homologous (corresponding) human HGD gene. And from there it was just a matter of sequencing the HGD gene in patients to show that loss-of-function missense variants were indeed the molecular genetic cause of alkaptonuria [3]. This century-long sequence of research accomplishments illustrates both the flow of classical (forward) genetics and the fruitful iterations of clinical investigation and molecular research using various model systems.

6.2.1.2 REVERSE GENETICS

Reverse genetics uses a different path, just as the term implies. Researchers want to learn about the functional role of specific gene sequences. For this, they can again use a wide range of cellular or whole-animal model systems to observe the consequence of a known sequence alteration. The crudest mode of exploration disables the gene permanently (knockout) or temporarily (knockdown). Site-directed mutagenesis and targeted DNA replacement (knock-in) are more specific techniques that can help us understand what common variants do in the controlled environments of model systems.

> **HIGHLIGHT**
>
> Forward genetics: starting from a phenotype to find the genotype.
>
> Reverse genetics: starting from a genotype to find the phenotype.

All experimental models are imperfect because none can capture the full range of human functionality. It is important to understand that the cell lines and animals used in nutrition research are not only different from humans because of their structural and functional differences but as importantly because the model systems lack the diversity that defines us. The uniformity that makes it easier to use these experimental models can be a major barrier to understanding human responses to our nutritional environment.

It is usually a mixture of different complementary approaches that can help us to understand the complex interactions of genome and nutrition. For each specific purpose, the most promising systems have to be chosen. It is important to understand that most of the time there is a multitude of functional solutions. The choice will often depend on availability and the researcher's familiarity with the model.

6.2.2 Genetic manipulation of cultured cells

In vitro culture of cells is very attractive because many aspects of their environments can be tightly controlled and there are no ethical concerns about their treatment. Genetic manipulation of the cells can take many different forms. DNA segments can be inserted randomly into the genome by transfection or, even better, the sequence can be inserted precisely in a desired location. It is particularly attractive to create cell lines with inserts that correspond to the variants found in humans. One group of investigators transfected a standard cell line (COS-1 cells) with 16 different constructs based on the most common *MTHFR* haplotypes [4]. This allowed them to express the different corresponding enzyme forms (allozymes) and investigate the precise impact of the genetic changes on protein quantity and quality (by measuring enzyme kinetics). Out of the many well-established cell lines, the ones that most closely reflect the tissue of interest are selected. Hepatocytes have to be used for investigating the behavior of a gene that is only expressed in liver, of course. Researchers most often use permanent cell lines, derived from cancers or developed by chemical or viral mutagenesis, because they are standardized and easiest to grow. Primary cells from surgical and autopsy samples are more difficult to obtain and to culture. Their advantage is that they more closely reflect the genetic diversity in the target population.

6.2.3 Genetic manipulation of animals

The sequence of the human genome is known and most of the 26,000 plus genes have been identified. But for many of them their exact roles in maintaining health is not fully understood. We know even less about the interaction of the genome with nutritional factors. We can change nutritional factors to some extent and then observe the changes, but we cannot change the genes in people. Of course, we can select people with specific genotypes to see whether they respond differently. This still leaves the possibility that people with the same local genotype differ at other positions. Gene manipulation in animals offers the opportunity to compare nutritional effects in animals that differ only in the places we want to investigate. Mice are a favored animal model because they are mammals and share many functional characteristics with humans. Their small size and rapid reproduction make it possible to conduct studies that require multigenerational breeding or interventions. We should not forget about the classical animal model of genetic research, the common fruit fly (*Drosophila melanogaster*), which continues to provide important insights into the inheritance of nutritional mechanisms. Another humble servant of genetic science is the roundworm *Caenorhabditis elegans*. This tiny worm has been helpful, for instance, for research into developmental biology and aging.

An illustration is research showing that the antioxidant resveratrol, which gives red wine its color, extends the lifespan of *C. elegans* [5]. Zebra fish (*Danio rerio*), particularly at the larval stage of development, have recently become another resource in the toolbox of molecular geneticists. This model has been used, for example, to investigate the expression of zinc transporters in response to zinc excess [6].

6.2.3.1 GENE KNOCKOUT

The classical experimental model explores the function of a gene by deleting it, hence the term *knockout*. Take the example of a mouse strain without a functional *ELOVL6* (OMIM 611546) gene [7], which encodes the elongase of long-chain fatty acids family 6 (EC 6.2.1.3). This elongase catalyzes the final step of the synthesis in the body of the saturated fatty acid stearic acid (C18:0), which appears to be an important signaling compound for fat and glucose metabolism. The knockout mice become obese and develop fatty liver on a high-fat diet. They have high blood sugar levels and too much insulin, just like people with metabolic syndrome. Being able to observe the consequences of this gene deficiency is very illuminating and helps us to understand its importance. This could motivate us, for instance, to focus our interest on genetic variants that reduce the activity of the ELOVL6 gene. Another line of research could be the investigation of nutritional (or pharmacological) interventions that target this gene.

6.2.3.2 TARGETED DNA SEGMENT REPLACEMENT

Investigators have gone a step further and can now replace specific stretches of the mouse genome with a piece of the equivalent (homologous) human genome [8]. This targeted replacement (knock-in) method immediately opened up opportunities to explore biological consequences of polymorphic variation. Investigators have, for instance, spliced the human *APOE* allele sequences E2, E3, and E4 into different lines of mice [9, 10]. A high-fat, high-cholesterol diet caused the development of extensive atherosclerotic plaques in mice with the *APOE* 2/2 genotype but not in the mice with the 3/3 genotype. The investigators found that the *APOE* 2/2 mice did not clear their partially metabolized very-low-density lipoprotein (VLDL) remnants well. The athero-sclerosis-causing remnants did accumulate in blood of the *APOE* 2/2 mice, similar to patients with type III hyperlipoproteinemia (OMIM 107741), but only when VLDL production was greatly increased by excessive fat intake. The *APOE* 3/3 animals handled the increased VLDL flux without an increase in harmful remnants [10].

Another research group wanted to test the hypothesis that the E4 allele protects against vitamin D deficiency [11]. They used mice with the humanized *APOE* genotypes and fed them the same diet with equal amounts of vitamin D (600 IU/kg chow). The serum concentrations of 25-hydroxy vitamin D after 3 months were much higher in animals with the *APOE* 4/4 genotype than in those with the *APOE* 3/3 genotype (Figure 6.2). With the transgenic mouse model we can directly compare the effects of the two different alleles in a way that would not be possible with humans. Because the mice are all from the same inbred strain and have been treated the same way, the remaining difference is due to the alleles. Of course, the anatomical and metabolic differences between mice and humans have to be considered. The investigators also measured 25-hydroxy vitamin D concentration in humans and found that carriers of the *APOE*4 allele had higher 25-hydroxy vitamin D concentrations than people with the *APOE* 3/3 genotype.

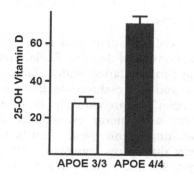

FIGURE 6.2
Transgenic mice with the humanized *APOE* 4/4 genotype have much higher 25-hydroxy vitamin D concentrations in serum than 3/3 animals [11].

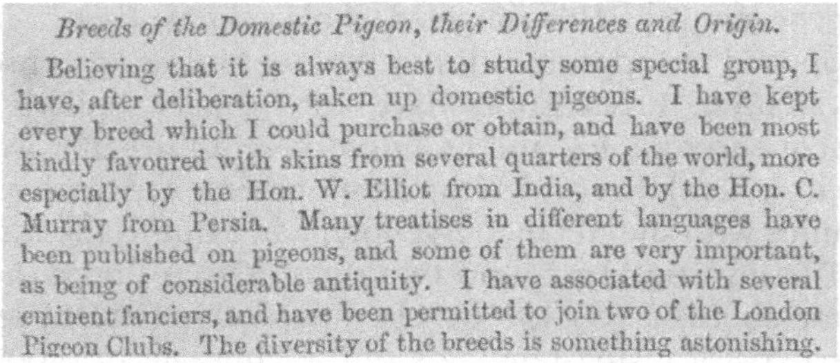

FIGURE 6.3

Shortly before completing his seminal work, Charles Darwin took up pigeon breeding, which helped him develop some of his concepts on evolution.

Charles Darwin, On the Origin of Species by Means of Natural Selection. John Murray, London, 1859, p.15.

> *Breeds of the Domestic Pigeon, their Differences and Origin.*
>
> Believing that it is always best to study some special group, I have, after deliberation, taken up domestic pigeons. I have kept every breed which I could purchase or obtain, and have been most kindly favoured with skins from several quarters of the world, more especially by the Hon. W. Elliot from India, and by the Hon. C. Murray from Persia. Many treatises in different languages have been published on pigeons, and some of them are very important, as being of considerable antiquity. I have associated with several eminent fanciers, and have been permitted to join two of the London Pigeon Clubs. The diversity of the breeds is something astonishing.

6.2.4 Animal cross breeding

The science of genetics owes much to the efforts of commercial and amateur breeders centuries ago, long before the efforts of Gregor Mendel and other founding fathers of the discipline. We need to remember that Charles Darwin based many of his ideas about evolution on his familiarity with the amazing power of carefully planned animal breeding programs (Figure 6.3).

Breeding programs as tools for genetic research have never gone out of style. They continue to be used to map specific phenotypes to genotypes. Where the study of inborn errors of metabolism has been limited by the small number of patients with the same form of a rare disease, animals can be selected and bred to similar specifications. The use of salt-sensitive rat strains, for instance, has greatly contributed to the understanding of the regulation of sodium excretion. Another classical example is the ob/ob mouse, which gets very obese. This mouse strain arose by chance and was preserved by selective breeding as a model for overeating, obesity, and the development of type 2 diabetes. It was eventually found to have a defective *LEP* (OMIM 164160) gene that fails to produce the satiety-signaling hormone leptin in adipose tissue. This poor animal does not know that it has already had enough to eat and is always hungry.

How animal breeding experiments have been brought into the twenty-first century is demonstrated by the Collaborative Cross project. This large-scale breeding project started with a large number of well-defined inbred mouse strains and is crossbreeding them systematically. The offspring are then characterized comprehensively. From a nutrigenetic perspective, traits such as obesity and energy expenditure are of particular interest [12]. Once an interesting phenotype is identified, it can be characterized further and the genetic differences compared with closely related animals without the phenotype. In many ways this works like family studies and overcomes rigid genetic uniformity, one of the key weaknesses of classical animal models.

6.3 NUTRIGENETIC STUDIES IN HUMANS

6.3.1 General considerations

Genetic studies always come down to finding a link between phenotype and genotype. We want to know which genetic differences are responsible for an inherited trait or outcome. To make it concrete, let us consider which sequences at specific loci on our chromosomes are responsible for the fact that one adult can drink considerable quantities of milk without having to worry about abdominal discomfort and another one responds like clockwork with stomach cramps or diarrhea.

A REMINDER
We speak of a recessive allele if a trait is observed only in people with two variant copies of the gene. Inheritance is codominant if both traits can be present at the same time or if there is an identifiable intermediate state.

In this example, the lactase enzyme activity measured in a duodenal biopsy sample, the hydrogen breath test following ingestion of 25 g lactose, and the abdominal symptoms in response to milk consumption are all distinct phenotypes. They describe observable appearances and responses. At the molecular level, we have the genotype that represents the information encoded in the DNA. This broader usage of the word genotype must be distinguished from the narrower usage of the term that refers to sequence information at a particular position (locus), usually on the two chromosome copies. We have seen in people of European ancestry that the allele (version) with a T in position LCT -13910 usually means that this chromosome copy is primed to express lactase when it gets the right signals in a cell of the duodenum. The allele with C in the same position will not respond to these signals in an adult duodenal cell and will not express lactase. Since healthy adults have usually two copies of the chromosome 6, and therefore two copies of the LCT 13910 locus, we need to know which allele is present on each of them. The usual convention is to say that someone has the genotype LCT -13910C/T when the C allele is present on one chromosome copy and the T allele on the other copy. Because one T allele is enough to trigger intestinal lactase expression in adults, we say that this allele is dominant.

Of course, the entire lactase gene between the upstream enhancer regions and downstream mRNA control region spans over 70,000 bases and any variants within that region can conceivably influence the regulation of lactase expression and the activity of the mature enzyme. We know that upstream of the LCT -13910 locus is another enhancer region. In Northern Europeans, LCT -22018 contains either a G or an A. The LCT -22018A allele provides an extra boost to intestinal lactase expression in adults, but the G allele does not. This raises the question whether the LCT -13910T allele and the LCT -22018A allele are situated on the same strand and therefore form a single haplotype or not. In someone with only one copy of each lactase-boosting allele, this will make a difference. If both are on the same strand, only that copy will get a boost. If the boosting alleles are on different strands, both copies will be expressed. The combination of two or more alleles is a haplotype. These haplotypes may be considered the words and sentences on our two chromosome copies, which are like two parallel banners with individual letters spelled out by the four bases. The banners would not be readable if the letters were scrambled back and forth across the two banner ribbons. A third consideration is shared genetic heritage that goes way beyond one local allele or even a haplotype combination of a few alleles. Let us not forget that it was presumably one particular individual a few thousand years ago who first had the LCT -13910C>T mutation with lactase persistence and that Northern Europeans with that allele are direct descendants. If that is true, then numerous unrelated genetic segments are shared by Northern

Genotype versus Allele Frequency

Database listings often refer to allele frequency and the question arises as to how that is different from genotype frequency.

Calculate each genotype frequency as a fraction of 1.0. For a biallelic variant (where there are two alternative alleles, a and A), you should get three different genotypes: a/a, a/A and A/A.

Remember that the Hardy-Weinberg equilibrium should apply in most cases:

$$\text{Genotype frequency}: \text{freq}(a/a) \times \text{freq}(a/a) + 2$$
$$\times \text{freq}(a/A) + \text{freq}(A/A)$$
$$\times \text{freq}(A/A)$$
$$= 1.0$$

Calculate the allele frequency of an autosomal biallelic variant a (where there are two alternative alleles and assuming the locus is in one of the numbered chromosomes) by adding twice the frequency of the homozygous genotype a/a (because the genotype contains two allele copies) plus the frequency of the heterozygous genotype a/A and divide it by two (because there are two chromosome copies).

$$\text{Allele frequency}: \frac{[\text{freq}(a/a) + \text{freq}(a/a) + \text{freq}(a/A)]}{2}$$

The diagram below is a handy graph that shows how genotype frequencies are related to biallelic frequencies, assuming that Hardy-Weinberg equilibrium applies.

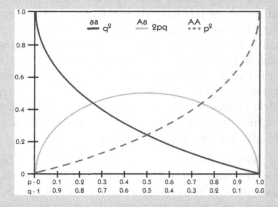

Europeans with the LCT -13910T allele. If the LCT -13910T allele arose independently in a different population, the trait of lactase persistence may be shared to some extent but other traits will be different. We know that the LCT -13910T allele is common among the Fulani people of the Sahel across North Africa [13], and the Fulbe and Hausa of the same region [14], and people in Northwest India [15]. Detailed analysis in Northern Indians with lactase persistence of all the variants near the LCT -13910 locus showed that they have the same haplotype as Northern European people with the LCT -13910T allele and lactase persistence. This means that they are all direct descendants of the first LCT -13910T carrier. It is as yet unknown whether African people with the LCT -13910T allele also share this Northern European ancestry.

6.3.2 Twin studies

We may suspect a genetic influence on many nutritional effects but don't know where to start investigating it. This is particularly true for complex events such as feeding behavior. One way to assess the potential contribution of genetic factors is the use of twin studies. This takes nothing more than a careful characterization of the nutritional effect or behavior and access to a lot of twins.

Let us start with a definition of our research subjects. Monozygotic (identical) twins come from the same fertilized egg and share almost their entire genetic sequence of about 6.6 billion bases. The reason we have to say "almost" is that 50–100 new mutations occur in every generation and those are not shared. Dizygotic (nonidentical or fraternal) twins come from two separate fertilized eggs and are therefore genetically the same as normal siblings born at about the same time. This means that they share about 50% of the genetic variants that occur in their parents. Note that this does not mean that 50% of the DNA sequence is the same (it is in fact well over 99% in unrelated people). The casual notion that monozygotic twins share 100% of their genes and dizygotic twins share 50% of their genes should be avoided, because it is misleading and conceptually inaccurate. It also important to recognize that self-reported sibship of twins (monozygotic vs. dizygotic) is wrong in a few percent of cases and should be ascertained analytically at a broad range of loci [16] or even better with whole-genome sequencing.

The rationale for a twin study is the assumption that twice the difference in the similarity between identical twins and between nonidentical twins is a measure of the genetic contribution to a phenotype. The traditional formula to indicate this assumption is

$$A = 2 \left(r_{mz} - r_{dz} \right)$$

where A is the estimated proportion of genetic influence on an observed variable and r_{mz} and r_{dz} are the correlation coefficients of the measured variable for monozygotic (mz) and dizygotic (dz) twins.

The justification for this formula is that all variation in a measured variable in people comes from a combination of genetic and nongenetic influences. The nongenetic influences include all environmental factors, aging, life events, and analytical errors. One nice aspect of twin studies is that the ages of these sib pairs are the same, though some may strongly disagree with that statement. An important assumption needed to justify the formula shown above is therefore that the impact of all nongenetic factors on the measured variable is the same in monozygotic and dizygotic twins. More complete models separate out the estimated contribution of shared and nonshared environmental influences where the information is available. Updated mathematical treatment of twin study data with multifactor models have been introduced [17] but their description goes beyond the scope of this text.

The original conceptual framework for the investigation of heritability of human traits with twin studies was developed well before the recognition of the digital nature of genetic inheritance. It is built on the implicit assumption that genetic traits are continuous and miscible. For outcomes that depend on a multitude of discrete genetic factors that is certainly a practical approach. But the closer we get to separating out the contribution of individual genes, the less viable the procedure becomes. As discussed in previous chapters, DNA variants are either present or not. Calculating an average variant score makes as little sense as calculating an average gender.

We should now be ready to review how researchers have used a twin study to assess the extent to which heritability of a complex dietary behavior may influence a measurable outcome. The investigators wanted to know whether the inherited eating rate of prepubertal children affects their body composition [18]. They videoed how many bites a minute the children took from standardized sandwiches and fruit at home. Multiple eating sessions were recorded to minimize measurement error and other extraneous variation. As initially hypothesized, overweight children ate about 12% faster than children with low-normal weight. The correlation of eating rates was greater in monozygotic twins (correlation coefficient $r_{mz} = 0.59$) than in dizygotic twins ($r_{dz} = 0.30$). This already points to a considerable genetic contribution to eating rate. Using the equation above we get:

$$A = (r_{mz} - r_{dz}) = 2\,(0.59 - 0.30) = 0.58$$

This calculation suggests that a very large part of the eating behavior contributing to overweight in children is inherited. Of course, this is only one study of one particular aspect of eating behavior and the significance of the finding should not be overestimated. Nonetheless, the results suggest that further research into the molecular mechanisms of this particular eating behavior is promising.

6.3.3 Family studies

The investigation of multiple generations of relatives is a very powerful approach for investigating genotype-specific responses to nutritional factors. The advantage

Twin Studies Assess Genetic Risk

Nutrigenetic questions naturally always want to know about the balance between nurture (food intake) and nature (genetic variation). Twin studies can be uniquely helpful for teasing out their relative contributions. Identical twins have nearly the same genetic sequence. The only difference will be about 60–100 newly arisen mutations and some divergent epigenetic modifications.

How do we know whether twins are identical (monozygotic) or nonidentical (dizygotic)?

Monozygotic twins have the same gender and share blood groups and DNA genotypes with extremely rare exceptions. A recent forensic conundrum illustrates how similar they are. Police found a glove with DNA traces left behind by burglars in a multimillion dollar jewelry heist (http://www.faz.net/artikel/C30176/kadewe-einbrecher-tatverdaechtige-wegen-zwillings-dna-entlassen-30142925.html). They quickly linked the DNA to monozygotic twins, but had to let them go free because they could not determine which of the two the DNA came from. Dizygotic twins, in contrast, differ from each other about as much as siblings born in different years, although they tend to share more environmental influences.

This means that we can compare the impact of nutrition on a condition or disease in monozygotic and dizygotic twins. The more similar a nutritional influence is in both types of twins, the less likely is the existence of a strong genetic influence. The degree of similarity is called *concordance*. If nutrition strongly influences an outcome in monozygotic twins and only moderately in dizygotic twins, a genetic interaction is very likely to exist.

Take this twin study of goiter on Crete [19].

	Concordant	Discordant	Statistics
Monozygotic twins	84/94 (89.4%)	10/94 (10.6%)	$\chi^2 = 8.97$
Dizygotic twins	102/139 (73.4%)	37/139 (26.6%)	$p = 0.005$

These data show that the Cretan monozygotic twins are concordant (in agreement) significantly more often than dizygotic twins are. The underlying assumption is that twins raised in the same family share much of their environmental factors. The only difference then is that monozygotic twins have a nearly identical genome and dizygotic twins do not. The smaller degree of concordance in the dizygotic twin suggests that both genetic and nutritional factors (intake of iodine and goiter-promoting isothiocyanates) influence the risk of developing goiter. Such twin data cannot show, however, whether there is a gene–nutrient interaction. For that we would need individual nutrition information.

is that just a few families may provide enough power to test for simple gene–nutrient interactions. You will remember the examples from Chapter 3 looking into lactose and fructose intolerance and discussing the inherited ability to perceive the bitter taste of white grapefruit. There will be less variation in nutrition and other environmental exposures in families, which makes it more likely to see a statistically significant result than in groups of unrelated individuals.

6.3.3.1 AGGREGATION ANALYSIS

The first step could be to simply ask whether first-degree relatives are more likely to be susceptible to a nutritional factor than unrelated family members. The genetic investigation of rare conditions usually starts this way. This is also a good start if it is unclear whether a trait is heritable or not. No gene sequence information is required for this kind of investigation; that comes later. At this stage, a careful assessment of the phenotype is the focus. Currently available software packages for genetic analyses can calculate estimated heritability.

For instance, investigators wanted to determine the heritability of 1,25-dihy-droxy-vitamin D concentration in Latin- and African-American populations [20]. They recruited more than a hundred large families and measured 25-hydroxyvitamin D and 1,25-dihydroxy-vitamin D concentrations. The concentration of the precursor 25-hydroxyvitamin D was used to adjust for different levels of sun exposure and therefore vitamin D production in skin. They found that heritability of 1,25-dihydroxy-vitamin D concentration was 0.48 in the African-American families, which means that it is influenced to a substantial degree by genetic factors.

6.3.3.2 SEGREGATION ANALYSIS

When we find that a genetic factor influences nutrition, we usually want to know the transmission pattern in the affected families. Segregation analysis of pedigrees can tell us whether the mode of inheritance is autosomal dominant, codominant, recessive, or maybe X-linked or non-Mendelian (Table 6.1).

A key metric is the segregation ratio: $R = \text{affected}/(\text{affected} + \text{non affected})$.

A codominant mode of inheritance does occur with nutrigenetic interactions, but is rarely perceived as such because we tend to distinguish between the presence and absence of a trait rather than quantitative differences. A typical example is lactase persistence in adults, which is usually seen as either present or not. In reality the two lactase gene copies in each intestinal cell act independently and enzyme activity correlates, for instance, with the number of LCT -13910T alleles.

Complex segregation analysis can also investigate the inheritance of multigenic predispositions and quantitative responses to nutrition—gene interactions.

Table 6.1	Quantitative Assessment of Segregation Ratios can Point to the Mode of Inheritance	
Mode of Inheritance	**Segregation Ratio (R)**	**Comments**
Autosomal dominant	0.50	—
Autosomal codominant	0.25 + 0.5 + 0.25	—
Autosomal recessive	0.25	—
X-linked dominant	1.00 daughters of affected males 0.00 sons of affected males 0.50 children of affected females	—
X-linked recessive	0.50 sons of female carriers 0.00 children of affected males 1.00 sons of affected females ≥ 0.00 daughters of affected females	All daughters of affected males are asymptomatic carriers; all children of an affected male and an affected female are affected
Mitochondrial	1.00 children of affected females 0.00 children of affected males	—

6.3.4 Association studies

Information about the influence of genetic variants on nutritional factors often comes from association studies that use opportunistic or systematic population data. Sometimes just a few students, hospital employees, or blood donors are assessed. At the other end of the spectrum are large representative studies like the National Health and Nutrition Examination Surveys (NHANES). These population samples are a great resource for examining the influence of genetic variants on measures of nutritional status.

Genetic association is expressed as the odds ratio of the proportion of trait carriers with the variant allele to trait carriers without the variant allele divided by the proportion of trait noncarriers with the variant allele to trait noncarriers without the variant allele. The significance of this odds ratio is calculated with a chi-square test or other suitable statistical test.

It gets really interesting when you can crawl along the DNA sequence and determine for one variant after another whether it is associated with the trait or phenotype of interest. When a very large number of variants are analyzed across all chromosomes we are talking about a genome-wide association study (GWAS). Conventionally, the results for each variant are plotted as the negative $\log_{10}$-transformed p-value. In other words, a more statistically significant association will have a greater value on the Y-axis. This form of display is called *Manhattan plot* because it tends to give the impression of a skyline. If there are significant associations, they tend to cluster around gene regions because variants are mostly inherited in haplotype blocks. The points from each chromosome are usually given different colors to signify their location. At the customary statistical significance threshold of $p < 0.05$, there would be thousands of positive results for a typical large-scale association study with 500,000 or more tested variants. To reduce such false-positive results to a more acceptable level, a Bonferroni correction is usually applied by dividing the statistical threshold by the number of tests. For a study with 500,000 variants, the threshold for statistical significance would be set at $p < 10^{-8}$, which is 8 on the negative $\log_{10}$-transformed Y-axis.

Let us look at an example of a GWAS. Results were drawn from three populations (women from the Nurses' Health Study and men and women from the Framingham Heart Study) and included 4763 adults [21]. Slightly over 2.4 million single nucleotide variants (SNP) were determined and their association with serum pyridoxal phosphate (PLP) measured. The results are displayed as a Manhattan plot (Figure 6.4). The only variants that were statistically significant were located around the liver alkaline phosphatase gene (*ALPL*; OMIM 171760). The variant rs1256335 gave the lowest p-value (shown in the Manhattan plot as the highest value because it is plotted as the negative $\log_{10}$); individuals with the G allele had lower PLP concentrations than people without it.

GWAS have become very sophisticated research instruments, with complex methods to optimize data collection and statistical analysis. A major challenge is the very high cost because large numbers of participants are needed to achieve

FIGURE 6.4
This Manhattan plot displays the statistical significance values ($\log_{10}$-transformed and inverted) for the association of 2.4 million variants with serum pyridoxal phosphate concentration in healthy American adults [21]. Results from across the entire genome are arranged from chromosome 1 on the left to chromosome 22 on the

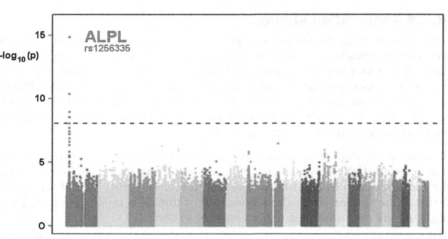

right; X and Y chromosomes are not included. Only a few associations near the alkaline phosphatase gene (ALPL) are below the statistical significance threshold of $p < 10^{-8}$ (indicated by the broken line).

acceptable statistical power. On the other hand, DNA from completed studies can be used, which can make GWAS more affordable and the results quickly available. Generally speaking, the robustness of the results depends very much on the quality of the input data. Careful phenotyping is probably the most important ingredient that goes into this multilayered pie because errors cannot be corrected at a later stage. If, for instance, participants were not fasting before a blood draw when they should have been, it will be very difficult to account for that at a later stage. Of course, the genetic analyses should be done with as low error rates as possible, but quality control measures and error detection algorithms are easier to implement with biological samples than with people.

What are the Odds

Genetic association studies ask how often someone with a genetic variant of interest has a certain detectable trait or outcome. Let's take the example of the LCT −13910T allele and lactase persistence in a healthy, nonsmoking Colombian population [22]. The listing is simplified in this case because only the genotypes C/C and C/T were observed. The third possible genotype T/T is rare in this population and was not observed.

First we set up this 2 × 2 contingency table for the presence or absence of the T allele by the presence or absence of lactase persistence.

The odds for someone with lactase persistence to have the T allele is $24/28 = 0.857$.

The odds for someone with nonpersistence to have the T allele is $2/74 = 0.027$.

We can now calculate the odds ratio, which is the ratio between the two odds we just calculated: $0.857/0.027 = 31.7$.

The p-value for this odds ratio is less than 0.001, as calculated with a chi-square test.

	T	no T
Lactase nonpersistence	2	74
Lactase persistence	24	28

6.3.5 Case-control studies

This type of study design aims to compare characteristics of individuals with an outcome (for instance, cancer) with the characteristics of otherwise similar people without that outcome. The major strength of this design is that it can generate results in a short time and with moderate effort because there is no need to wait for an outcome to occur. This is particularly important for cancer, cardiovascular disease, and other outcomes that take many years to develop. Case-control studies should be followed up with targeted intervention studies whenever possible, since the expected effect be distorted by bias or require intervention during a limited window of opportunity, such as during early childhood or adolescence.

Let's look at a typical example that examines a nutrigenetic effect in obese Spanish children and nonobese controls [23]. The question was whether saturated fat promotes being overweight and obesity more in carriers of the A allele in a common FTO variant (rs9939609) than in those without the variant. Both overweight and normal weight children got an average of about 13% of their total energy intake as saturated fat. This suggests that higher saturated fat intake promotes weight gain no more than other fats do. However, when we separate by genotype, we can see that children carrying the A allele are distinctly more likely to be obese when they have higher saturated fat intake than with lower intake (Figure 6.5).

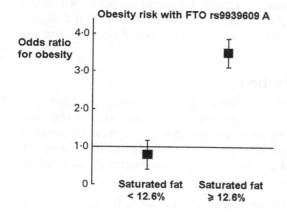

FIGURE 6.5
The FTO rs993960 A allele was associated in children with a higher risk for obesity only in those with higher than average consumption of saturated fat.

In contrast, higher saturated fat intake did not make children without the A allele more prone to obesity (Figure 6.6).

The take-home message of this example is that bringing both nutrition and genetics into this case-control comparison finds a genotype-specific dietary difference when no effect of saturated fat is detected in the group as a whole.

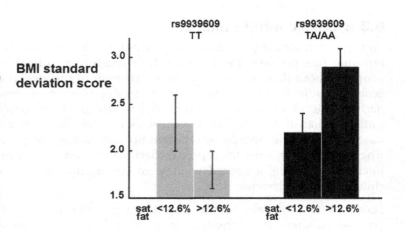

FIGURE 6.6
Obese children carrying one or two FTO rs9939609 A alleles (dark bars on the right) with higher than average saturated fat (sat. fat) intake deviated more from the average body mass index (BMI) of their peer group than A allele carriers with lower saturated fat consumption [23]. This obesity-promoting effect of saturated fat was not detectable in children without the A variant. If anything, the effect of saturated fat was the opposite in the absence of the A variant.

6.3.6 Prospective cohort studies

Following a large group of individuals of similar age and other personal characteristics over extended periods of time, sometimes lifelong, can give us insight into the effects of nutrition on health outcomes. Prospective cohort studies are in many cases the only practical design for the relatively bias-free investigation of nutritional effects on cancer, cardiovascular disease, and other very slowly developing conditions. A major weakness of this type of study is that it can only observe what happens spontaneously. It is rare that enough members of a cohort will use a particular unusual diet pattern of interest on their own. Well-designed intervention trials (as discussed below) are necessary as follow-up to test the effect of a specific combination.

6.3.7 Mendelian randomization

Mendelian randomization (MR) is a very powerful genetic strategy that explores the effect of genetically fixed long-term exposure on an outcome [24]. In other words, this strategy uses suitable genetic markers to estimate the long-term nutritional exposure in place of unavailable dietary factors. MR can be used in conjunction with cross-sectional, case-control, and cohort studies.

The concept was first suggested to determine whether lowering blood cholesterol concentration increases cancer risk [24]. This was of great concern at the time and provided an easy excuse for many not to switch to a healthier, cholesterol-lowering diet. People are not very good at sticking to lifestyle recommendations anyway, particularly for the many years it takes for cancer to become detectable. So how could we ever know with reasonable certainty whether we would just exchange one demon, heart disease, with another, cancer? The answer was to look for a gene for which one allele lowers cholesterol throughout life, while the other one does not. It was a long time before

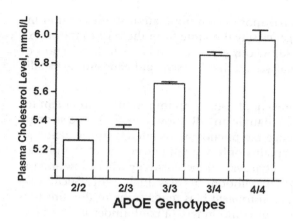

FIGURE 6.7
APOE genotypes generate a stepwise distribution of plasma cholesterol concentrations in untreated populations [25].

a MR study actually tested the question directly [25]. This study made use of the fact that common variations in *APOE* have a predictable impact on blood cholesterol concentration. People carrying the minor allele E2 have about 7% lower cholesterol levels than people who are homozygous for the most common form, E3 (Figure 6.7). Carriers of the other minor allele, E4, have 3−4% higher cholesterol concentrations than people with two copies of the E3 allele. This means that there is a nice stepwise relationship between the various *APOE* genotypes, exactly spanning the difference to blood cholesterol concentrations that dietary change can readily achieve.

As had been observed in earlier studies, participants initially in the lower third quartile for blood cholesterol concentrations were about twice as likely to develop cancer and die from it. But people with genetically low cholesterol (E2 carriers) tended to have a lower cancer risk (Figure 6.8) than those with genetically fixed cholesterol concentrations (E3 and E4 carriers). If low blood cholesterol had increased cancer risk, the genotype-linked data should have shown a higher risk for E2 carriers. This means that the cancer could not have been caused by the low cholesterol concentrations. Probably the causality is the other way around: early cancer lowers blood cholesterol concentration. This elegant and extremely important example shows the power of leveraging

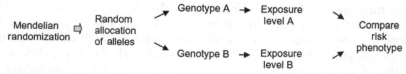

FIGURE 6.8
Mendelian randomization studies use the fixed metabolic or nutritional difference caused by particular genotypes. If someone is born with a predisposition to a lower plasma cholesterol concentration, the effect is life long and changes long-term health risks.

knowledge about fixed genetic predisposition in population studies. If nothing else, we can be certain that our genes stay the same from the start to the end of our life. Since we usually don't know our genotypes, there is little selection or compliance bias. The genetic analyses are inexpensive and exposure is usually lifelong.

There are limitations to this approach, of course. An important requirement for using a gene variant in an MR study is that it should strongly influence the phenotype of interest. If the genotype explains only a small percentage of the phenotype variation, sensitivity will be low and the likelihood of bias high [26]. Such *weak instrument* bias tends to be in the direction of confounders (a confounder is a variable that influences both the exposure and the outcome). A biased result can thus reinforce the erroneous assumption of causality by a confounder and undermine the whole purpose of the exercise. To put it in the framework of the cholesterol—cancer example, diabetes is a potential confounder, because it raises both blood cholesterol concentration and cancer risk [27]. Diabetes was twice as common in study participants with low cholesterol concentration as in those with high cholesterol concentration. If the influence of *APOE* genotype on cholesterol levels was weaker, the confounding influence of diabetes might have mimicked a cholesterol—cancer link and the MR results may have falsely supported a cholesterol—cancer link.

KEY CONCEPT
Mendelian randomization compares outcomes between people who have different long-term nutritional exposures due to a genetic variant.

The main point to remember from this discussion is that genetic variants should be selected and their effect on the intermediary phenotype fully tested before designing an MR study. Avoiding reliance on a weak predictor will minimize the risk of erroneous conclusions due to bias and low sensitivity.

Another interesting genotype-specific association study was used to examine the previously suggested link between high iron intake and increased colon cancer risk. Iron is known to be retained better in carriers of common variants in the hemochromatosis gene (*HFE*; OMIM 613609) controlling iron balance, and more likely to cause accumulation to excessive levels [28]. Investigators used a large study of colon cancer cases and matched controls in North Carolina to explore the suggested iron—cancer link [29]. They found that high iron intake was associated with a doubling of colorectal cancer risk in carriers of HFE variants, but not in noncarriers. A later investigation into the same question in the large Nurses' Health Study found a much weaker effect [30], highlighting the difficulty of getting conclusive answers to such questions. Nonetheless, the investigators would not have found any effect at all, not even a weak one, if they had examined all study participants together without separating them into genotype-specific groups.

Let's look at a last example that relates a strong genetic predictor of alcohol intake to esophageal cancer risk. It is difficult to know how much alcohol people consume in the long term, because they often don't like to admit to their intake level. So it would be nice to have a genetic variant that pegs carriers to a particular amount. The ALDH2 variant 2 (rs671) does that to some extent. Asians with the ALDH2 genotype 1/1 do not mind drinking alcohol; some of them actually like it a whole lot, while those with the ALDH2 2/2 genotype tend to consume little [31]. The reason is, as discussed earlier, that people with *ALDH2*2* metabolize the ethanol metabolite acetaldehyde to acetate only very slowly and even a single drink rapidly raises acetaldehyde concentrations in blood and the brain to very uncomfortable levels, causing facial flushing and typical hangover symptoms, including headache, hypersensitivity to light, and nausea. We can, therefore, assume that in the Japanese population with a mixture of genotypes the 2/2 genotype carriers get little alcohol exposure and the 1/1 carriers get significantly more. The final step then is to link ALDH2 genotypes to esophageal cancer risk. And there the findings are very consistent [32]: The available cohort and case-control studies all point to a two to three times higher risk of cancer of the esophagus attributable to the consumption of alcoholic beverages.

6.3.8 Intervention trials

Studies that attempt to determine what happens when we actually change dietary habits are the rarely achieved gold standard of nutrigenetic evidence. All the usual rules for intervention trials apply to nutrigenetic investigations as well. To be meaningful, an alternative intervention must be included and assignment to either treatment must be randomized. Neither test subjects nor investigators should be aware of the assignments. Suitable comparisons often will contrast the presence or absence of a target compound. In this type of design, capsules or pills may be used that either contain the ingredient or don't (placebo). Color, odor, taste, and other detectable characteristics must be hidden with great care to avoid breaking of blinding. Obviously this can complicate the investigation of materials like garlic or fish oil with their unmistakable odor.

There are two challenges that are specific to genotype-specific intervention trials. The first one is the fact that most variants that might be able to discriminate between responders and nonresponders occur in a minority of people. Let us assume that the prevalence of the less common (minor) genotype is 10% and that we need 20 individuals in each group. We have then to screen at least 200 candidates to find 20 potential participants with the minor genotype. Of course, only some of them will pass all inclusion and exclusion criteria and then actually participate. That means that compared to a standard intervention trial we have to screen at least five times as many people for a genotype-specific study. For less common variants, the ratio would be even more unfavorable. The second barrier is that compared to a simple nutrition hypothesis there are usually many more

Table 6.2		Previously Suggested Gene Variants that Might Determine Whether Individuals Gain Weight with a Higher Proportion of Saturated Fat in their Usual Diet		
Gene	Variant	Genotype Frequency	Effect	Reference
APOA2	rs5082	15% −265C/C	1–2 BMI point difference	[33]
CLOCK	rs1801260	49% 3111C	8.0 vs. 10.4 kg weight loss	[34,35]
FTO	rs1121980	17% T/T	2–3 BMI point difference	[36]
FTO	rs9939609	18% A/A	2–3 BMI point difference	[36]
STAT3	Several	> two risk alleles	2 cm waist circumference	[37]

BMI, body mass index.

hypotheses to test. Take, for example, the question of whether consuming a higher percentage of fat as saturated fat will promote weight gain. We already know from observational studies of at least five candidate variants that may contribute to a genotype-specific effect of saturated fat on weight gain (Table 6.2). This means that we may already have to do five times as many studies to resolve that originally simple question about the contribution of saturated fat intake to weight gain.

To date, there are very few published genotype-directed nutrition intervention trials with a fully developed a priori (a priori means at the outset) variant-based design and adequate power. An excellent example is an investigation [38] that aimed to determine how much folate it takes to achieve a normal plasma homocysteine concentration (9 µmol/L). 5,10-Methylene tetrahydrofolate reductase (*MTHFR*; OMIM 607093) provides the active 5-methyl-tetrahydrofolate precursor for the homocysteine remethylation reaction. The *MTHFR* variant 677T (rs1801133) encodes an enzyme with lower activity than the more common 677C allele. Carriers of the 677T variant may require a higher dietary folate intake to compensate for their lower enzyme activity.

The first step was to find a sufficient number of participants for each of the three genotype groups: C/C, C/T, and T/T. They were then fed four different levels of folate, each level for 4 months. The homocysteine concentrations measured at the end of each feeding period painted a pretty clear picture (Figure 6.9): Carriers of the T/T genotype came close, but were slightly above, the targeted 9 µmol/L homocysteine level with about 800 µg dietary folate equivalents (DFE). Participants with the C/C genotype had no trouble even staying below the target level with about 200 µg DFE.

Homocysteine concentration may not be the best or most important indicator of folate adequacy. Nonetheless, this study shows unequivocally that the same amount of dietary folate has quantitatively different effects depending on genotype. The strength is that a sufficient number of people

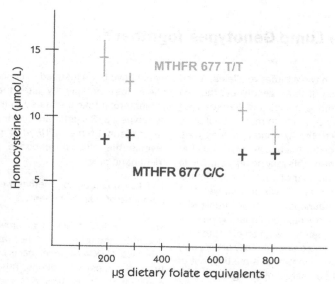

FIGURE 6.9
Dose finding for dietary folate equivalents to achieve 'normal' plasma homocysteine concentration (9 μmol/L).

with the critical T/T genotype were studied. This was obviously quite a challenge, because only about 10% of the study population carries this genotype.

6.4 MOST FIRST REPORTS OF NUTRIGENETIC INTERACTIONS ARE FALSE

6.4.1 Really, false

Many researchers viscerally resist the notion that their conclusions may be wrong. Such sentiments have existed since the earliest times of science and across all science disciplines [40]. Newly emerging fields like nutrigenetics are particularly vulnerable to false discoveries [41] because they are treading on untested ground. This presents its practitioners with the conundrum of needing to enthusiastically embrace new paradigms while remaining appropriately skeptical of their findings. It is important to face this challenge head on and understand why *most* initially reported nutrigenetic interactions will not hold up to later investigations, even if everything is done right.

KEY QUESTION
What we want to know, in the end, is whether a future edition of this book will still describe a particular nutrigenetic interaction in roughly the same terms as now after it has been thoroughly tested by follow-up studies.

Should we Lump Genotypes Together?

Biallelic variants (those with two alternative alleles, A and a) allow for the occurrence of three genotypes, the two homozygote forms A/A and a/a, and the heterozygote combination A/a. It is very common to read in published reports that individuals with the A/a and a/a genotypes were grouped together because the a/a genotype was too rare to be analyzed on its own. This may be unavoidable in many cases because the number of study participants is too small. We need to be aware, however, that we often miss an important effect that happens only in people with the homozygous genotype. It would be equally wrong, on the other hand, to inappropriately assign an effect to both heterozygotes and homozygotes when it really occurs only in homozygotes. This is not a matter of statistics, but of proper interpretation and optimal study design.

Study the previously discussed (in Chapter 4) example of the dietary cholesterol effect on serum cholesterol concentration by *APOE* genotype [39]. Had they grouped the *APOE* genotypes 3/4 and 4/4 together, the authors

would have been justified in concluding that the *APOE* 4 allele does not greatly alter the response to dietary cholesterol intake compared to the most common *APOE* genotype 3/3 (left-hand side). Only the specific comparison of the *APOE* 4/4 group with the others reveals the distinct response to dietary cholesterol (right-hand side).

The key question is whether we are interested in the response of a small minority.

The study design needs to provide sufficient power to determine the response of the least common genotype, in this case *APOE* 4/4, which occurs with a frequency of less than 2% in most populations. This study could answer the question about the *APOE* 4/4-specific response because the investigators specifically recruited equal numbers of participants with the *APOE* 3/3, 3/4 and 4/4 genotypes. A lot of people have to be screened to find enough study subjects with a rare genotype!

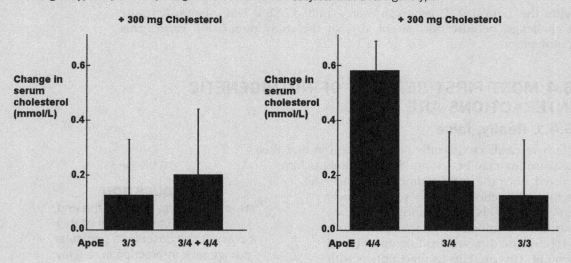

6.4.2 Positive predictive value

The relevant measure of how well we can expect a reported research finding to hold up to further investigations is its positive predictive value (PPV). This post-study probability that the finding is true depends, of course, on the rate of false-positive (alpha, type I) and false-negative (beta, type II) error rates. In addition, the PPV is strongly dependent on the a priori likelihood, R, that a positive result is to be found at all [41]. In other words, R is the ratio of ultimately confirmed to ultimately unconfirmed findings in the probed research area. Note that we are using the term *confirmed* rather than *true* because science history gives us enough examples of 'true' conclusions that were eventually upended by later investigations.

PPV = (1 − beta)R/(R − betaR + alpha)   alpha = false-positive error rate
PPV = positive predictive value   beta = false negative-error rate
                                          R = a priori likelihood

Let's look at a genome-wide association study of plasma homocysteine concentrations in three populations [21]. Associations with three *MTHFR* variants were significant below the threshold of the test statistic of $p = 3 \times 10^{-7}$, taking into account multiple testing for 347,000 or so genetic variants. Univariate regression models with adjustments were statistically significant for all three variants. The association with the *MTHFR* 677T variant (rs1801133) was expected, since it had been observed in numerous previous studies. The Bayesian a priori likelihood constant R could be reasonably set at 0.95 in light of the extensive prior knowledge. The variants rs12085006 and rs1999594, which are located 106 kb upstream of the *MTHFR* transcribed gene region, seemed to be much more robustly associated with homocysteine concentrations, both with test statistics of around $p = 5 \times 10^{-10}$. What was the prior evidence for ultimately true associations of these two variants with plasma homocysteine concentration? As far as the authors of the study were aware, there was none. If we assume that there are actually ten true associations (just to be optimistic) among all of the possible ones between homocysteine concentration and 347,000 variants, it follows that the Bayesian a priori likelihood constant R is approximately 1:34,700 [41]. The prestudy probability for any association, which can be calculated as R/(R + 1), may then be estimated as 2.9×10^{-5}. Let's assume for the sake of this illustration that this study had statistical power (1 − beta) of 0.60. We can then calculate the post-study probabilities (PPV) for each of the three variants using the formula shown above (Table 6.3).

The calculation shows us that we can continue to have confidence in the association with the rs1801133 variant. The probabilities of the associations with the previously unexplored variants, on the other hand, were barely higher after the study than before it, leaving their reliability very much unresolved. Thus, the prestudy odds can radically change the meaning of a finding from a seemingly confident conclusion to a tantalizing observation with modest hope for confirmation.

	Genome-wide Association Study of Plasma Homocysteine Concentrations Identifies Statistically Significant Associations with Three *MTHFR* Variants. But how do the Positive Predictive Values Compare?			
Table 6.3				
MTHFR variants	Uncorrected *p*	Multivariate *p*	Prestudy probability	Post-study probability (PPV)
rs1801133	5.8×10^{-8}	0.002	0.95	0.997
rs12085006	5.8×10^{-10}	0.02	0.000029	0.00058
rs1999594	6.2×10^{-10}	0.02	0.000029	0.00058

The practical challenges from this Bayesian view of evaluating findings in the largely unexplored field of nutrigenetics are undeniable. An ever-growing number of individual genetic variants and haplotype combinations are just waiting to be linked to one of too-many-to-count specific nutritional conditions to see whether this pair does any of a nearly endless number of wonderful things. Obviously, only some of this astronomical (or should we now coin a new expression, nutrigenetic) number of possible combinations will be true. This means that there is an a priori likelihood that R is going to be a very small number for all newly discovered nutrigenetic interactions. We can only guess how small it really is in a specific instance, but it is usually much smaller than we would like. And if we just blindly explore these kinds of relationships as is typical for GWAS, the value of R will be very low. We just have no a priori reason to assume that there is anything to be found. This means that even some apparently robust associations will eventually disappoint us.

6.4.3 Bias in nutrigenetic science

A bias is any influence that distorts results. The battle against bias is a constant effort in all branches of science, but there are a few particular types of bias that practitioners of nutrigenetics must be especially aware of and guard against.

6.4.3.1 PUBLICATION BIAS

Publication bias is common in all emerging science disciplines. Consider that in the course of their work thousands of investigators generate a range of study data. Which of the results and associated conclusions are most likely to be published? It is easy to see that it is very difficult to publish a detailed report of a novel interaction between a genetic variant and a nutrient that is not statistically significant. Think quickly: when did you last check the most current issue of the Journal of Negative Results in Biomedicine? Before you even ask, this is indeed a peer-reviewed public-access journal. No scientific career has been built on the basis of research that indicated the absence of a nutrigenetic effect. Both the apparent effect size and strength of newly discovered associations tend to be inflated because this is what it takes to get them noticed [42]. Publication bias

A Jar of Fake Bills

Many eyes will glaze over while trying to plow through explanations why the a priori likelihood should have any bearing on the strength of a study finding. How could it be that precisely the same study data mean very different things depending on the hypothesis in question? Does my nice low *p*-value count for nothing? But it is really a fairly simple concept.

Imagine the associations between an outcome and all possible combinations of genetic markers and nutritional factors as little envelopes in an enormous lottery jar, with one envelope for each possible association. Imagine further that each association that holds true is an envelope with a cash prize and all other envelopes are empty. Unfortunately, a certain percentage of the envelopes contain counterfeit bills. Those are the associations that don't hold up to follow-up investigations. It is then easy to see that the likelihood of finding good money (an association that stands the test of time) depends on the percentage of real bills in the jar. If there are no real bills in the jar, no amount of trying will find one. And it should also be obvious that we can always get counterfeit prizes just by opening up a lot of envelopes even if there is no good money in the jar at all. It is just a matter of effort and resources. Unfortunately some of the fake money will not be recognized immediately and using it might get us into trouble, particularly if we don't check each bill very carefully.

Compare the following two fictional scenarios. There are two lottery jars full of envelopes, some of which contain money. Unbeknownst to us, the jar on the left contains no envelopes with good money. Ten of the one thousand envelopes in the jar on the right contain good money. One percent of the envelopes in each jar contain counterfeit bills. The more envelopes you examine from either jar, the more false positives you will get. Only a detailed follow-up analysis may show that all of the bills you recovered from the jar on the left are counterfeit, but that a few from the jar on the right appear to be good money. But you will never know for certain; they may just be particularly good fakes. The Bayesian a priori likelihood R would have helped you not to rely too heavily on your winnings from the jar on the left and not to start spending your newfound fortune (which would be like using the nutrigenetic finding in practice right away). Unfortunately, we need prior evidence to estimate the percentage of good money in each jar. If we had considered that none of the bills from the jar on the left and half of the bills from the jar on the right seemed to be good, we could have adjusted our expectations accordingly.

does not mean that anybody was acting deceptively. It is a matter of human nature as much as of market forces.

An interesting corollary of favoring interesting reports over mundane ones is that a report may be selected for publication because of the controversy it

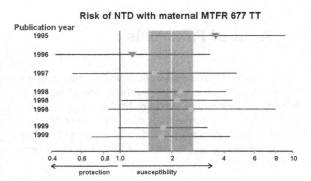

FIGURE 6.10
The Proteus phenomenon relates to the observation that extremely contradictory results are published immediately following an original report, often based on retrospective studies carried out in response to the earlier publication. This figure shows the extreme divergence of the reported impact of the maternal *MTHFR* 677T/T genotype on the child's risk of a neural tube defect. After the early pendulum swings, the reports tend to cluster more tightly [43].

generates [43]. Original publications that are rapidly followed by retrospective studies are particularly likely to generate contradictory results. Such extreme swings in findings have been dubbed the *Proteus phenomenon* (Figure 6.10) after the shape-shifting behavior of the original carrier of the name. Proteus was a son of the Greek god Poseidon and could assume any form he wanted. If you could catch him while he was asleep, he would tell you the future. Wouldn't it be nice, if we could hold on to a report of a gene—nutrient interaction and make it tell us whether it will stay for good?

6.4.3.2 SELECTION BIAS

Selection bias comes about when specific kinds of people are more likely to be recruited into studies than others. For instance, studies may narrowly focus on only one ethnic group. Low activity of intestinal lactase in adulthood was seen as an abnormality and called *lactose intolerance*. This had a lot to do with using Northern Europeans as a reference and most of them just happen to carry the LCT −13910T variant. Now that we know that two-thirds of the adult world population does not express much intestinal lactase, we call the condition *lactase persistence* instead. Selection bias can also occur when exclusion criteria in studies remove people with particular genotypes, for instance because they are more likely to use blood pressure medication.

6.4.3.3 OBSERVATIONAL OR INFORMATION BIAS

Observational or information bias means that some people are misclassified or observations in them are more likely to be wrong than in other people. This could happen, for instance, if people with one allele are falsely assigned to the group with the other allele. This could happen due to a technical genotyping problem that occurs only with one allele and not in the alternative one. More commonly, it will happen because of a mistaken assumption. The LCT -13910T variant is now increasingly used to determine lactase persistence status. This works quite well as long as the practice is restricted to people of Northern European ancestry. Others may not have a LCT -13910T allele but have lactase persistence, nevertheless. How? All it needs is the presence of one of the other variants known to cause lactase persistence. Those variants may be uncommon in Americans of Northern European ancestry but it's possible that one got mixed in a few generations ago.

6.4.3.4 POPULATION STRATIFICATION

Population stratification may not exactly count as a bias (it has a potential confounding effect), but it is an important potential cause for distortion of interactions. This distortion can happen when there are two or more genetically distinct groups in a population and each is exposed to different nongenetic risk factors. An obvious example might be a population with some people of Caucasian ancestry and others of African ancestry. A recent study of women in Birmingham, Alabama, investigated the impact of polymorphisms of uncoupling protein 2 (*UCP2*; OMIM 601693) on insulin sensitivity [44]. Prior studies had suggested that African-Americans have a greater acute insulin response to ingested glucose than European-Americans. The investigators used multiple genetic markers to quantify each participant admixture from the two racial groups. By controlling quantitatively for racial admixture and removing genetic and nongenetic racial effects, they could clarify the previously inconsistent findings and show that women with the UCP2 164C/C genotype (rs660339) secrete more insulin in response to glucose than women with the other genotypes.

Populations often contain genetically distinct groups that can distort results, not just people of Caucasian and African ancestry. This is as good a time as any to remember that one African ethnic group can be more distinct from another African group than from any other population in the rest of the world. The geographical proximity (ignoring the huge size of the continent) can be deceptive and wrongly suggest close genetic similarity.

Case-control studies and population-wide linkage studies are most likely to suffer from this problem because they are built on the assumption of random mating. In contrast, family studies are not likely to be affected.

6.4.4 Replication

The ultimate question in research will always be whether findings hold up to further investigations over time. A recent review by a number of nutrigenetics experts emphasized the high priority of replicating and validating reported genotype-specific responses to nutrition [45]. But what constitutes sufficient replication? Simply put, the same association needs to be found with adequate statistical significance at least once by other investigators who were not involved in the original investigation. Ideally, this follow-up investigation would be undertaken specifically to test the originally reported association. Unfortunately, this does not usually happen.

A comprehensive review of genetic association studies found that follow-up replication studies were not available for the vast majority (72%) of 603 initially reported findings [46]. Only 16% of the associations were observed at least once again, although they were often balanced by studies that did not find the same association. Only 1% was replicated at least twice.

Even if the finding is replicated, the size of the difference between responder and nonresponder tends to be smaller than originally reported [42]. This phenomenon has a lot to do with publication bias as explained above. Other potential problems can come from choices made for the statistical analysis and for displaying results in the reports. None of these are fatal flaws. They just make it less likely that the originally reported effect will be exactly duplicated by others.

The main take-home message here is that we cannot assume that an association is real or as strong as reported just because it has been published by a reputable source. We need to look for genotype-specific responses to nutrition that have staying power. Meta-analysis is a method for the quantitative analysis of results from multiple investigations of the same question. Unfortunately, we need a good number of studies before we can combine any data. In other words, (much) more research is needed.

Practice Questions

How can we know whether elderly women with the *MTHFR* 677T/T genotype need more food folate than other women?

What model system would you use to find out how the effect of saturated fat on apolipoprotein A-II impacts body composition?

Dizygotic (17.3%) and monozygotic (33.4%) twins were found to be discordant for kidney stones in an investigation of US veterans. What does this tell us about the heritability of the condition?

Why is the first report on the investigation of a genotype-specific nutrient effect on children's body mass index more likely to be false than the sixth report about research with a comparable design?

What is the likely mode of inheritance in a family in which the mother, both daughters, and all three granddaughters have hypomagnesemia, but none of the three sons or five grandsons do?

What is the likelihood that the daughter of a male with G6PD deficiency, an X-linked recessive condition, will suffer from hemolysis after eating broad (fava) beans?

Select one of the nutrigenetic interactions presented in this book and evaluate, based on the available literature, the robustness of the claim.

SUMMARY AND SEGUE TO THE NEXT CHAPTER

Studies using a wide range of cultured cells and animal models help to explore nutritional mechanisms and functional effects of genetic variants, but the results cannot reliably tell us what this means for humans. Human studies are indispensable and we should never draw conclusions about gene–nutrient interactions without them. But human studies have their own pitfalls. We need to investigate responsiveness to nutritional factors in sufficiently large populations to identify the responsible genetic variants. As in all population studies, analytical, statistical, and interpretational errors are always conspiring to confuse us. An awareness of these potential problems will help us to keep reported findings in perspective. Most important is patient reassessment and replication of the first findings before relying on them for making critical decisions.

Now, let us assume that the evidence is strong enough, but is it practical? The next chapter will explore the potential for the actual use of promising nutrigenetic combinations.

References

[1] Sofi F, Conti AA, Gori AM, Eliana Luisi ML, Casini A, Abbate R, et al. Coffee consumption and risk of coronary heart disease: a meta-analysis. Nutrition, metabolism, and cardiovascular diseases: NMCD 2007;17(3):209−23.

[2] Cornelis MC, El-Sohemy A, Kabagambe EK, Campos H. Coffee, CYP1A2 genotype, and risk of myocardial infarction. JAMA 2006;295(10):1135−41.

[3] Fernandez-Canon JM, Granadino B, Beltran-Valero de Bernabe D, Renedo M, Fernandez-Ruiz E, Penalva MA, et al. The molecular basis of alkaptonuria. Nat Genet 1996;14(1):19−24.

[4] Martin YN, Salavaggione OE, ECkloff BW, Wieben ED, Schaid DJ, Weinshilboum RM. Human methylenetetrahydrofolate reductase pharmacogenomics: gene resequencing and functional genomics. Pharmacogenet Genomics 2006;16(4):265−77.

[5] Zarse K, Schmeisser S, Birringer M, Falk E, Schmoll D, Ristow M. Differential effects of resveratrol and SRT1720 on lifespan of adult Caenorhabditis elegans. Horm Metab Res 2010;42(12):837−9.

[6] Zheng D, Feeney GP, Kille P, Hogstrand C. Regulation of ZIP and ZnT zinc transporters in zebrafish gill: zinc repression of ZIP10 transcription by an intronic MRE cluster. Physiol Genomics 2008;34(2):205−14.

[7] Shimano H. Novel qualitative aspects of tissue fatty acids related to metabolic regulation: Lessons from Elov16 knockout. Prog Lipid Res 2012;51(3):267−71.

[8] Sullivan PM, Mezdour H, Aratani Y, Knouff C, Najib J, Reddick RL, et al. Targeted replacement of the mouse apolipoprotein E gene with the common human APOE3 allele enhances diet-induced hypercholesterolemia and atherosclerosis. J Biol Chem 1997;272(29):17972−80.

[9] Sullivan PM, Mezdour H, Quarfordt SH, Maeda N. Type III hyperlipoproteinemia and spontaneous atherosclerosis in mice resulting from gene replacement of mouse Apoe with human Apoe*2. J Clin Invest 1998;102(1):130−5.

[10] Mihovilovic M, Robinette JB, DeKroon RM, Sullivan PM, Strittmatter WJ. High-fat/high-cholesterol diet promotes a S1P receptor-mediated antiapoptotic activity for VLDL. J Lipid Res 2007;48(4):806−15.

[11] Huebbe P, Nebel A, Siegert S, Moehring J, Boesch-Saadatmandi C, Most E, et al. APOE epsilon4 is associated with higher vitamin D levels in targeted replacement mice and humans. FASEB J 2011;25(9):3262−70.

[12] Mathes WF, Kelly SA, Pomp D. Advances in comparative genetics: influence of genetics on obesity. Br J Nutr 2011;106(Suppl. 1):S1−10.

[13] Lokki AI, Jarvela I, Israelsson E, Maiga B, Troye-Blomberg M, Dolo A, et al. Lactase persistence genotypes and malaria susceptibility in Fulani of Mali. Malar J 2011;10:9.

[14] Mulcare CA, Weale ME, Jones AL, Connell B, Zeitlyn D, Tarekegn A, et al. The T allele of a single-nucleotide polymorphism 13.9 kb upstream of the lactase gene (LCT) (C-13.9kbT) does not predict or cause the lactase-persistence phenotype in Africans. Am J Hum Genet 2004;74(6):1102−10.

[15] Gallego Romero I, Basu Mallick C, Liebert A, Crivellaro F, Chaubey G, Itan Y, et al. Herders of Indian and European Cattle Share Their Predominant Allele for Lactase Persistence. Mol Biol Evol 2012 Jan;29(1):249−60.

[16] Hannelius U, Gherman L, Makela VV, Lindstedt A, Zucchelli M, Lagerberg C, et al. Large-scale zygosity testing using single nucleotide polymorphisms. Twin Res Hum Genet 2007;10(4):604−25.

[17] Wood AC, Neale MC. Twin studies and their implications for molecular genetic studies: endophenotypes integrate quantitative and molecular genetics in ADHD research. J Am Acad Child Adolesc Psychiatry 2010;49(9):874−83.

[18] Llewellyn CH, van Jaarsveld CH, Boniface D, Carnell S, Wardle J. Eating rate is a heritable phenotype related to weight in children. Am J Clin Nutr 2008;88(6):1560−6.

[19] Malamos B, Koutras DA, Kostamis P, Rigopoulos GA, Zerefos NS, Yataganas XA. Endemic goitre in Greece: a study of 379 twin pairs. J Med Genet 1967;4(1):16−18.

[20] Engelman CD, Fingerlin TE, Langefeld CD, Hicks PJ, Rich SS, Wagenknecht LE, et al. Genetic and environmental determinants of 25-hydroxyvitamin D and 1,25-dihydrox-yvitamin D levels in Hispanic and African Americans. J Clin Endocrinol Metab 2008;93(9):3381−8.

[21] Hazra A, Kraft P, Lazarus R, Chen C, Chanock SJ, Jacques P, et al. Genome-wide significant predictors of metabolites in the one-carbon metabolism pathway. Hum Mol Genet 2009;18(23):4677−87.

[22] Mendoza Torres E, Varela Prieto LL, Villarreal Camacho JL, Villanueva Torregroza DA. Diagnosis of adult-type hypolactasia/lactase persistence: genotyping of single nucleotide polymorphism (SNP C/T-13910) is not consistent with breath test in Colombian Caribbean population. Arquivos de Gastroenterologia 2012;49(1):5−8.

[23] Moleres A, Ochoa MC, Rendo-Urteaga T, Martinez-Gonzalez MA, Azcona San Julian MC, Martinez JA, et al. Dietary fatty acid distribution modifies obesity risk linked to the rs9939609 polymorphism of the fat mass and obesity-associated gene in a Spanish case-control study of children. Br J Nutr 2011:1−6.

[24] Katan MB. Commentary: Mendelian Randomization, 18 years on. Int J Epidemiol 2004;33(1):10−1.

[25] Trompet S, Jukema JW, Katan MB, Blauw GJ, Sattar N, Buckley B, et al. Apolipoprotein e genotype, plasma cholesterol, and cancer: a Mendelian randomization study. Am J Epidemiol 2009;170(11):1415−21.

[26] Burgess S, Thompson SG. Bias in causal estimates from Mendelian randomization studies with weak instruments. Stat Med 2011;30(11):1312−23.

[27] Khandekar MJ, Cohen P, Spiegelman BM. Molecular mechanisms of cancer development in obesity. Nat Rev Cancer 2011;11(12):886−95.

[28] McLaren CE, Garner CP, Constantine CC, McLachlan S, Vulpe CD, Snively BM, et al. Genome-wide association study identifies genetic loci associated with iron deficiency. PLoS One 2011;6(3):e17390.

[29] Shaheen NJ, Silverman LM, Keku T, Lawrence LB, Rohlfs EM, Martin CF, et al. Association between hemochromatosis (HFE) gene mutation carrier status and the risk of colon cancer. J Natl Cancer Inst 2003;95(2):154−9.

[30] Chan AT, Ma J, Tranah GJ, Giovannucci EL, Rifai N, Hunter DJ, et al. Hemochromatosis gene mutations, body iron stores, dietary iron, and risk of colorectal adenoma in women. J Natl Cancer Inst 2005;97(12):917−26.

[31] Higuchi S, Matsushita S, Muramatsu T, Murayama M, Hayashida M. Alcohol and aldehyde dehydrogenase genotypes and drinking behavior in Japanese. Alcohol Clin Exp Res 1996;20(3):493−7.

[32] Oze I, Matsuo K, Wakai K, Nagata C, Mizoue T, Tanaka K, et al. Alcohol drinking and esophageal cancer risk: an evaluation based on a systematic review of epidemiologic evidence among the Japanese population. Jpn J Clin Oncol 2011;41(5):677−92.

[33] Corella D, Tai ES, Sorli JV, Chew SK, Coltell O, Sotos-Prieto M, et al. Association between the APOA2 promoter polymorphism and body weight in Mediterranean and Asian populations: replication of a gene-saturated fat interaction. Int J Obes (Lond) 2011;35(5):666−75.

[34] Garaulet M, Corbalan MD, Madrid JA, Morales E, Baraza JC, Lee YC, et al. CLOCK gene is implicated in weight reduction in obese patients participating in a dietary programme based on the Mediterranean diet. Int J Obes (Lond) 2010;34(3):516−23.

[35] Garaulet M, Lee YC, Shen J, Parnell LD, Arnett DK, Tsai MY, et al. CLOCK genetic variation and metabolic syndrome risk: modulation by monounsaturated fatty acids. Am J Clin Nutr 2009;90(6):1466–75.

[36] Corella D, Arnett DK, Tucker KL, Kabagambe EK, Tsai M, Parnell LD, et al. A High Intake of Saturated Fatty Acids Strengthens the Association between the Fat Mass and Obesity-Associated Gene and BMI. J Nutr 2011;141(12):2219–25.

[37] Phillips CM, Goumidi L, Bertrais S, Field MR, Peloso GM, Shen J, et al. Dietary saturated fat modulates the association between STAT3 polymorphisms and abdominal obesity in adults. J Nutr 2009;139(11):2011–17.

[38] Ashfield-Watt PA, Pullin CH, Whiting JM, Clark ZE, Moat SJ, Newcombe RG, et al. Methylenetetrahydrofolate reductase 677C—>T genotype modulates homocysteine responses to a folate-rich diet or a low-dose folic acid supplement: a randomized controlled trial. Am J Clin Nutr 2002;76(1):180–6.

[39] Sarkkinen E, Korhonen M, Erkkila A, Ebeling T, Uusitupa M. Effect of apolipoprotein E polymorphism on serum lipid response to the separate modification of dietary fat and dietary cholesterol. Am J Clin Nutr 1998;68(6):1215–22.

[40] Kuhn TS. The structure of scientific revolutions. 3rd ed. Chicago, IL: University of Chicago Press; 1996.

[41] Ioannidis JP. Why most published research findings are false. PLoS Med 2005;2(8):e124.

[42] Ioannidis JP. Why most discovered true associations are inflated. Epidemiology 2008;19(5):640–8.

[43] Ioannidis JP, Trikalinos TA. Early extreme contradictory estimates may appear in published research: the Proteus phenomenon in molecular genetics research and randomized trials. J Clin Epidemiol 2005;58(6):543–9.

[44] Willig AL, Casazza KR, Divers J, Bigham AW, Gower BA, Hunter GR, et al. Uncoupling protein 2 Ala55Val polymorphism is associated with a higher acute insulin response to glucose. Metabolism: clinical and experimental 2009;58(6):877–81.

[45] Fenech M, El-Sohemy A, Cahill L, Ferguson LR, French TA, Tai ES, et al. Nutrigenetics and nutrigenomics: viewpoints on the current status and applications in nutrition research and practice. J Nutrigenetics Nutrigenomics 2011;4(2):69–89.

[46] Hirschhorn JN, Lohmueller K, Byrne E, Hirschhorn K. A comprehensive review of genetic association studies. Genet Med 2002;4(2):45–61.

CHAPTER 7
Practical Uses of Nutrigenetics

Terms of the Trade

- DRI: The Dietary Reference Intakes summarizes intake recommendations for healthy people.
- EAR: Estimated average requirement is the intake level for assessing adequacy of groups.
- Multimodal distribution: Non-normal distribution generated by subgroups with different means.
- RDA: Recommended dietary allowance, the intake level that meets the needs of most people.

ABSTRACT

There are many different uses of nutrigenetic science and practice. The most important consideration is to become familiar with the concepts and consider individual responsiveness to nutrition. Nutrigenetic tests have to be selected with careful attention to the expected health benefit, performance of the required laboratory analyses, and cost. Ethical concerns often relate to autonomy and privacy. Patients and clients must understand what they are getting themselves into at the outset. All genetic tests require explicit and clearly documented informed consent. Healthcare providers need to know that genetic analyses can give false results, although not as often as many other laboratory tests, and that the evidence base for the interpretation of results can change with new research.

Interest in the practical application of personalized nutrition continues to grow. It is important to understand that average consumers cannot change their nutrition profile without in-depth guidance. This may be done with individualized text messages or with more comprehensive computer-based nutrition guidance.

A better understanding of the different nutritional needs of individuals must find its way into the development of population-wide guidelines and all levels of

Nutrigenetics. http://dx.doi.org/10.1016/B978-0-12-385900-6.00007-1

nutrition research. It is no longer acceptable to assume that most people will respond in more or less the same way to a given nutrition factor. Development of nutritional guidelines has to make use of the new analytical and interpretative tools and frameworks. A similar challenge for research is to actively embrace genetic diversity and tap into the lessons of human diversity.

7.1 GENETIC INHERITANCE IS JUST A FIRST OFFER

Nutritionists and other healthcare providers are facing a growing wave of questions about the risks and benefits of personal genetic information as billions of data bits become readily available. What does it all mean? What can I do about it? Don't expect to find all the answers here—just food for thought to nurture your professional expertise.

It is very important to understand that common genetic variants exist because they have served some of our ancestors well. Since many polymorphisms are adaptations to particular nutritional circumstances (nutritopes), we can often improve health outcomes by adjusting lifestyle and treatments to achieve a closer fit with the individual genome's functional settings.

The challenge is then to develop strategies that provide the best nutrition to the largest number of people. This can be done for one person at a time using genetic information to guide personal lifestyle choices and tailor clinical interventions. Opportunities for the use of nutrigenetic strategies do not stop at the individual level. You should always consider the implications of an individual client's genetic information for his or her close relatives, and not least for current or future children. Knowledge of individual responsiveness to specific nutritional factors can direct research toward achieving deeper insights into the mechanisms of health maintenance and disease processes. Recognition of the diverse nutritional needs of groups or populations should also guide the development of dietary guidelines, nutritional recommendations, treatment strategies, and food programs. We do not always need to know everybody's genetic predisposition to achieve better health outcomes. Just taking into account known differences in response to nutritional exposures can improve our decision-making processes at all levels, from the design of intervention strategies to the development of nutrition policies.

7.2 PERSONAL NUTRITION

7.2.1 Criteria for selecting genetic tests

The first question about any clinical test has to be whether it provides a demonstrable benefit that is worth the cost and significantly greater than potential harms. This is no different for nutrigenetic tests than for other tests, such as a blood cholesterol test or an HIV test.

First, we have to understand the nature of the test. Does the test detect just the presence or absence of a specific genetic allele or genotype? This question is important because genetic tests are increasingly doing much more, such as providing the DNA sequence around the variant position, or maybe even the entire gene. Will this added information be just ignored or can it be used in some way? What will we do if we discover new variants or variants with unknown biological impact? We have to know about the sensitivity (i.e., the percentage of correctly identified people with the allele or genotype of interest) and specificity (i.e., the percentage of correctly identified people without the allele or genotype of interest) of the test.

Then we have to define and understand the group of people for which the test will be used. Few will question that it makes sense to analyze the *ornithine transcarbamoylase* (*OTC*; OMIM 300461) gene of a young woman who just barely survived an episode of excessive ammonia accumulation due to OTC deficiency, as described in Chapter 1. Knowing the precise nature of the gene defect will help evaluate her risk and simplify necessary follow-up testing of relatives. We can weigh the benefits and potential harms from testing because we can find out everything we need to know about the circumstances of this patient. Contrast this targeted use of genetic testing with the proposition to screen all women for genetic OTC variants. Here, we encounter mostly unknowns and have to infer needed information from surveys and other information sampling. Remember that we would have to test about 14,000 women for each individual with a serious *OTC* gene defect, one that presumably works just like the defect in known cases. This seems simple enough. Find an *OTC* variant and you have a new patient. Of course, you know enough by now to know that this is not how it works. There will be some variants that have occurred in previous patients with known OTC deficiency. But we don't necessarily know whether some other variant nearby or far away epistatically changes the impact of that observed variant. It is important to remember that many different parts of the genome (genotype) working together in the context of anatomy and metabolism of a particular person generate the physical presentation (phenotype). This means that all kinds of genomic variation may be important, including copy number variation, epigenetic modifications, and the local and extended haplotype. This makes it very difficult to predict rare phenotypes with reasonable sensitivity and specificity based on genetic information, not least because there are few cases to learn from. The other side of the nutrigenetic coin is the usual nutritional state of the group. Some genetic variants make carriers more vulnerable to low or high intake. If there is no increased risk for these responders at current intake recommendations, we don't need to know about the genetic difference because the advice will be the same for carriers and noncarriers.

A REMINDER

Epistasis refers to an effect of a DNA sequence on the function of a distant gene.

Once the scope of the test and the target population are understood, we can tackle the question of modifiable risk. The question is, as indicated before, how much benefit the proposed genetic test provides for the specified target group. It is a common misunderstanding that risk associated with a variant has to be high to make it important. It is quite possible that at the same initial risk level, responders may have a significant risk reduction and nonresponders don't. At low soy intake, for example, men with and without the ESR2−13950C variant (rs2987983) have about the same prostate cancer risk. Eating tofu a few times a week appears to reduce risk for men with the ESR2−13950C variant, but raise it for men without the variant. Determining the ESR2−13950 genotype might be a very helpful test, then, at least for men who consider using soy or an isoflavone supplement for keeping their prostate cancer risk in check.

The only remaining item for supporting practical use would be confirmation in a few other large cohorts that the test can reliably (i.e., with adequate sensitivity and specificity) distinguish between men who have a lower prostate cancer risk with high isoflavone intake and those who do not have this benefit. Ideally, this ability to discriminate responders from nonresponders would come from prospective intervention studies. Of course, none of the current recommendations in the Dietary Reference Intake (DRI) publications meet this high bar, nor are they likely to in the foreseeable future. The generally ignored question is whether genotype-specific guidance needs to be supported by a stronger evidence base than that of current official nutrition guidelines.

7.2.2 Ethical considerations

Decisions about test use involve more than just test performance or expected benefit. Many questions about tangible and intangible risks of the tested individual have to be asked. Then there are potential implications for relatives and for society as a whole. Actually, everyone considering such tests needs to ponder such questions and realize that genetic testing is not a game. The entertainment value may be more than they bargained and paid for. Healthcare professionals need to recognize the scope of the issues and be prepared to provide information and guidance.

7.2.2.1 CHOICE

There are very few circumstances that justify compulsion for a genetic test. One of the conceivable exceptions is paternity testing and this always requires a court order. Another reasonable exception is newborn screening, which is required by law in many jurisdictions. Employers or insurers should not require genetic testing, particularly as it relates to nutrigenetic predispositions. Laws in some countries already impose restrictions on such required testing but are rarely stringent and proactive enough to prevent abuse. It is the personal obligation of healthcare providers and of all staff at diagnostic facilities to resist unethical or illegal practices.

7.2.2.2 SELF-DISCLOSURE

The average consumer does not realize the potential burden of knowing too much about their own fate, particularly when it comes to risks they have not even thought about. How would they feel about knowing that they have a particularly high risk of cancer, heart disease, or premature dementia? Would they mind the dietary constraints compelled by the results? What about finding out about inherited personality traits, unexpected paternity issues, or unpopular ancestry? Will they have the inner strength and external support to deal with potentially difficult or even shocking information?

7.2.2.3 INFORMATIONAL AUTONOMY

People's attitudes about sharing personal information vary greatly and change over time. A starting point for a conversation about their feelings could be whether they would be comfortable if their genetic information was posted on a website or sent to them on a postcard. Would they mind if their detailed genetic information was known to their family, friend, neighbors, coworkers, or employers? How about government agencies or insurance companies? Which parts of the information would they want to keep to themselves? Do they know that information cannot be made private again after it has been released? Would they rather not have their genetic information circulate in unknown files for the next 50 or more years? Are they comfortable with transferring ownership of their genomic information to the diagnostic laboratory or another entity, even when all identifying personal information is removed?

7.2.3 Analytical implementation

Methods for the detection of genetic variants continue to evolve very rapidly and will not be discussed here in detail. Suffice to say that whole-genome or targeted sequencing will play an increasingly important role in routine testing. Tests directed at single base polymorphisms will be used less often because there are just too many of them with an important impact on any outcome of interest. Narrowly targeted tests will become more important for the detection of deletions, insertions, copy number variants, and epigenetic signatures. There are a few principles that apply to virtually all tests, regardless which methodologies and targets are used for testing. The Centers for Disease Control and Prevention (CDC) have published a comprehensive description of current best practices in molecular genetic testing [1].

7.2.3.1 ACCREDITATION

Accreditation of laboratories ensures minimum standards for clinical or commercial testing services. In the USA, the Clinical Laboratory Improvement Amendments of 1988 (CLIA) define the procedures and obligations of diagnostic facilities. These regulations define many details for the facilities and procedures of genetic laboratories, including the qualifications of personnel, quality control procedures, and external oversight. New York State and California have additional rules for genetic testing. Many other countries have analogous and often even more stringent regulations.

7.2.3.2 INFORMED CONSENT

Informed consent is obligatory for all genetic tests. Consent by proxy is extremely problematic for all but the most consequential nutrigenetic conditions, such as phenylketonuria or urea cycle defects. The client or patient needs to understand the implications of the test and have realistic expectations about typical risks and benefits. Due to the nature of genetic tests, they may reveal more about personal circumstances and other sensitive information than desired. There also needs to be education about potential errors in the analytical assessment and in the genotype-specific advice given. The client or patient needs to know that information about the significance of results can change in unforeseeable ways. No test should be carried out without documentation of explicit consent on file.

7.2.3.3 SPECIMEN HANDLING

Specimen handling is a perennial source of error, more often when clients do the sampling on their own instead of having it done in a professional laboratory. Comprehensive training of staff and step-by-step instructions in writing are essential. Just as important is seamless oversight and rigorous enforcement of compliance with the instructions. Sample mix-ups in molecular genetic laboratories may occur as often as once per thousand submissions [2].

7.2.3.4 EVALUATION AND VALIDATION

Evaluation and validation of all components of a test result have to be done properly. This means that we need to know the error rate of an assay and whether there are any circumstances that make an analytical error more likely. Sensitivity (rate of the correctly classified variant genotype) and specificity (rate of the correctly classified normal genotype) need to be documented and should be available on request. It is important to assess the performance in diverse populations because there might be interfering sequence variants that occur in one population but not in another. This evaluation has to be carried out for every individual variant, regardless of how many thousands are used for nutrigenetic testing.

7.2.3.5 ERROR RATES

Proficiency testing evaluates the ability of genetic laboratories to correctly classify the factor V (FV) Leiden variant (F5 1691G>A; rs6025), to highlight a commonly tested example. The false-negative rate is about 1.2%; the false-positive rate about 0.7%. Such error rates for a single, well-characterized variant are much higher than many expect and may have to do with testing by less experienced and less specialized laboratories. Part of the problem reflected in high error rate is the vulnerability of tests to unanticipated genetic interference as previously mentioned. In the case of the FV Leiden variant, it is the failure of some test systems in samples with nearby variants [3]. When proficiency tests contain such 'difficult' samples, several laboratories using the same test system will fail and we get the reported high error rates. The same laboratories might have gotten zero errors with 'easy' test samples.

Error rates with genome-wide testing in specialized facilities appear to be much lower, but some may still give the wrong result for 0.3% of the variants [4]. Such an error rate is lower than what is reported for other clinical laboratory data, but genotyping errors can have long-lasting consequences for misdiagnosed individuals. It is important to understand likely error rates when ordering tests and discussing them with clients or patients.

Error rates can be reduced to a minimum with uncompromising attention to the entire diagnostic process from sample intake to reporting [5]. An important approach to the elimination of seemingly intractable sporadic errors is duplicate analysis through independent processing streams and in separate runs. One might even argue that samples should be obtained on two separate occasions. Duplicate measurements reduce error rates for the individual and they can identify weaknesses in the analytical process. The higher cost is justified in light of the lifelong consequences of false assignment of a genetic variant and to resulting nutritional instructions. The goal is to reduce laboratory-related error rates to well below one in 100,000, which is called the *six sigma range* by business management experts. When a large number of genetic variants are to be integrated into nutritional guidance, a high error rate would add up to unacceptable levels and undermine confidence in the usefulness of nutrigenetic testing.

KEY CONCEPT
Samples for personal nutrition should always be analyzed twice in separate runs.

7.2.3.6 QUALITY CONTROL AND EXTERNAL PROFICIENCY TESTING

Quality control and external proficiency testing help to maintain performance at an acceptable level. Negative and positive controls have to be included with each analytical run. In addition, amplification-based tests have to include at least one suitable template-free control with each run to demonstrate that the test did not suffer from contamination. The performance of quality controls has to be documented and any deviations researched and their cause resolved before any further runs.

Cumulative quality control records from the previous months and external proficiency results should be available to healthcare professionals on request. They are the laboratory's calling card and consistently good performance inspires confidence. This is comparable to the sanitary rating that restaurants have to display.

External quality assessment of clinical genetic laboratories in the UK goes further than schemes in many other countries by querying both the knowledge of key staff and laboratory performance. Inadequate performance is actively followed up with further testing and monitoring [6]. The only nutrigenetic test included so far in this scheme, however, is for cystic fibrosis. The lack of attention and oversight over other nutrigenetic tests in the UK reflects the reality in the rest of the world.

7.2.3.7 INTERPRETATION AND REPORTING

Interpretation and reporting are integral parts of the diagnostic testing process. For nutrigenetic analyses, the interpretation becomes paramount because the effect size tends to be small and the evidence base is often thin. Starting with identical raw data, the interpretations given to requesters of direct-to-consumer tests have sometimes been very different. Where one provider suggested that the risk of rheumatoid arthritis was lower than average, another provider predicted a much higher risk [4]. This gives just a hint of the challenges that lie ahead with more widely adopted nutrigenetic testing.

There are also important considerations about confidentiality and inappropriate disclosure to the patient or client, as discussed in Chapter 8. The practical implications of the test result should get priority over the simple listing of test data that are not understood or helpful for healthcare professionals or patients/ clients. This becomes particularly obvious if dozens of variants are tested. Many of the variants may make sense only in context because they interact with other variants and factors.

7.2.4 Targeted nutrition interventions

Clinical genetic counseling has traditionally dealt with one defective gene or chromosome region at a time because severe genetic disorders are rarely caused by variants in multiple regions. We find an equivalent scenario with a few nutrigenetic variants, usually those with obvious and severe health conse-quences. In many ways, these severe, monogenic nutrigenetic diseases have served as a bridge between clinical genetics and the more recently emerging multigenic risk factors. Clinical genetics builds on a small number of highly trained specialists and provides careful evaluation of individual patients and their immediate families. Many hours are spent providing the best available guidance for rare and often even unique diseases. An important part of a clinical geneticist's work is to provide counseling to couples about their risk of having a child with a significant genetic disorder. Multigenic risk factors predispose to a wide range of serious diseases and conditions such as cancer, cardiovascular disease, or diabetes. But, because they do so in concert with many other genetic and nongenetic factors and are very common, they do not fall into the usual domain of the already overtaxed clinical geneticists. Besides, until the advent of nutrigenetics, therapeutic options did not usually depend on a full genetic evaluation. Fortunately, very few people need the resource-intensive attention of clinical genetics, but every single one of us has many genetic variants that have relevant and often preventable health consequences. In other words, the difference between clinical genetics and more broadly conceived nutrigenetics reflects the contrast between retail patient care and population-wide lifestyle guidance.

The best examples of nutrigenetic diseases in the domain of clinical genetics are all the rare inborn errors of metabolism, such as biotinidase deficiency, maple syrup disease, phenylketonuria (PKU). The incidences of the most common of these are well below one in ten thousand. We screen for them, affected patients get personal attention from a clinician, and healthcare resources are usually up to the task. Celiac disease (CD) and hemochromatosis occupy an intermediate position both in terms of severity and incidence. There is no systematic screening mechanism in place and their detection is delayed until symptoms make themselves felt. Carriers of the risk genes receive individual care, although not always by specialists. Genetic variants that shift the individual intake optimum for a nutrient or food, such as the *MTHFR* 677T allele does for folate and riboflavin intake, are at the other end of the spectrum. More than 10% of many populations are homozygous for this variant and there is simply not enough competent counseling capacity to explain the health significance and reproductive implications to each of the genotype carriers. Care for people with such common variants with a modest health impact is rarely provided in a predictable and effective manner. And this is just one of so many genetic variants that deserve equal or even more attention. The individual healthcare model, that is, one genetic variant at a time, has a hard time dealing with such conditions.

7.2.4.1 HEMOCHROMATOSIS

This iron retention disease is very common and important for health in many countries, for instance accounting for nearly 1% of all mortality in Denmark [7]. First of all, it needs to be understood that family members of individuals with known familial hemochromatosis must be educated about screening. The reason is that early lifestyle adjustment can reduce excess risk. Helpful measures include bloodletting (blood donations) several times a year, tempering alcohol consumption, doing away with iron-containing dietary supplements and iron-fortified cereals, and avoiding eating or handling raw seafood. Regular medical controls monitor iron accumulation but also need to ensure that iron levels do not fall below desirable levels in young women, particularly during pregnancy.

While most experts will agree that these simple preventive measures would reduce disease burden with modest cost (mainly arising from more frequent check-ups), there is not a concerted effort to detect carriers at the earliest possible stage. When it comes to screening, most discussion focuses on the question of whether to use transferrin saturation (the percentage binding capacity of the transferrin molecule filled by iron) or ferritin concentration. Both tests become abnormal only after iron has accumulated to an unhealthy level. Transferrin saturation values often become false positives due to diurnal variation, inflammation, recent food intake, or unrelated liver disease, which means suboptimal sensitivity. Both measurements have limited sensitivity, missing 15–30% of patients at risk, depending on cut-offs. Specificity is also imperfect, which means that a much larger number of false positives (typically 5–20%) than true positives (about 0.4%) are identified. This is less concerning, since most of the false positives are classified properly with later confirmation tests.

The use of a comprehensive genetic scan and assessment of all potential gene variants that could predispose to iron retention is rapidly coming within our grasp. Such a scan, most likely distilled from whole-genome sequence data, will provide very early information that can be used for preventive measures before significant iron accumulation has occurred. The costs would be modest if sequence data are already available.

7.2.4.2 CELIAC DISEASE

More than 1% of adults in North America and Europe have CD (OMIM 212750), a genetic predisposition to respond with intestinal and systemic inflammation to gliadin (gluten) from barley, rye, wheat, and closely related grains. More details have been already discussed in Chapter 3. As a reminder, the potential health consequences are infertility, immune-mediated disease (Grave's disease, Hashimoto thyroiditis, type 1 diabetes, and uveitis), kidney disease, liver damage, migraines, intestinal lymphoma (T-cell non-Hodgkin lymphoma), osteoporosis, and poor iron status and other nutrient deficits due to malabsorption. The measurement of immunoglobulin A-anti-endomysial antibodies identifies individuals with currently active disease and predicts future development of the disease in asymptomatic individuals [8]. Inheritance of susceptibility to the disease (involving HLA DQ2 or DQ8 and variants in over 40 additional genes) is a necessary requirement but is not sufficient on its own. Exposure to gluten is another necessary requirement. Recent episodes of infectious gastroenteritis may constitute an additional trigger for disease manifestation [9].

CD is clearly a nutrigenetic condition that needs to be searched out proactively because treatment is unquestionably effective. There are strong arguments for population-based screening, though it may not be sufficiently cost effective. Not to initiate blood-based testing of people with potentially related conditions (anemia, immune-mediated disease, infertility, migraine, or type 1 diabetes) or in close relatives of individuals with known CD is nothing short of a healthcare failure. Determined nutrition guidance and persistent follow-up health assessments of patients with CD help to prevent new CD-related disease and reduce the severity of existing disease [10].

7.2.5 Comprehensive lifestyle tailoring

We have already seen numerous examples of genotype-specific responses to a wide range of nutritional factors. In the end, all of those hundreds or thousands of variants with relevant nutritional impact generate a unique profile of personal nutrition targets. In one person this may be folate and vitamin C intake targets slightly above average, sodium and iron intake targets below average, and so forth. Another individual may be able to tolerate a bit more sodium, and need only an average amount of folate but extra iron and vitamin C. A third one may have CD shaping most other individual nutrition needs. Not many people are likely to have the same or even closely similar profiles. How will we be able to reconcile all these different profiles and make the right food selections every day?

The usual approach up to now has been to inform individuals of their various genotypes and explain to them how to optimize their nutrition for each genotype. Different combinations of genetic tests are marketed direct to consumers or to healthcare providers. One of the expectations is that the genetic tests increase interest in good nutrition and this may be true in some instances. For instance, a randomized controlled study compared the effect of genotype-specific advice on nutrition behavior [11]. They analyzed five common genetic variants (CYP1A2,

GSTM1, GSTT1, TAS1R2, and ACE) and prepared a personalized list of dietary recommendations for caffeine, sodium, sugar, and vitamin C. This is a scheme that is commercially marketed to dietitians.

The following texts illustrate genotype-specific (left-hand side) and generic recommendations [11]:

Health Canada's recommendation for caffeine is at most 300 mg/day for women of child-bearing age and at most 400 mg/day for other adults. Since you have the CC version of the CYP1A2 gene, you might benefit from limiting your caffeine intake to no more than 200 mg/day. Caffeine is found in coffee, tea, cola beverages, and energy drinks. One small (8 oz) cup of coffee contains about 100 mg of caffeine, while an 8 oz cup of tea contains about 50 mg of caffeine. One can (355 mL) of cola contains about 30 mg of caffeine, while the caffeine content of energy drinks can range from 80 to 200 mg depending on the serving size and brand.

Health Canada's recommendation for caffeine is at most 300 mg/day for women of child-bearing age and at most 400 mg/day for other adults. Caffeine is found in coffee, tea, cola beverages, and energy drinks. One small (8 oz) cup of coffee contains about 100 mg of caffeine, while an 8 oz cup of tea contains about 50 mg of caffeine. One can (355 ml) of cola contains about 30 mg of caffeine, while the caffeine content of energy drinks can range from 80 to 200 mg depending on the serving size and brand.

Most participants liked the more detailed, genotype-specific recommendations. Recipients of the genotype-specific messages were slightly more likely than recipients of the generic recommendations (88% vs. 72%) to respond that the recommendations would be useful when considering their diet. Unfortunately, there is no evidence that they will do so or actually adjust their dietary pattern according to directions in the short or long term.

A major limitation with this approach is that the complexity of food choices increases exponentially with every additional nutrient or food group. It is hard enough to work with clients and patients on one or two targets, for instance cholesterol and saturated fat. But try yourself for just a few days to match your food selections to the individual nutrition targets in Table 7.1. You will have to look up your own age- and gender-specific recommended daily allowance (RDA) from the most current DRI tables and then make the indicated adjustments.

This exercise requires adjusting intakes of only six nutrients and total energy. With truly a comprehensive scope, the majority of nutrients and food groups would have to be adjusted in some individuals.

When you come back from this exercise you will have found out how difficult it is to modify food selections to meet personal targets. There are no templates or easy rules (such as 'have at least five fruits and vegetables a day') that can serve for guidance. Everybody has to find out on their own which combinations work for them.

An even more serious problem relates to the necessary disclosure of raw genetic information to the user. Usually this is done without the critically important

Table 7.1 Genotype-Specific Nutrient Intake Targets for a Personalized Nutrition Experience

Food/ Nutrient	Enter your RDA here	Multiply the RDA by this Adjustment Factor	*Genetic Variant that Drives this Adjustment Factor*	Enter how much you Need here
Energy		0.925	*IL-6−174C/C* (rs1800795) [12]	
Sodium		0.8	*AGT−6A/A* (rs5051) [13]	
Total folate		1.5	*MTHFR 677T/T* (rs1801131) [14]	
Folic acid		0.33	*DHFR 19 bp del/del* [15]	
Vitamin C		1.5	*XRCC1 26304T/T* [16]	
Saturated fat		0.5	*APOA2−265C/C* (rs5082) [17]	
Calcium		Calcium/magnesium	*TRPM7 4727G/A*	
Magnesium		ratio under 2.6	(rs8042919) [18]	

RDA, recommended daily allowance.

filter and guidance of a genetic counselor or other comparably trained healthcare professional. The assumption is that the genetic information is so inconsequential that it can do no harm. Unfortunately, new scientific advances and findings can show genetic information to be harmful long after it has been released. What was considered inconsequential at one time may prove to be an untreatable risk factor for a particularly nasty form of cancer or turn out to predispose to a stigmatizing personality trait.

7.2.6 Computer-assisted nutrition guidance

Since we cannot really find food combinations that fit our personal nutrition targets, we need a little bit of help. Computer-assisted nutrition guidance will be indispensable for any practical personal nutrition scheme. A prototype of such a support tool (PONG, Personal Online Nutrition Guidance; Figure 7.1) has been developed (accessible at http:/www.nutrigen.com).

PONG uses polynomial distance functions to calculate how well daily menus meet an individual user's nutritional targets. The nutritional targets consist of a full set of nutrients and food groups that can be individually configured depending on the primary nutrition concerns. The targets for this sample individual can be seen at the bottom left of the display. The user starts with a number of menus that already meet his or her personal nutrition needs (Figure 7.1) and can then make adjustments by adding, deleting, or replacing, items (Figure 7.2).

An important feature of this program is the support for anonymization, as illustrated in Figure 7.3. The scheme for anonymization is called *double masking* in analogy to double-blind testing because critical information is withheld from both the client and the genetic testing facility.

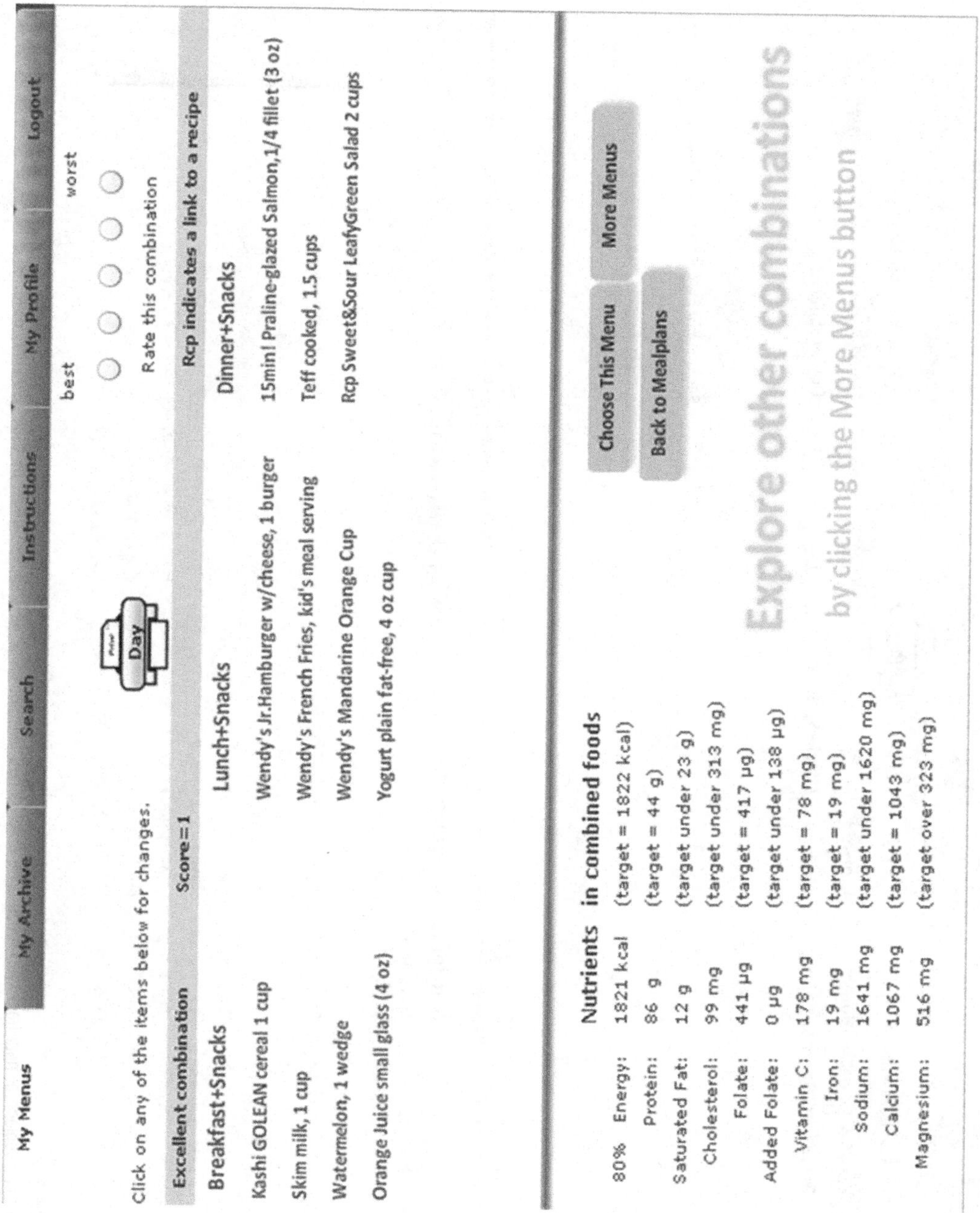

| My Menus | My Archive | Search | Instructions | My Profile | Logout |

Click on any of the items below for changes.

best ○ ○ ○ ○ worst

Rate this combination

Excellent combination   Score=1

Rcp indicates a link to a recipe

Breakfast+Snacks

Kashi GOLEAN cereal 1 cup

Skim milk, 1 cup

Watermelon, 1 wedge

Orange Juice small glass (4 oz)

Lunch+Snacks

Wendy's Jr.Hamburger w/cheese, 1 burger

Wendy's French Fries, kid's meal serving

Wendy's Mandarine Orange Cup

Yogurt plain fat-free, 4 oz cup

Dinner+Snacks

15min1 Praline-glazed Salmon,1/4 fillet (3 oz)

Teff cooked, 1.5 cups

Rcp Sweet&Sour LeafyGreen Salad 2 cups

Choose This Menu   **More Menus**

Back to Mealplans

Explore other combinations

by clicking the More Menus button

	Nutrients	**in combined foods**
80% Energy:	1821 kcal	(target = 1822 kcal)
Protein:	86 g	(target = 44 g)
Saturated Fat:	12 g	(target under 23 g)
Cholesterol:	99 mg	(target under 313 mg)
Folate:	441 µg	(target = 417 µg)
Added Folate:	0 µg	(target under 138 µg)
Vitamin C:	178 mg	(target = 78 mg)
Iron:	19 mg	(target = 19 mg)
Sodium:	1641 mg	(target under 1620 mg)
Calcium:	1067 mg	(target = 1043 mg)
Magnesium:	516 mg	(target over 323 mg)

FIGURE 7.1
Personal online nutrition guidance helps with the creation of daily menus that meet individual intake targets.

My Menus   My Archive   Search   Instructions   My Profile   Logout

Now choose a substitute below on the right:

best   worst

○ ○ ○ ○ ○   Rate this combination

Excellent combination	Score = 1		Rcp indicates a link to a recipe
Breakfast+Snacks	Lunch+Snacks	Dinner+Snacks	

Day

Replace Rcp Sweet&Sour LeafyGreen Salad 2 cu

Nutrients in combined foods

80%	Energy:	1821 kcal	(target = 1822 kcal)
	Protein:	86 g	(target = 44 g)
	Saturated Fat:	12 g	(target under 23 g)
	Cholesterol:	99 mg	(target under 313 mg)
	Folate:	441 µg	(target = 417 µg)
	Added Folate:	0 µg	(target under 138 µg)
	Vitamin C:	178 mg	(target = 78 mg)
	Iron:	19 mg	(target = 19 mg)
	Sodium:	1641 mg	(target under 1620 mg)
	Calcium:	1067 mg	(target = 1043 mg)
	Magnesium:	516 mg	(target over 323 mg)

Select an item below (rated from *** Excellent to -- Poor)

*** Rcp Sweet&Sour LeafyGreen Salad 2 cups

*** Rcp Romaine Salad w/Balsamic Vinaigr.., 2.5 cups

*** Rcp Romaine Salad w/Balsamic Vinaigr.., 3 cups

** Fresh Spinach leaves, 3 cups

* Rcp Romaine Salad w/Balsamic Vinaigr., 2 cups

-- Rcp Cranberry-Spinach Salad, 1 cup

-- Romaine lettuce, shredded 2 cups

-- Fresh Spinach leaves, 2 cups

-- Rcp Caesar Salad, 1.5 cups

FIGURE 7.2

The user can replace a meal item with foods on a color-coded list that presents choices ranked by their goodness of fit.

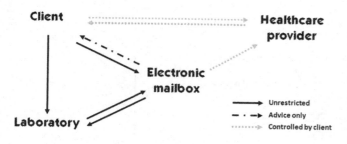

Client

Healthcare provider

Electronic mailbox

Laboratory

→ Unrestricted
— · —▶ Advice only
········▷ Controlled by client

FIGURE 7.3
Anonymization with a double-masking scheme. The laboratory does not know the identity of the client because transmitted samples are only identified with an alphanumeric code. The client gets only nutritional advice, but no raw genetic information. The client can authorize access by a qualified healthcare professional, who can provide additional counseling and may disclose genetic information.

The DNA-containing sample is submitted with a numeric code from the program. The genetic laboratory never knows the identity of the client. The genetic information is only used to adjust individual intake targets and is never made directly available to the client. After the genotype-specific adjustments have been entered, the client can focus on the task at hand, which is putting together a meal plan for the next few days. The client never sees the genetic results because this information is not needed at this stage. The client controls the transfer of any information to a healthcare provider. If the client wants to get more detailed genetic counseling or desires to see the actual genetic information, an accredited healthcare professional can get client-authorized access and then provide guidance and appropriate counseling. This scheme permanently secures genetic information and avoids self-disclosure as well as any unintended disclosure to third parties. Because genetic information is never disclosed, there is no need for intensive genetic counseling. General online information will be adequate to help clients deal with the information.

7.2.7 Nutritional pharmacogenetics

Sometimes we deliberately cause a nutritional imbalance to achieve specific health effects. This may be done with the use of pharmaceutical compounds that antagonize physiological functions of a nutrient or alter its normal metabolic processing. Other applications use much larger doses of a nutrient than is available from regular food sources or use derivatives that are not found in foods. These applications are at the intersection of pharmacogenetics and nutrigenetics because individual genetic predispositions influence the bioavailability and effectiveness of both drug and nutrient.

7.2.7.1 VITAMIN K ANTAGONISTS

Warfarin (Coumadin) and similar compounds put a brake on vitamin K-dependent blood clotting factors in order to prevent the formation of dangerous blot clots and thrombosis. The problem with this therapy is the risk of unchecked bleeding, which can easily be fatal.

Genetic variation in the *VKORC1* gene has great therapeutic relevance. As shown in Figure 7.4, one of the vitamin K-epoxide reductases can be inhibited by

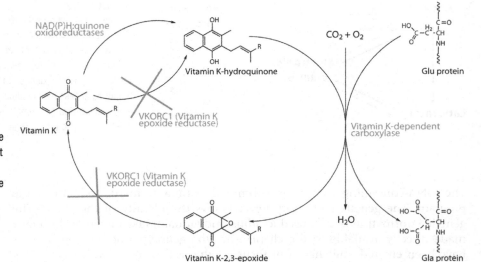

FIGURE 7.4
Vitamin K antagonists like warfarin (Coumadin) limit the availability of active vitamin K by inhibiting the vitamin K-epoxide reductase VKORC1. CO_2, carbon dioxide; H_2O, water; O_2, oxygen molecule.

warfarin (Coumadin) and related drugs, which dramatically reduces the amount of active vitamin available for the gamma-carboxylation of vitamin K-dependent coagulation factors. Anticoagulation is a risky form of therapy, because the dosing should be high enough to prevent blot clots, but no higher, because then the patient might suffer dangerous bleeding. The effect of the medication on coagulation (prothrombin time) is measured frequently and the dose adjusted accordingly. It usually takes months of trial and error to get the dose right for a patient. Because people differ so much in their handling of vitamin K, it is difficult to know at what dose to start. Because the initial dose could be far too high for the patient, the usual approach has been to start with a low dose and then slowly increase the medication incrementally over a period of many weeks and months. Wouldn't it be nice to have some better guidance that estimates a reasonable starting dose without putting the patient at risk? This is what the International Warfarin Pharmacogenetics Consortium has come up with. They developed an algorithm for deciding on the warfarin dose that is most likely to be appropriate for a given patient. The current scheme takes into account a common variant (-1639G>A; rs9923231) in the promoter of the VKORC1 gene [19]. Let's assume we want to start an elderly, slightly overweight (178 cm; 88 kg) Caucasian 72-year-old male on an anticoagulant regimen. The algorithm would then give us for the G/G genotype an estimated weekly warfarin dose of 21 mg, but 39 mg for the A/A genotype. Go to the Supplementary Appendix of the reference [19], open up the MS Excel file, and try it out! By the time you read this text, the algorithm may take into account additional variants.

7.2.7.2 LIPID-LOWERING THERAPY WITH NIACIN

Next, a word about the effect of nicotinic acid on lipoprotein metabolism. It has been known for a long time that large amounts of nicotinic acid lower elevated concentrations of low-density lipoproteins (LDL) and very-low-density lipoproteins by inhibiting the release of fatty acids from adipose tissue. Nicotinic acid also lowers concentrations of the very atherogenic lipoprotein (a) (LPA; OMIM 152200) and raises the concentration of high-density lipoprotein. Nicotinic acid has this effect largely through its action on the *hydroxycarboxylic acid receptor 2* (HCAR2 or GPR109A; OMIM 606039) and its very closely related homolog, the slightly larger *G protein-coupled receptor 109B* (HM74 or GPR109B; OMIM 606039). The difference between the two genes is subtle, mainly causing a higher affinity of nicotinic acid to the GPR109A protein than to GPR109B. Several common genetic variants in the two niacin receptors predict different individual responsiveness to lipid-lowering therapy with nicotinic acid [20]. GPR109A in particular has been linked to the very cumbersome side effects of flushing and itching [21], which affect some people much more than others. This use of nicotinic acid for lipid-lowering purposes and the extensive genetic variation provides us with a good opportunity to reflect on the extensive overlap of nutrigenetics and pharmacogenetics. There is no doubt that most of the common variants with effects on the response to medications have arisen as adaptations to different nutritopes.

7.2.7.3 ANTIFOLATE THERAPIES

Treatment of various forms of leukemias and cancers has relied for many decades on methotrexate, a powerful inhibitor of dihydrofolate reductase (DHFR; EC 1.5.1.3; OMIM 126060). DHFR is a key enzyme for dividing cells because it reactivates the dihydrofolate that synthesis of thymidine generates. Unless this dihydrofolate is reverted without delay into tetrahydrofolate, a dividing cell will run out of folate-bound one-carbon building blocks for the synthesis of DNA (adenine, guanine, and thymidine) and RNA (adenine and guanine). The typical amount of folate would be insufficient to duplicate 1% of its DNA. Because the cell division cycle cannot turn back once it has started, cells then die as incomplete ruins.

Chemotherapy is only helpful if the methotrexate dose is high enough to kill the malignant cells while allowing healthy cells to survive. The therapeutic window for effective and survivable therapy is very narrow. Genetic variability in the response adds considerably to the already daunting challenges of methotrexate chemotherapy. Multiple *DHFR* genetic variants strongly influence an individual response to treatment [22]. Simply put, greater DHFR activity increases dose requirements. The *DHFR* variants with known effect on methotrexate efficacy are located across the gene. There are several each in the 5′ upstream promoter region, in introns and exons, and in the 3′ untranslated region sequence that controls the stability of the circular mRNA for enzyme synthesis.

7.2.7.4 EXCESSIVE WEIGHT GAIN WITH ANTIPSYCHOTICS

Second-generation medications for the treatment of schizophrenia and other mental disorders have the potential to induce significant weight gain. This side effect is of great concern because the medications are used over long periods of time and the health burden from obesity-related diabetes, cardiovascular disease, cancer, and other chronic disease is significant. Not everybody is equally at risk. A genome-wide association study (GWAS) in a large number of patients in American and European institutions [23] pointed to variants in the *melanocortin 4 receptor* gene (*MC4R*; OMIM 155541). Patients with two copies of the rs489693 A allele appear to be most consistently at risk. Three months of treatment with aripiprazole, quetiapine fumarate, or risperidone induced more than twice as much weight gain in patients with the rs489693 A/A genotype than in other patients. This means that drug-induced weight gain could be anticipated and preventive measures taken.

7.3 TRANSCENDING PERSONAL INTERESTS

KEY CONCEPT
When a genetic syndrome with grave and preventable consequences is suspected or diagnosed, always consider the relevance for close relatives and follow-up with them as far as is practical.

Consider this case. A 7-year-old girl is diagnosed with type 1 diabetes mellitus. Her pediatrician orders a test for CD because this condition is often associated with childhood-onset diabetes. Blood tests come back positive and follow-up biopsy confirms the diagnosis. The doctor explains to her parents that the child should never again have any exposure to gluten in foods, drinks, and even in cosmetics for the rest of her life. An appointment with a dietitian is made to help with the transition to a gluten-free lifestyle. Everything goes well and the family learns to cope with the everyday challenges of insulin injections, carbohydrate counting, and gluten avoidance. But did anybody discuss with the parents the possibility that they or their other children also might have CD? What about the grandparents or the parents' siblings?

7.4 FOCUS ON THE WEAKER LINK

7.4.1 Dietary reference intakes

Current nutritional recommendations aim to define nutrient intake levels that meet the needs of at least 97–98% of the healthy population. The Food and Nutrition Board of the Institute of Medicine, which authors the Dietary Reference Intakes for the USA and Canada, uses 22 subgroups based on age, gender, and pregnancy/lactation status. This is done by estimating the median nutrient requirement (EAR) of a target group and then adding two estimated standard deviations to the recommended daily intake (RDA). Since standard deviations usually cannot be derived from actual data, estimates of 10–20% are used instead. If the individual nutrient requirements are normally distributed (meaning that they can be described by a bell curve) and the estimate for the standard deviation is correct, the intake targets will indeed cover 97.5% of the population. The current folate intake recommendations for men and women set

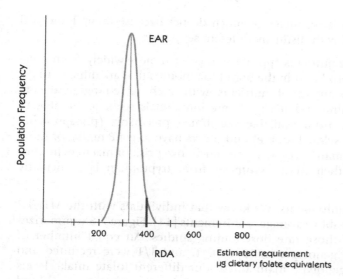

FIGURE 7.5
Nutrient intake recommendations are assumed to be normally distributed and accurately characterized by the median (estimated average requirement; EAR) and standard deviations. The recommended daily allowance (RDA) is set at two standard deviations above the EAR in order to meet the requirements of 97–98% of the healthy target group.

an EAR of 320 µg/day and assume a standard deviation of 10%, which means that the RDA is 400 µg (Figure 7.5). You should note that intakes are given as dietary folate equivalents (DFE) to account for the fact that folic acid has higher bioavailability than food folate.

The EAR is the measure to which the nutritional adequacy of comparable groups is compared. Let's say that we want to assess whether a particular group of American women get enough folate. We would then estimate folate intakes of a sufficient number of representative members of our target group and determine the folate intake level that is exceeded by half of the individuals. We then compare this median folate intake level with the relevant EAR. If the estimated median folate intake level of the group is well below the EAR, we conclude that the investigated group gets less folate than recommended.

7.4.2 But we are not all the same

Obviously, everybody will agree that requirements have to be determined separately for men and women. But why stop there? We know that requirements for many nutrients are not normally distributed within DRI groups due to genetic and other categorical differences.

There is a fundamental incompatibility of the current DRI model with the emerging understanding of genetically diverse individual requirements. The necessary assumption of the DRI model is that all members of the group have more or less the same nutrient requirements. Deviations from the median can only come from various continuous variables such as height and weight. It would make no sense to include men and women in the same group because we know that their requirements are fundamentally different. It is not just that women tend to be smaller and have a higher fat mass as a percentage of their total body mass. The discrepancies go much deeper because many biological processes are regulated with distinctly different set points.

Requirements for some nutrients appear to range even more widely, from none (because they can be produced in the body) to amounts that are difficult to get from foods alone. Examples of nutrients with such wide-ranging dietary requirements are choline and niacin. Some individuals seem to be able to make enough choline from available metabolite precursors (phosphoethanolamine and S-adenosylmethionine) and others have to get 8 mg/kg or more from foods. Similarly, many people have a much lower risk of niacin deficiency than others because their niacin synthesis from tryptophan is particularly effective.

Let's turn to folate requirements. We know that individuals with the *MTHFR* 677T/T genotype have above average requirements [14]. Figure 7.6 summarizes the results of one of those rare dose-finding studies. An equal number of participants with the genotypes 677C/C, C/T, and T/T were recruited and participants of each genotype maintained four different folate intake levels for an extended period of time. Homocysteine concentration in blood was chosen as a measure of folate adequacy because folate (to be precise, 5-methyl-tetrahydrofolate [5-mTHF]) is needed for homocysteine remethylation to methionine. If there is not enough 5-mTHF, homocysteine cannot be cleared and its concentration rises. Comparing the effect of all folate intake levels within the same person is very important (and hard to do) for eliminating variation other than the genotype. It is very clear that adding folate on top of the lowest intake amount (221 µg DFE/day) changes homocysteine concentration very little in participants with the 677C/C genotype. One can easily conclude that daily folate intake of 200–300 µg/day achieves already the optimal result in terms of homocysteine metabolism. Participants with the 677T/T genotype, on the other hand, need several times as much folate to bring their homocysteine concentration into the same low concentration range. Those with the C/T genotype (not shown) have a somewhat intermediate requirement.

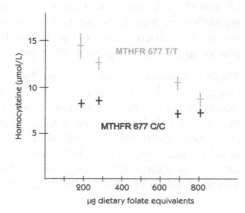

FIGURE 7.6
Individuals with the *MTHFR* 677CC (C/C) genotype did not have significantly lower homocysteine concentrations (shown are means and standard errors) when they increased their folate intake over the minimal level of 221 µg/day dietary folate equivalents per day. Individuals with the 677T/T genotype clearly need much higher folate intakes to achieve the same low homocysteine concentration.

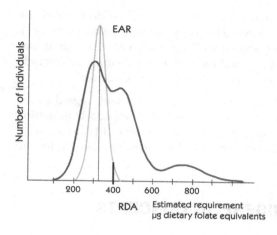

FIGURE 7.7
Multimodal distribution of predicted folate requirements based on the observed requirements of individuals with the three *MTHFR* 677 genotypes. EAR, estimated average requirement; RDA, recommended daily allowance.

These data make it very clear that many more people than the intended 2–3% are not covered by the current folate RDA. Figure 7.7 illustrates the discrepancy. The narrow bell curve around the EAR of 320 µg DFE is the range of folate requirements according to the DRI. The multimodal distribution with three peaks is calculated based on the observed folate requirements, assuming a low 12% frequency for the T/T genotype. Latin-American populations are known to have 677T/T genotype frequencies in excess of 30%, which shifts the center of the distribution curve much more toward higher folate requirements. These individuals are not served well when genotype-specific requirements are swept under the carpet.

7.4.3 An alternative model for setting intake recommendations

If we accept that a parametric model cannot work for a population composed of individuals with fundamentally different response characteristics, an alternative is needed. Just to be clear, the goal is to search for the intake level that meets the needs of at least 97–98% of the target population. We have learned that the behavior of someone with one genotype cannot possibly predict the needs of someone with a different genotype. The problem is, of course, that most of the time we don't even know what complex genotypes are out there, much less what impact they have on intake requirements. The following solution may sound paradoxical, but the most reliable way is to simply ignore all genetic variation and to directly determine the intake level that works for most people. Let's say that we investigate folate requirements in 100 people. We will then simply rank their individual estimated folate requirements and calculate the 97.5th percentile. This would be the average of the intake levels of the 97th and 98th highest ranked individuals. We can worry later about which genetic or other factors are responsible for their high requirements. More importantly, we would then know the elusive 97.5th percentile for certain without having to make wild assumptions based on dubious averages and immeasurable standard deviations.

A major barrier to the implementation of this approach is the need for a much more serious exploration of nutrient requirements than has ever been attempted. The current recommendations are mostly based on dose–response investigations in just a handful of individuals (only five in the case of folate!). The only reason that this number was considered sufficient was that it was assumed that all individuals would respond more or less the same. We certainly know now that such an assumption is contradicted by the evidence.

A more fundamental concern is that people may also differ with regard to the intake level that may cause harm. A more sophisticated approach to setting recommendations will have to take this upper tolerable level (UL) into account, threading the needle by truly finding the amount that provides a benefit for the greatest percentage of people.

7.5 LEVERAGING NUTRIGENETIC CONCEPTS IN RESEARCH

7.5.1 Separating the wheat from the chaff

Important relationships that affect a significant number of individuals in an entirely predictable way often get lost when we investigate groups as a whole. The fact that there are many unaffected individuals (nonresponders) does not diminish the benefit or harm a nutritional factor brings to affected individuals (responders). If we can show that a minority of individuals with a specific predisposition respond in a desirable way to a simple and affordable nutrition intervention, we should use the intervention with everybody as long as the majority is not harmed.

Take the example of the genotype-specific benefit of generous niacin intake in people with type 2 diabetes discussed in Chapter 5. Carriers of a common *SIRT1* (OMIM 604479) haplotype appear to be particularly vulnerable to suboptimal niacin supplies [24]. The reason is that *SIRT1* encodes a nicotinamide adenine dinucleotide (NAD)-dependent histone deacetylase, which is known to be strongly linked to longevity in organisms from microscopic worms to mammals. The approximately 16% of diabetes patients with two copies of the susceptible *SIRT1* haplotype 1 (consisting of the alleles A-G-G at rs7895833-rs1467568-rs497849) have a survival benefit of more than 7 years with optimal versus suboptimal niacin intake, while there is no such benefit for the majority of patients without haplotype 1. Patients with only one copy have a much smaller benefit. The question is very simply whether we can detect a niacin benefit for the entire group. The answer is no because the vast majority of patients experiences either no benefit (36% of the patients; the ones without haplotype 1) or only a small benefit (48% of the patients; the ones with a single copy of haplotype 1). Once we know that optimal niacin intake provides a major benefit for patients with two copies of haplotype 1, we can ask whether there is evidence that patients with fewer copies of haplotype 1 are harmed by a niacin intake of 20% above current recommendations (14 mg for women; 16 mg for men). The answer is quite clearly no. This means that we can recommend this slightly increased niacin intake level for all diabetes patients without having to know about their genotypes first. We can be reasonably confident that some will benefit and the others will suffer no significant harm. Of course, it would be nice to see this genotype-specific niacin effect studied again to test this interaction. But it would make little sense to do such a niacin study without taking into account these haplotypes.

7.5.2 Finding kernels of truth

All genetic studies have the potential to draw attention to previously unknown or ignored molecular mechanisms. The exploration of genotype-specific responses to nutritional factors may identify previously unknown pathways and mechanisms. It is not unusual that a GWAS of responses to a nutritional factor identifies a new player. For instance, a GWAS of vitamin B12 concentration in plasma identified a gene that had not been linked to vitamin B12 metabolism before [25]. This gene encodes *galactoside 2-alpha-L-fucosyl transferase 2* (*FUT2*; EC 2.4.1.69; OMIM 182100), which adds a fucose sugar to the sugar chains of various glycolipids and glycoproteins of the cell membranes. Linking *FUT2* variants to vitamin B12 may have seemed like a long shot at the time, but a plausible mechanistic link was proposed and has since been confirmed in several studies. People with the *FUT2* 461G allele (rs601338) are more accommodating to bacteria (e.g., *Helicobacter pylori*) and viruses (e.g., the Norwalk virus) than people without the allele. In particular, infection of the stomach lining with *H. pylori* leads to a loss of the cells that produce intrinsic factor (*GIF*; OMIM 609342), without which vitamin B12 is not well absorbed. Less efficient absorption due to the infection explained the effect of the *FUT2* 461G variant quite elegantly and at the same time established a novel interaction of the gastric microbiome with vitamin B12 metabolism. Without this interest in nutrigenetic interactions, we might have missed the connection of gastric bacteria with vitamin B12-dependent outcomes such as anemia and cognitive decline.

7.5.3 Nutrigenetic research approaches

Now that we understand that gene–nutrient interactions are common and functionally important, we have to design nutrition research with genetic variation in mind. While there is no simple answer for how to do this, there may be at least a few practical tips.

7.5.3.1 DIVERGENT FINDINGS

Unexplained inconsistency of nutritional effects in several high-quality studies may hint at an underlying gene–nutrient interaction. For example, the effect of soy consumption on prostate cancer has been repeatedly investigated with promising, but highly variable results [26]. We now know that increasing soy intake lowers prostate cancer risk in the many individuals (one-third of Americans) with an *ESR2* −13950C allele (rs2987983), but may actually increase risk in men without this allele [27]. This is certainly not the only factor responsible for the variable results. We also know that the composition of the intestinal microbiome (bacteria and other microorganisms), which strongly influences the processing of ingested soy isoflavones, makes a big difference [28]. As we have seen in the example of *H. pylori* and vitamin B12 absorption, genetic factors determine to some extent which kinds of microbes colonize our intestines.

7.5.3.2 OUTLIERS

Somewhat related is the approach of taking individual results seriously even when they are more than two standard deviations from the average. Sometimes such results are even discarded because they are considered implausible. There is no doubt that poor compliance, misclassifications, and analytical errors are often responsible for outliers. Even so, it is still worth studying previous reports and other available data collections to see whether there might be a pattern of outliers due to a genetic difference.

7.5.3.3 NON-NORMAL DISTRIBUTIONS

Distributions of individual responses to nutritional factors are normally distributed less often than commonly thought. Deviations from normal distribution are often missed because there are relatively few data points. In other cases, attempts are made to 'normalize' distributions by using a logarithmic or other nonlinear scale. The important question is whether the deviation from a nice bell-shaped curve is due to subgroups. Distribution curves resulting from more than one subgroup with different responses are called *multimodal* (referring to more than one peak or *mode*). Curves that come from such multimodal distributions do not necessarily have distinctly separate peaks, particularly when the medians for the subgroups are within about one standard deviation. Even with large sample sizes there may not be much more to see than a broadening of the peak or a shoulder in the curve.

Recognizing multimodal distributions becomes even more difficult when one subgroup is much smaller than the other.

Take a look at the distribution of protein requirements per kilogram body weight on which current protein intake recommendations are based (Figure 7.8). These

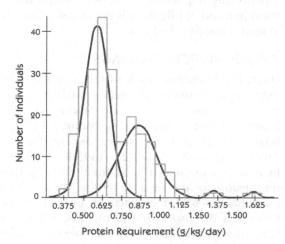

FIGURE 7.8
Current Dietary Reference Intakes for protein are based on a clearly multimodal, non-normal distribution of intake requirements.
Dietary Reference Intakes for Carbohydrate, Fiber, Fat, Fatty Acids, Cholesterol, Protein, and Amino Acids (Macronutrients), National Academic Press, Washington, D.C., 2005, ISBN 978-0-309-08525-0, p.637.

data were used to set the recommended daily allowance (RDA) at 0.8 g/kg body weight. The RDA is intended to meet the dietary needs of 97—98% of healthy people. The data strongly suggest that there are at least four subgroups with average requirements of about 0.6, 0.9, 1.3, and 1.6 g/day. Note that the distribution of the apparent subgroup with an average requirement around 0.9 g/day just generates a shoulder to the right of the largest (apparent) subgroup with average requirement of 0.6 g/day. It is not known what might cause this clearly multimodal distribution, but thinking about genetic variation could lead to a better understanding of actual needs of individuals.

Practice questions

What is the likelihood that the next naturally conceived child of a couple will have PKU if they already have a child with that condition?

How would you go about investigating whether higher folate intake reduces symptoms of clinical depression?

You are dealing with a middle-aged woman recently diagnosed with autoimmune thyroiditis. What are your next steps to determine the personal nutritional needs of this patient?

What information do you have to provide to a young woman before getting her consent to have her *MTHFR* 677 genotype tested?

How can genetic testing improve the clinical use of anti-coagulants in a 72-year-old man with atrial fibrillation?

SUMMARY AND SEGUE TO THE NEXT CHAPTER

Nutrigenetic tests can only be useful if they provide actionable information based on strong evidence. It is important that the evidence comes from a comparable target population and that the genetic analyses are reliable.

Remember that close relatives of someone with an identified genetic risk could have the same predisposition and may have a right to know about that.

Ethical considerations should always ask whether testing is truly the individual's choice, the potential burdens from learning about unfavorable genetic information are properly understood, and the tested person retains adequate control over the genetic information.

So now you might want to benefit from this new science, but worry about what a future employer might think and whether you will still be able to get life insurance. The final segment will review such risks and the legal and practical protections that are available.

References

[1] Chen B, Gagnon M, Shahangian S, Anderson NL, Howerton DA, Boone JD. Good laboratory practices for molecular genetic testing for heritable diseases and conditions. MMWR 2009;58(RR-6):1—37.

[2] Hofgartner WT, Tait JF. Frequency of problems during clinical molecular-genetic testing. Am J Clin Pathol 1999;112(1):14−21.

[3] Mahadevan MS, Benson PV. Factor V null mutation affecting the Roche LightCycler factor V Leiden assay. Clin Chem 2005;51(8):1533−5.

[4] Imai K, Kricka LJ, Fortina P. Concordance study of 3 direct-to-consumer genetic-testing services. Clin Chem 2011;57(3):518−21.

[5] Plebani M. The detection and prevention of errors in laboratory medicine. Ann Clin Biochem 2010;47(Pt 2):101−10.

[6] Ramsden SC, Deans Z, Robinson DO, Mountford R, Sistermans EA, Grody WW, et al. Monitoring standards for molecular genetic testing in the United Kingdom, the Netherlands, and Ireland. Genet Test 2006;10(3):147−56.

[7] Ellervik C, Tybjaerg-Hansen A, Nordestgaard BG. Total mortality by transferrin saturation levels: two general population studies and a metaanalysis. Clin Chem 2011;57(3):459−66.

[8] Kurppa K, Rasanen T, Collin P, Iltanen S, Huhtala H, Ashorn M, et al. Endomysial antibodies predict celiac disease irrespective of the titers or clinical presentation. World J Gastroenterol 2012;18(20):2511−6.

[9] Riddle MS, Murray JA, Porter CK. The incidence and risk of celiac disease in a healthy US adult population. Am J Gastroenterol 2012;107(8):1248−55.

[10] Gibson PR, Shepherd SJ, Tye-Din JA. For celiac disease, diagnosis is not enough. Clin Gastroenterol Hepatol 2012;10(8):900−901.

[11] Nielsen DE, El-Sohemy A. A randomized trial of genetic information for personalized nutrition. Genes Nutr 2012 Mar 11 [Epub ahead of print].

[12] Kubaszek A, Pihlajamaki J, Punnonen K, Karhapaa P, Vauhkonen I, Laakso M. The C-174G promoter polymorphism of the IL-6 gene affects energy expenditure and insulin sensitivity. Diabetes 2003;52(2):558−61.

[13] Watkins WS, Rohrwasser A, Peiffer A, Leppert MF, Lalouel JM, Jorde LB. AGT genetic variation, plasma AGT, and blood pressure: An analysis of the Utah Genetic Reference Project pedigrees. American Journal of Hypertension 2010;23(8):917−23.

[14] Ashfield-Watt PA, Pullin CH, Whiting JM, Clark ZE, Moat SJ, Newcombe RG, et al. Methylenetetrahydrofolate reductase 677C—>T genotype modulates homocysteine responses to a folate-rich diet or a low-dose folic acid supplement: a randomized controlled trial. Am J Clin Nutr 2002;76(1):180−6.

[15] Xu X, Gammon MD, Wetmur JG, Rao M, Gaudet MM, Teitelbaum SL, et al. A functional 19-base pair deletion polymorphism of dihydrofolate reductase (DHFR) and risk of breast cancer in multivitamin users. Am J Clin Nutr 2007;85(4):1098−102.

[16] Shen J, Gammon MD, Terry MB, Wang L, Wang Q, Zhang F, et al. Polymorphisms in XRCC1 modify the association between polycyclic aromatic hydrocarbon-DNA adducts, cigarette smoking, dietary antioxidants, and breast cancer risk. Cancer Epidemiol Biomarkers Prev 2005;14(2):336−42.

[17] Corella D, Peloso G, Arnett DK, Demissie S, Cupples LA, Tucker K, et al. APOA2, dietary fat, and body mass index: replication of a gene-diet interaction in 3 independent populations. Arch Intern Med 2009;169(20):1897−906.

[18] Dai Q, Shrubsole MJ, Ness RM, Schlundt D, Cai Q, Smalley WE, et al. The relation of magnesium and calcium intakes and a genetic polymorphism in the magnesium transporter to colorectal neoplasia risk. Am J Clin Nutr 2007;86(3):743−51.

[19] Klein TE, Altman RB, Eriksson N, Gage BF, Kimmel SE, Lee MT, et al. Estimation of the warfarin dose with clinical and pharmacogenetic data. N Engl J Med 2009;360(8):753−64.

[20] Zellner C, Pullinger CR, Aouizerat BE, Frost PH, Kwok PY, Malloy MJ, et al. Variations in human HM74 (GPR109B) and HM74A (GPR109A) niacin receptors. Hum Mutat 2005;25(1):18−21.

[21] Soudijn W, van Wijngaarden I, Ijzerman AP. Nicotinic acid receptor subtypes and their ligands. Medicinal Res Rev 2007;27(3):417−33.

[22] Owen SA, Hider SL, Martin P, Bruce IN, Barton A, Thomson W. Genetic polymorphisms in key methotrexate pathway genes are associated with response to treatment in rheumatoid arthritis patients. Pharmacogenomics J 2012 Mar 27 [Epub ahead of print].

[23] Malhotra AK, Correll CU, Chowdhury NI, Muller DJ, Gregersen PK, Lee AT, et al. Association between common variants near the melanocortin 4 receptor gene and severe antipsychotic drug-induced weight gain. Arch Gen Psychiatry 2012 May 7 [Epub ahead of print].

[24] Zillikens MC, van Meurs JB, Sijbrands EJ, Rivadeneira F, Dehghan A, van Leeuwen JP, et al. SIRT1 genetic variation and mortality in type 2 diabetes: interaction with smoking and dietary niacin. Free Radic Biol Med 2009;46(6):836—41.

[25] Bernstein IL, Longley A, Taylor EM. Amiloride sensitivity of chorda tympani response to NaCl in Fischer 344 and Wistar rats. Am J Physiol 1991;261(2 Pt 2):R329—33.

[26] Hwang YW, Kim SY, Jee SH, Kim YN, Nam CM. Soy food consumption and risk of prostate cancer: a meta-analysis of observational studies. Nutr Cancer 2009;61(5):598—606.

[27] Hedelin M, Balter KA, Chang ET, Bellocco R, Klint A, Johansson JE, et al. Dietary intake of phytoestrogens, estrogen receptor-beta polymorphisms and the risk of prostate cancer. Prostate 2006;66(14):1512—20.

[28] Akaza H. Prostate cancer chemoprevention by soy isoflavones: Role of intestinal bacteria as the "second human genome." Cancer Sci 2012;103(6):969—75.

CHAPTER 8
Keeping Genetic Information Safe

Gabrielle Kohlmeier and Martin Kohlmeier

Terms of the Trade

- Certificate of confidentiality: Letter issued by the Department of Health and Human Services to shield from all compelled inquiry.
- CLIA: Clinical Laboratory Improvement Amendments regulate laboratories in the USA.
- FDCA: Federal Food, Drug, and Cosmetic Act (USA).
- Genetic exceptionalism: The idea that genetic information should be treated differently from other kinds of medical data.
- GINA: The Genetic Information Nondiscrimination Act of 2008 (USA).
- HIPAA: Health Insurance Portability and Accountability Act (USA).
- IRB: Independent review boards; their approval is needed for all medical research.

ABSTRACT

Clients and patients have important rights that protect their interests and genetic service providers have significant obligations. The right to confidentiality, information, and privacy cover individuals who have genetic analyses done. Service providers must always obtain informed consent, have the duty to warn of health-relevant irregular results, and might have to warn affected relatives. A number of laws and regulations in the USA cover the analysis, transfer, and use of genetic information, but many gaps remain. Consumers and providers should always consider genetic information to be sensitive. Since undesired information transfer is always possible, genetic testing requires a heightened awareness of the consequences. Permanent anonymization is the ideal, particularly when it comes to nutrigenetic information.

Nutrigenetics. http://dx.doi.org/10.1016/B978-0-12-385900-6.00008-3

8.1 WHAT IS THERE TO LOSE?

The digital age has heightened concerns about privacy, while simultaneously creating many enticing opportunities to distribute personal information without much thought through participation in social media such as Facebook, online gaming, and online shopping. The technological advances that make it possible to find out intimate details about ourselves through our genetic information create even greater such opportunities and potential pitfalls. As we open the door to find out all of the things we can learn from our genetic information, we need to think as well about what there is to lose, including what rights we have regarding our genetic information and privacy. One of the biggest things we have to lose is control—once genetic information is released, an individual may have little command over who has access to it and what they do with it.

The focus of this chapter is the same as that of most legal and ethical regulations: patient safety and rights. Two key questions from a public policy perspective are whether the information that patient-consumers receive is accurate and what control patient-consumers have over their information. For those providing genetic testing services, the flipside of these issues is important: what standards of quality must be met and what duties are owed to the patient, in terms of information, privacy, and quality.

This chapter therefore examines protections (or the lack thereof) for genetic information. We begin with an introduction to various relevant legal and regulatory concepts followed by an overview of legal protections that exist—and highlight what these legal protections leave vulnerable.

8.2 KEY LEGAL TERMS AND CONCEPTS FOR UNDERSTANDING GENETIC INFORMATION

8.2.1 Overview

Before discussing the legal issues arising from genetics and genetic privacy, a brief overview of key terms is helpful. Remember that legal definitions may sometimes differ from biological reality. This section therefore breaks down some of the key terms and concepts from a legal perspective. First, it is important to understand what exactly we mean by genetic information and the different forms that this information can take. Next, we explore a few terms related to how we can get that genetic information, including direct-to-consumer services and other testing services. We then touch on some of the legal issues that may arise from testing, including fraud, misrepresentation, and product defects. Finally, we define terms and concepts at the heart of genetic privacy, including access, disclosure, and rights.

8.2.2 Different types of genetic information

Genetic information is the hereditary information about genes, gene products, or other inherited characteristics contained in chromosomal DNA or RNA that are derived from an individual, families, or populations. Legally, genetic information of an individual has been defined as 'information about: (a) such individual's genetic tests; (b) the genetic tests of family members of such individual; and (c) the manifestation of a disease or disorder in family members of such individual' 42 USC § 300 gg-91(d)(16) [1]. The legal definitions do not distinguish between raw genetic information and derived genetic information. *Raw* genetic information is the particular genetic variants accompanied with a brief global summary explaining what the genetic information means. Thus, the raw genetic information would include an actual gene sequence such as 'CCCATAGCTATAAAAAAC . . .' along with a description of the meaning of this genetic information. For example, it would show that the presence of a Taq1 t (rs731236) allele was detected—which indicates not only that the individual has a higher vitamin D requirement but also several other conditions or predilections that the individual may not need or want to know, such as increased risk of cancer, osteoporosis, and other dire health conditions. In contrast, *derived* genetic information consists only of the meaning of genetic information and not the specific genetic variants. So you get just the information you were testing for, and not any extras that may not have any meaning to them or that reveal more than one wants to know. When people think of genetic information, they generally think of raw genetic information even though only the derived information actually has meaning to nonprofessionals.

Whether in raw or derived form, genetic information includes information about an individual's genetic tests and the genetic tests of an individual's family members, as well as family medical history—that is, information about the manifestation of a disease or disorder in an individual's family members. Family medical history is included in the definition of genetic information because it is often used to determine whether someone has an increased risk of getting a disease, disorder, or condition in the future. Genetic information also includes an individual's request for, or receipt of, genetic services or participation in clinical research that includes genetic services by the individual or a family member of the individual; the genetic information of a fetus carried by an individual or by a pregnant woman who is a family member of the individual; and the genetic information of any embryo legally held by the individual or family member using an assisted reproductive technology (Genetic Information Nondiscrimination Act of 2008, USA).

A related concept is that of *genetic exceptionalism*: the idea is that genetic information is inherently unique, should receive special consideration, and should be treated differently from other kinds of medical data. Genetic information stands apart from other types of information because it reveals the health of family members, parentage, reproductive options, and future health risks; goes to the essence of who and what an individual is; and is regarded as unique by individuals and third parties.

8.2.3 Testing and testing services

There are several different types of testing services. The differences between testing service providers has become particularly relevant as testing has expanded.

Traditionally, testing involved diagnostic, screening, and predictive tests for a disease. Nonpathological testing, however, is now becoming increasingly available, and expectations of the future demand for such nonpathological tests are high. Nutrigenetic testing, for example, is an emerging area of nonpathological genetic testing that looks at the influence of interactions between diet and genotypes on health and disease. This new type of nonpathological testing has increased the popularity of less traditional testing services as well.

8.2.3.1 TRADITIONAL TESTING SERVICES

Pathological genetic testing looking for Alzheimer disease, breast cancer, or prenatal diagnostics for Down syndrome is firmly grounded in the medical context, which provides a structured framework. Prior to testing, patients go to their doctor's office or to a genetic counselor, where the trained practitioner advises them of the implications, meaning, and risks of tests. The health professional can also highlight limitations of the tests and explain why the patient may not want to undergo the test. If the patient chooses to proceed, the professional obtains consent and sends the patient's specimen to federally certified laboratories. The doctor then receives a test report of the genetic variations and often a brief interpretation (essentially a short summary saying that these results point to a specific probability of the tested disease or condition, for example). The patient returns for an intensive, detail-oriented session, where the professional explains the findings of the tests and the risks of proceeding based on this information. If patients insist, they will be given a copy of the report with the raw genetic information and the global evaluation. Genetic testing provides only one component of information about a person's health. Other genetic and environmental factors, lifestyle choices, and family medical history also affect a person's risk of developing many disorders and should generally be discussed.

8.2.3.2 COMMERCIAL TESTING

Commercial testing just means that the testing is done for profit, whether for pathological or nonpathological purposes. Commercial testing may give rise to conflicts of interest, for example when commercial testing services have incentives to hype their test to persuade the consumer to buy it. Tests ordered through more traditional routes such as physicians' offices are presumably less susceptible to such conflicts. Some commercial testing services employ genetic counselors or physicians to help deliver their testing service and provide counseling to customers. The greatest risk and potential problems arise in the context of direct-to-consumer testing.

8.2.3.3 DIRECT-TO-CONSUMER TESTING

Many people like the idea of taking control of their own destiny. They constantly hear about exciting new scientific advances in the media and what it will do for them. Reports about nutrigenetic advances are especially appealing for the lay public because they concerns nutrition, which is something they unavoidably engage in on a daily basis. They do not want to miss out on better health just because an insurer or government agency has not signed off on a genetic procedure or considers it not to be cost effective. Their natural interest and curiosity make them an attractive target for direct marketers that can bypass the usual medical hierarchy and healthcare system.

The process can be simple enough. The genetic testing company sends a simple sample collection device to the consumer. The consumer takes a few painless, noninvasive swabs from the inside of the cheek and mails the samples to the designated address for testing. A few weeks later, the company sends the test results as a printed or electronic report and provides some advice. The lack of comprehensive service standards means that the consumer cannot know what to expect. Even assuming that the genetic analyses are carried out in a Clinical Laboratory Improvements Amendments (CLIA)-compliant facility with high performance standards (which is not a given), they will know nothing about the validity of the test interpretations or the credibility of the advice. Consumers are confronted by inscrutable genetic information that should be disclosed only after careful counseling. The nutritional advice they receive tends to be generic and rarely meets their actual need for information and emotional support. Another concern is that their genetic and personal information may not be adequately protected over their lifetime, even if assurances are given by the company.

8.2.4 Traditional legal claims: product and product-marketing defects

Flawed genetic tests may give rise to several legal concerns. When the information delivered is inaccurate or representations about the genetic testing system are false, this may give rise to different legal claims, depending on the legal jurisdiction. The problem in the commercial context, particularly the direct-to-consumer context, is that the likelihood of information being inaccurate may be higher, as it is unclear whether many of the regulations currently covering genetic testing even apply in this situation. Profit incentives and uneven regulatory oversight may create incentives for commercial entities to offer questionable or unproven tests. In fact, the US Government Accountability Office has already exposed numerous websites offering nutrigenetic testing as scams based on the inaccurate information they have provided. In the USA, when representations about the genetic testing system are false, this may give rise to product-marketing defect claims, including misrepresentation and fraud. Under US law, misrepresentation occurs when a person, by their words or acts, creates in the mind of another person an impression not in accordance with facts. The fact is material if the other party would not have entered into the transaction or contract had he or she known about the misrepresentation. In that case, the consumer may have a viable tort claim. Fraud goes a step further, since it is a *deliberate* misrepresentation or nondisclosure of a material fact made with the intention that the other party will rely on it. If the party does in fact rely upon such a misrepresented statement, and this causes injury, then the person may bring an action to rescind the contract.

Product defects can include manufacturing defects, which occur in the manufacturing process and usually involve poor-quality materials or shoddy workmanship. Product design defects occur where the product design is inherently dangerous or useless (and hence defective) no matter how carefully manufactured. This may be demonstrated either by showing that the product fails to satisfy ordinary consumer expectations as to what constitutes a safe product, or that the risks of the product outweigh its benefits.

It seems that genetic testing that misrepresents the accuracy or efficacy of the tests or which presents—whether purposely or accidentally—false results should be subject to any or all of these theories of product liability. Arguably, direct-to-consumer testing provides a service, not a product, and therefore is not subject to product liability regimes [2].

8.2.5 Potential legal issues: disclosure and exposure of genetic information

Even if the genetic information provided by a testing service is *accurate*, problems may arise for patients or consumers. Two issues that enshrine these problems are exposure and disclosure. Exposure denotes information that the individual receives about his or her own genetic information. Disclosure relates to information that is revealed to third parties.

Exposure to genetic information may harm, rather than benefit, the individual. Other medical test results may show a disease or condition where there are a variety of options for treatment, such as dietary changes, exercise, prescription medicine, or even surgical intervention. In contrast, genetic tests may just show a possibility of someday developing a disease, with little that can be done in the meantime to prevent it. There is no current cure, for example, for Alzheimer and Huntington diseases, two diseases for which genetic testing exists. And even where there are certain treatments available, a genetic test does not actually indicate when, if ever, the patient or consumer will get the disease. Nonetheless, patients may take serious steps based on the genetic information they receive. Patients who find they have a gene related to breast cancer may opt for an elective double mastectomy rather than waiting to see if the cancer ever materializes. The results of prenatal genetic screenings may lead some women to terminate a pregnancy.

A second even more concerning issue related to exposure is that the individual may misinterpret the results of his or her genetic tests. A patient facing a positive *BRCA1* test result may not comprehend what a 60% chance of developing breast cancer means. In particular, the absence of quality genetic counseling to accompany the test may lead to misunderstandings that have dire consequences. On the flipside, a patient or consumer whose test result comes back negative may erroneously believe they are not susceptible to that particular disease.

A third exposure problem occurs when the individual gets more information than necessary. Overexposure may reveal more information than the individual was seeking, including distressing information about potential disease susceptibility. Although on first thought it may seem that a patient or consumer should obviously have the right to obtain all of their personal genetic information, this must be balanced against the patient's right for privacy and the right *not* to have to learn information they do not necessarily want or understand, or both.

Disclosure of genetic information to third parties may be harmless, especially when not tied to a particular individual. For example, genetic information may be disclosed to medical professionals for medical use, and may even enhance the medical care provided to the individual. It may be disclosed to medical researchers who are never directly connected to the individual, in which case the information disclosure might be considered harmless. Yet there are various other disclosures, including for nonmedical uses. These may include disclosure for commercial purposes, criminal law, domestic relations, education, employment, forensics, insurance, and personal-injury litigation. Genetic information may also be used for identification and in such contexts as immigration, kinship, paternity, schools, and the settlement of estates.

Disclosure and exposure may give rise to tort claims, such as the tort of negligence. Genetic test providers may have a duty to inform about the availability of tests and the results of tests, and they may be found negligent for misdiagnoses. Informed consent may be another area that gives rise to legal liability.

All of these issues give rise to the question of what rights patients or consumers have over their genetic information. The next section discusses rights, including the right to be informed, the right to privacy, and the right to control one's own information. It also discusses how far legal protections of these rights extend—and their limits. It then discusses what duties may arise for those dealing with genetic information.

8.3 THE LEGAL REGIME AND LAWS RELATING TO GENETIC INFORMATION

8.3.1 Risks and regulation: what regulation is and why we need it

Regulation may take the form of official legal structures, such as courts and government entities. It may also take the form of self-regulation, through professional and standard-setting organizations. Or it can take the form of consumer organizations that use formal or more informal channels, such as community pressure, to regulate and change practices. Regardless of the form that the regulation takes, ethical and legal structures are intended to mitigate the risks to patients/consumers in obtaining genetic information.

As discussed, obtaining genetic information, especially in a direct-to-consumer, nonmedical context, may seem harmless but may pose significant risks. For example, a genetic test result may affect not only a person's mental well-being but can also affect a person's ability to obtain disability, health, life, or long-term care insurance. It could also affect the ability to obtain or keep a job.

Like other diagnostic laboratory tests, genetic tests in the USA and many other jurisdictions are subject to some federal regulatory oversight. The most important requirement in the USA is that human genetic tests for clinical or commercial purposes can only be carried out in CLIA-certified laboratories. Beyond this, there is currently no uniform or comprehensive system in the USA to assess the analytic and clinical validity of tests before they are offered to patients, and there are no laboratory standards that specifically address molecular or biochemical genetic testing or require laboratories to enroll in proficiency testing programs that assess their ability to perform the tests correctly [3].

Contrary to common expectation, genetic tests are not necessarily subject to Food and Drug Administration (FDA) regulation, as long as they can be considered a professional service. However, the FDA considers direct-to-consumer testing to constitute medical devices that *may be* subject to FDA oversight under section 201 of the Federal Food, Drug, and Cosmetic Act (FDCA). This continues many years of regulatory uncertainty. It is not only the laboratory analyses themselves that are under contention but also the sampling devices for collecting saliva for the DNA analyses.

Although there have been repeated calls to have all genetic tests go through physicians, only New York and California require that a physician order clinical genetic tests. The Office of In Vitro Diagnostic Evaluation and Safety at the FDA uses external experts, the Molecular and Clinical Genetics Panel, to advise on issues relating to genetic tests. This panel could not find agreement on the question of required referral from physicians. This means that the issue remains contentious and unresolved.

Calls for increased oversight of genetic tests have been made for nearly a decade, yet federal and state regulation remains patchy. The Secretary's Advisory Committee on Genetic Testing set forth its findings nearly 7 years ago that '[b]ased on the rapidly evolving nature of genetic tests, their anticipated widespread use, and extensive concerns expressed by the public about their potential for misuse or misinterpretation, additional oversight is warranted for all genetic tests.' Although several federal agencies and laws, as well as state regulations, do touch on genetic information, they provide incomplete safeguards of genetic information, both in terms of accuracy and privacy. In 2008, the US Secretary of Health and Human Services charged an Advisory Committee on Genetics, Health, and Society with identifying the problems related to the adequacy and transparency of current oversight for genetic testing. The committee identified gaps in five main areas: the regulations governing clinical laboratory quality; the oversight of the clinical validity of genetic tests; the transparency of genetic testing; the level of current knowledge about the clinical usefulness of genetic tests; and meeting the educational needs of health professionals, the public health community, patients, and consumers, along with providing tools to assist these groups with the interpretation and communication of genetic test results [4].

8.3.2 US Federal agencies

Several federal agencies deal with protection of consumers and of health-related issues. The US Federal Trade Commission (FTC), for example, has a Consumer Protection Bureau that often deals with healthcare claims and issues. The FTC's mandate under the Federal Trade Commission Act is to prevent 'unfair or deceptive acts or practices in or affecting commerce.' The issues of accuracy, informed consent, and privacy could arguably fall under the FTC's mandate, but the agency has not taken steps in that direction yet, nor is it clear that it would be willing or able to provide the necessary resources to do so.

The Department of Health and Human Services (DHHS) and the FDA, which falls under DHHS's purview, are tasked with protecting the public health. Oversight of all laboratory tests and their components falls within the purview of the FDA, pursuant to the Federal Food, Drug, and Cosmetic Act. Some commentators have suggested that the FDA is in the best position to protect the public from genetic test problems, and that the profound dangers of genetic information 'warrant the public delegating the resources and mandate to the FDA to ensure that the troubling issues and agonizing choices occasioned by genetic testing are not compounded by poorly developed or even misleading information.' [5] Despite its arguably broad powers to act in the area, current FDA regulation of genetic testing is minimal. Though there are 'genetic tests available for close to 1000 diseases or conditions . . . only about a dozen genetic tests have been reviewed and approved . . . to ensure their safety and effectiveness.' Furthermore, FDA regulation is incoherent and vacillating. For example, the FDA deems genetic testing *kits*—classified as medical devices—to be subject to premarket approval, yet genetic testing *services* (including nutrigenetic testing) are not. In fact, '[c]linical laboratories that plan to market tests as services and that have not received federal funds are under no requirement to consult [independent review boards (IRB)]. . . . [and] few have sought IRB approval or consulted the FDA.' Because certain genetic tests do not involve testing for the presence of a single marker, but rather involve complex evaluation of numerous different genetic variants, test *kits* for these types of services are unlikely to be offered any time soon. Those types of tests therefore remain a genetic testing *service* and thus part of the market that the FDA has barely acknowledged, much less regulated. Overall, FDA policy and statements indicate an unwillingness to regulate genetic testing services, whether because of lack of political will or lack of resources. Aside from the unwillingness or inability of the FDA to regulate genetic testing services, real concerns remain regarding the FDA's competence to address the complex social issues attached to genetic testing that go beyond mere product performance concerns.

These federal agencies do not have a clear mandate or resources to address genetic testing services, largely because of the lack of comprehensive, uniform laws governing genetic testing services. The next section discusses the types of federal laws that currently exist in order to highlight what is shielded and what remains uncovered by legal protections.

8.3.3 US Federal laws

First, the CLIA amendments of 1988 establish minimum quality levels in laboratory testing practices. CLIA may theoretically extend to federal oversight of DNA analyses but coverage is quite patchy. Although various governmental advisory bodies found that "a smooth transition of genetic testing from research to practice" would require 'creat[ion of] regulations under CLIA that focused specifically on genetic tests,' CLIA has no specific category or requirements for genetic tests. Because genetic tests are broadly included as part of all laboratory tests, there are no specific personnel, quality control, or proficiency testing requirements for the vast majority of genetic tests. As a result, nonmedical genetic tests are 'sometimes performed in laboratories that have not been approved under CLIA.' The effect of delays in implementing CLIA, not to mention its gaps, means that 'neither healthcare providers nor consumers can be confident in the oversight mechanisms in place to ensure genetic tests are accurate and reliable. While genetic science and genetic technologies have leapt into the 21st century, the agency entrusted with ensuring laboratory quality is stuck in the past.' Most importantly, CLIA does not address the serious issues relating to genetic counseling or informed consent. Finally, CLIA arguably does not fit the paradigm of genetic testing because of the huge number of tests involved in nutrigenetics. The 50K to up to 1M tests (that analyze 50,000 to up to 1,000,000 genetic variants for one person) used for ancestry, pharmacogenetic, and other developmental testing are already orders of magnitude higher than what can reasonably be quality-controlled under CLIA.

The Health Insurance Portability and Accountability Act (HIPAA) in the USA is also frequently cited as a protection in the context of genetic testing. HIPAA provides comprehensive protection to individually identifiable health information. The statute is not focused on genetic information specifically, but it does provide that genetic information may not be treated as a condition 'in the absence of a diagnosis related to such condition.' Yet this indirect and minimal approach to protecting genetic information is inapposite to deal with the large number of concerns presented by genetic information. The statute is full of gaps that cut down the limited protections afforded. HIPAA is designed to protect confidentiality of health records generally and is not focused on the special issues surrounding genetic information. In fact, genetic information is only protected if it falls under the definition of protected health information, and there is no recognition of the differences that distinguish genetic information from other types of personal information.

The Genetic Information Nondiscrimination Act of 2008 (GINA) in the USA demonstrates some efforts by legislators in both the House and Senate to introduce federal genetic information protections. GINA seeks to protect against genetic discrimination both by employers and insurance companies. Title I deals with the use of genetic information in health insurance and falls under the purview of the Departments of Labor, Health and Human Services and the Treasury; Title II deals with genetic discrimination in employment and is enforced by the US Equal

Employment Opportunity Commission. After several years stuck in legislative limbo due to strong opposition from powerful special interests, both titles took effect in 2009. Under the Act, it is illegal to discriminate against employees or applicants because of genetic information. Title II prohibits the use of genetic information in making employment decision; restricts employers and other entities covered by Title II (employment agencies, labor organizations, and joint labor-management training and apprenticeship programs—referred to as *covered entities*) from requesting, requiring, or purchasing genetic information; and strictly limits the disclosure of genetic information.

There are still significant gaps uncovered by GINA. The provisions of the law do not apply to disability insurance, life insurance, or long-term care insurance. GINA also does not apply to members of the military, veterans obtaining health care through the Department of Veterans Affairs, health benefits plans for federal employees, and the Indian Health Service. GINA also does not apply to educational or athletic programs.

It is important to understand that GINA has not been fully tested and that we should not rely on enforcement of the law as an adequate protection at this point in time.

8.3.4 US State laws

State efforts to provide genetic testing oversight are spreading. More than a dozen states have prohibited direct consumer access to any laboratory testing. As of 2008, 31 states had genetic privacy laws, 47 states had genetic nondiscrimination in health insurance laws, and 34 states had genetic nondiscrimination in employment laws. But state regulations remain far from uniform and share many of the problems of federal regulations. First, the state laws contain different definitions of the relevant terms, such as discrimination, genetic information, and genetic tests. Second, the state statutes contain significant gaps that leave genetic information unprotected from exploitation or nefarious purposes. Although some states have implemented measures that indirectly regulate some aspects of genetic testing services—such as quality assurance requirements beyond those mandated by CLIA and genetic counselor licensing requirements—nonmedical genetic testing falls largely outside the realm of those regulatory efforts. Third, despite their good intentions and efforts, none of the state laws afford individuals the autonomy to fully control who gets access to their genetic information. Finally, as jurisdictional issues involving services offered through the internet remain judicially unresolved, even the most stringent state laws may not offer the necessary measures to protect information affecting individuals and their families because whether a web-based genetic testing service is subject to a particular state's jurisdiction remains uncertain.

8.3.5 Professional and ethical guidelines

Some professional organizations provide a certain degree of oversight over the provision of services, including genetic testing. A number of organizations help assure the quality of laboratory practices and assist in developing clinical practice guidelines to ensure that genetic tests are used appropriately. Ethical guidelines also cover medical professionals providing genetic testing services. Patient advocacy groups and families with genetic conditions can likewise influence the development of standards and guidelines for genetic testing. Yet the extent that any of these ethical and professional guidelines covers nonmedical genetic testing is unclear. Genetic testing services for ancestry or nutrition, for example, may not feel obligated to abide by professional responsibilities to the same extent as medical professionals.

This section, like much of this chapter, focuses primarily on US regulations, agencies and laws, which may differ considerably from those in other countries. Several EU countries, for example, offer greater protection for genetic information. One particularly useful resource for more specific European regulations and protections is a compendium published by the European Union entitled *Genetic Testing: Patients' Rights, Insurance and Employment; A Survey of Regulations in the European Union* (http://ec.europa.eu/research/biosociety/pdf/genetic_testing_eur20446.pdf).

8.4 PATIENT RIGHTS

8.4.1 What we are talking about

US legal precedent regarding genetic information is sparse, as few courts have dealt with genetic information suits and the relevant legal regimes are still in their nascent stages. The rights and duties owed to patients are largely theoretical based on arguably analogous rights and duties in other contexts. This section discusses the types of patient rights that apply in the medical context generally. Commentators have suggested these rights apply to different degrees in the genetics information context. As these debates demonstrate, the degree to which any of these rights applies is unclear; they are discussed here to lay out the types of rights and duties that may apply, not to state any particular mandate.

8.4.2 Patients' rights

Patient rights have been a hot topic in the medical and bioethics field for quite some time. There has been a shift away from a patriarchal approach to patients—that medical professionals know best and patients cannot make decisions for themselves—and toward requiring greater patient input into decisions affecting their care. Patient rights now generally enshrine the ideas of autonomy, individual liberty, and self-determination. Yet the exact contours of patient rights are amorphous and undefined. One issue to consider in the shifting context of genetic testing from medical to nonmedical genetic testing is whether the person receiving the test is still a patient, or if the person is now classified as a consumer. This shift may not be merely semantic, but may entail less or different protections. Since this book is geared toward healthcare professionals, however, this section will assume that patient rights apply and discuss these in further detail.

Theoretically, patient rights may include the right to receive information from physicians and to discuss the benefits, risks, and costs of appropriate treatment alternatives; the right to make decisions regarding the health care that is recommended by the physician; the right to courtesy, dignity, respect, responsiveness, and timely attention to health needs; the right to confidentiality; the right to continuity of health care; and the basic right to have adequate health care. This section focuses on the right to privacy, the right to information, and the right to confidentiality.

8.4.2.1 THE RIGHT TO BE INFORMED AND THE RIGHT TO INFORMATION

A clear right enshrined both in law and ethical standards is the patient's right to informed consent. For a long time, patients had the right to give or withhold permission for medical treatment. It is only more recently, however, that this came to mean that if the patient needs a treatment, her healthcare provider should give her the information needed to make a knowledgeable decision about that treatment. The right to be informed about one's health status and treatment is closely tied to this idea of giving the patient adequate information and thereby giving them a say in medical decisions.

The shift toward greater autonomy also has to do with the realities of chronic disease in an aging population. Their lifestyle and health behavior drives many of the common and very costly diseases. A higher information level is known to increase motivation for adopting healthier choices. In a time of rising healthcare costs, this economic aspect cannot be ignored. In the end, everybody wins because health care becomes more affordable and a healthier public has a higher quality of life.

The genetic information context, especially in terms of genetic information that is not intended for the treatment of any medical conditions, raises the question of whether there is a right to information beyond that needed for treatment. And there is not only the right to know to consider, but also the right *not* to know or to limit the information received.

Beyond the right of the patient to be informed (or not informed), a further issue is the right of third parties to receive or limit information. Information about the patient may have key significance to relatives, for example. One thing for medical professionals and other genetic services providers to consider is the extent to which there exists a duty to inform these third parties, as discussed further below.

8.4.2.2 THE RIGHT TO PRIVACY

Privacy is defined in terms of a person having control over the extent, timing, and circumstances of sharing oneself (behaviorally, intellectually, or physically) with others. Privacy refers to the right of individuals to limit access by others to aspects of their person that can include thoughts, identifying information, and even information contained in bodily tissues and fluids. Even though privacy is not explicitly mentioned in the United States Constitution, many consider privacy a basic human right and maintaining confidentiality a professional obligation (http://ccnmtl.columbia.edu/projects/cire/pac/foundation/index.html#1). In 1993, the Council for International Organizations of Medical Sciences and the World Health Organization published the *Ethical Guidelines for Biomedical Research Involving Human Subjects*. These guidelines provide explicit provisions for respecting the privacy of research participants and maintaining the confidentiality of their personal information. The right to privacy involves limiting access to a person, the right to be let alone, and the right to keep certain information from being disclosed to other individuals. As such, it is clear that the right to privacy may lie in tension with the right to information that third parties may claim. Genetic information has enormous implications to an individual and his or her family. The privacy of that information is a major concern to patients and, in particular, who should have or needs access to their information.

8.4.2.3 THE RIGHT TO CONFIDENTIALITY

Closely related to the right to privacy is the right to confidentiality. Confidentiality is the practice of protecting an individual's privacy. Information disclosed in the context of a trusted relationship, such as one that may arise in the context of genetic testing, may carry the expectation that the information will not be divulged to others without permission.

In short, confidentiality involves the right of an individual to prevent the redisclosure of certain sensitive information that was originally disclosed in the confines of a confidential relationship. The confidentiality of genetic information may need to be guarded even more stringently than other medical information because of the type of immutable information genetic test results reveal.

Protecting confidentiality can be difficult because others may think they should have the right to see an individual's information or use an individual's information for other purposes. Entities with a legitimate interest in genetic information include research facilities and biobanks. In all instances, subjects have to agree in writing to the information transfer.

The predictive power of genetic testing may also make genetic testing particularly liable for misuse. Employers and insurance companies have been known to deny individuals essential health care or employment based on knowledge of genetic predisposition. This type of discrimination can be socially debilitating and have severe socioeconomic consequences. It is important, therefore, to ensure the confidentiality of test results and to establish legislation permitting only selective access to this information. This confidentiality extends to all relatives, which can create significant conflicts of interest. Particularly with the advent of widespread whole-genome sequencing, there is an increasing likelihood that variants are seen that indicate a potential risk for close relatives. Procedures for such instances are not fully formed, particularly in regard to the threshold when disclosure becomes urgent.

8.5 DUTIES OF MEDICAL PROVIDERS

8.5.1 What we are talking about

On the other side of the coin from patient rights are the duties owed by medical professionals. This section discusses the types of duties imposed on medical service providers dealing with genetic information. The relationship of trust between a medical professional and a patient entails the duty to act competently and non-negligently. Part of this entails balancing the duty to inform and obtain informed consent with the duty to protect the patient, including confidentiality, respecting the patient's privacy, and warning the patient—and in some cases even third parties—about potential harm. Two areas of particular concern are therefore the duty to obtain informed consent—and what exactly informed consent entails in the genetic testing context—and the duty to inform or warn. The latter is a hot issue that requires consideration not only of the duty to inform the patient or consumer but also potentially affected relatives. Another area of growing concern is the intentional or unintentional sharing of genetic information with third parties, sometimes compelled by legal action. It is important that service providers know about their obligations to protect sensitive data and about mechanisms that protect them and their clients against unwarranted intrusions.

8.5.2 Duty to obtain informed consent

As discussed above, patients have a right not only to consent to medical procedures but also to understand the implications and risks of those procedures. The landmark case on informed consent, Canterbury v. Spence, set forth that physicians have a duty to reasonably inform a patient about all the risks that may affect the patient's decision, but that 'the physician discharges the duty when he makes a reasonable effort to convey sufficient information although the patient, without fault of the physician, may not fully grasp it' (Canterbury v. Spence, 464 F.2d 772, 780; D.C. Cir. 1972) [6]. Medical professionals have a legal duty to help ensure that patients understand the risks and benefits of health care choices. Different federal agencies and laws take different approaches to informed consent in different contexts. In the genetic testing context, the issue is to what extent a patient or consumer must be advised of the ramifications and risks of obtaining the test—especially when the risks are not physical but rather entail employment, insurance, psychological, and other nonmedical risks.

Although it is unclear that explaining such risks is necessary from a legal standpoint, a prudent practitioner or genetic services provider should make sure that patients or consumers have carefully discussed and understood the following before undergoing testing.

Testing is voluntary;
Risks, limitations, and benefits of testing or not testing;
Alternatives to genetic testing;
Details of the testing process (accuracy of test, sampling, turn-around time, etc.);
Privacy/confidentiality of test results;

Potential consequences related to results, including:
Impact on health;
Possible emotional and psychological reactions;
Treatment/prevention options; and
Ramifications for the family.

The goal of imposing a duty to provide complete information and obtain truly informed consent on medical providers is to protect and promote patient autonomy. There are therefore calls for courts to impose the same duty on any other genetic testing providers, whether that testing occurs in a medical context or otherwise [2].

8.5.3 Duty to warn the patient (recontact)

Advances in medicine, but particularly in the realm of genetic testing, have given rise to two different duties to warn: first, the duty to recontact, or warn the patient about newly discovered risks, and second, the duty to warn third parties that may also be affected by the genetic testing.

The issue of recontacting patients about past genetic test results arises because of the uncertainty in the results of many genetic tests. Although the extent of any legal duty to recontact remains unclear at present, several commentators have proposed helpful guidelines for consideration. These considerations include:

The degree to which current *variants of unknown significance* will become interpretable in the future;
Whether incidental findings will proliferate with multiarray or whole-genome testing;
How clear it is whose responsibility it would be to recontact patients about new interpretations of genetic tests;
Which healthcare providers might bear responsibility for test results included in the medical record;
The specific patient's expectations regarding being recontacted;
Implications that genetic test results or reinterpretations may have for the patient's relatives;
Whether genetic test results can be interpreted in isolation, or if a variation's effect may depend on many interacting factors;
Whether the patient is directly ordering genetic tests and inserting their results into the medical record;
Whether an otherwise short-term physician—patient relationship might be extended if genomic test results are subject to reinterpretation; and
The extent to which clinicians and consumers require education about the promise and pitfalls of genomic testing [7].

8.5.4 Duty to warn others

On the issue of a duty to warn third parties, genetic service providers may face the conundrum of balancing a duty to warn with the duty to protect patient confidentiality. If the counselor takes it upon himself to warn relatives, he faces a possible lawsuit from the counselee based on a violation of privacy. However, if the counselor does not warn a relative, he might be sued by the relative based on the failure to prevent a medical issue, such as the birth of a genetically defective child. Although commentators are split, current precedent suggests that since the counselee has no duty to inform relatives, the counselor has no such duty and no business to warn the relatives. The duty to inform varies by state, and courts have ruled differently in different cases. Individuals and families often seem opposed to doctors informing at-risk members without their consent, even in cases where the disease is easily preventable. The idea that a nonmedical genetic services provider would have such a duty—even if it were practicable—seems unlikely under the current legal framework.

8.5.5 Data protection

Genetic data may be released to insurers and others who pay for medical care subject to HIPAA regulations. It is important to recognize that genetic data may be transmitted inadvertently with other patient files. Sometimes unauthorized and illegal transfer can occur. Information may even end up in the files of employers when they have a self-funded insurance plan. Individual health information may be held by entities such as the Medical Information Bureau (MIB), which serves the needs of insurance companies in the area of fraud detection and deterrence. Such entities are not subject to the HIPAA. Individuals will rarely know when such data 'leakage' occurs and have little redress.

8.5.6 Release of genetic information to third parties

Courts can compel the disclosure of genetic information to third parties against an individual's wishes or interests. This could come about, for instance, in paternity suits. Issuers of disability, life, and long-term care insurance may also compel disclosure because these forms of insurance are not covered by GINA.

8.5.6.1 CERTIFICATE OF CONFIDENTIALITY

There is a procedure that can shield the genetic information of individuals from compelled disclosure under some circumstances [8]. Institutions can apply to the DHHS for a certificate of confidentiality under the Public Health Service Act 42 USCA 241([d]).

The certificate is designed to permanently prevent the release of sensitive genetic information of participants in 'biomedical, behavioral, clinical, or other research' [8]. DHHS considers data sensitive when 'information that if released could reasonably be damaging to an individual's financial standing, employability, or reputation within the community' or 'information that normally would be recorded in a patient's medical record, and the disclosure of which could reasonably lead to social stigmatization or discrimination.' The certificate allows the researcher to refuse to reveal the identity of a research subject to 'any Federal, State, or local civil, criminal, administrative, legislative, or other proceedings.' The protection does not end with a subject's death: it is permanent.

It is important to note that the certificate does not prevent a researcher from disclosing the genetic information voluntarily, depending on the specifications of the informed consent form that the subject signed.

8.6 DOUBLE MASKING

As explained in the previous chapter, it would be best not to disclose nutrigenetic information to consumers at all because they do not really need it. Many consumers are intrigued by their own genetic information but they rarely have the scientific education to put it into perspective. What they really need is nutritional guidance that helps them make the food choices that fit their personal needs. An online computer program (PONG [Personal Online Nutrition Guidance]; accessible at http://www.nutrigen.com) can do precisely that without ever disclosing raw genetic information to the consumer. Because this program only provides personal nutrition guidance based on combinations of normal foods and recipes, without the need for any proprietary formulations or products, there is no stigmatizing or otherwise disturbing effect that might be associated with the disclosure of raw genetic data. This means that there is no need for specific genetic counseling. After all, the guidance is directed at more or less healthy people, not at patients. The daily menus are based on general nutrition guidelines with slight modifications as directed by the user's individual genome.

The program also permanently shields the user's identity from the laboratory or other third parties that are not explicitly authorized by the user. This permanent anonymization scheme provides failsafe protection against unintended disclosure of genetic information linked to identifiable individuals. Since the laboratory does not know the user's identity, not even insider access or a court order can penetrate this privacy shield.

Practice questions

Is a provider of a direct-to-consumer nutrigenetic test panel required to provide evidence of a health benefit?
What legal qualifications does a genetic test provider have to meet?

Can a health insurance carrier deny coverage when an applicant refuses to release genetic information from a direct-to-consumer genetic test?

SUMMARY

The legal framework regarding genetic information and genetic testing services, both in the USA and in other countries, is still developing and in many instances is quite sparse and patchy. Patients or consumers obtaining genetic testing must be cognizant of the risks to themselves and to their privacy that such testing may entail. It is critical to understand that there is no comprehensive legal protection for their genetic information and that even if the protections existed they may not remedy the harm that the exposure of genetic information may have engendered.

At the same time, providers of genetic testing services should carefully consider their potential obligations and duties to their patients and customers. These duties, though they may appear tenuous given the lack of current judicial precedent, may give rise to significant legal liabilities. Following ethical guidelines and carefully considering the rights of patients may significantly reduce legal risks. Ultimately, obtaining comprehensive and considered informed consent and protecting patient confidentiality appear to be the most prudent courses for avoiding legal liability.

References

[1] Title 42, US Code, Section 300, gg-91 (d)(16). Published on Thomas (the Congressional online system) and in the printed US Code.
[2] Kilgore D. Test at your own risk: Your genetic report card and the direct-to-consumer duty to secure informed consent. Emory Law Journal 2010;59:1553−604.
[3] Huang A. Who regulates genetic tests? Genetics & Public Policy Center, Washington, DC; 2006. Available from: http://www.dnapolicy.org/policy.issue.php?action=detail&issuebrief_id=10.
[4] Secretary's Advisory Committee on Genetics HaS. System of Oversight of Genetic Testing: A Response to the Charge of the Secretary of Health and Human Services. In: Services UDoHaH; 2008.
[5] Huang AFDA. Regulation of genetic testing: Institutional reluctance and public guardianship. Food & Drug Law Journal 1998;53(3):555−91.
[6] Ekberg M. Governing the risks emerging from the non-medical uses of genetic testing. Aust Emerg Tech Soc 2005;3:1−16.
[7] Pyeritz RE. The coming explosion in genetic testing—is there a duty to recontact? N Engl J Med 2011;365(15):1367−9.
[8] Earley CL, Strong LC. Certificates of confidentiality: a valuable tool for protecting genetic data. Am J Hum Genet 1995;57(3):727−31.

Genome Glossary

Adenine: Purine base (A) that is a constituent of nucleotides (e.g., ATP) and the polynucleotides RNA and DNA; in DNA is usually paired with thymine (T) in the complementary strand.

Adenosine: Nucleoside containing the base adenine and the sugar ribose.

Alleles: DNA variation in a specific sequence (locus). The combination of the allelic forms inherited from the parents is the individual's genotype.

Allozyme: Enzyme with altered amino acid sequence corresponding to a variant allele.

Alu sequence: Very common repeating DNA sequences (comprising more than 10% of the entire human genome) containing the motif 5′-AGCT-3′, which is cleaved by the *AluI* restriction endonuclease.

Amino acid: Compounds that contain both an amino group (NH2-) and a carbonyl or other acidic group.

Amplicon: The DNA or RNA segment resulting from polymerase chain reaction (PCR) amplification of a polynucleotide sequence.

Amplification: An increase in the number of copies of a specific DNA fragment in vivo or in vitro.

Analog: A protein or DNA that shares with another protein or DNA functional characteristics, but not sequence or ancestry.

Anaphase: The phase during cell division when the duplicated chromosomes separate.

Aneuploidy: Presence of an abnormal number of chromosomes.

Annotation: Contextual information about the normal and abnormal behavior of the protein(s) encoded in a gene.

Anticipation: Increasingly earlier manifestation of a genetic disease with each successive generation (for instance in Huntington disease).

Autosomes: The 22 numbered chromosomes that are not involved in sex determination.

Avuncle: The sibling of a parent of a person of interest (i.e., an aunt or uncle).

Base pair (bp): Two complementary nucleotides (adenine and thymine, or guanine and cytosine).

Base sequence: The order of nucleotide bases in DNA or RNA molecules.

Carrier: Person with one copy of a recessive allele that on its own does not cause disease or a trait; in X-linked disease, female carriers have a 50% chance of transmitting the pathological allele to their sons.

cDNA: Complementary DNA that is generated by copying a messenger RNA segment, either in vivo by a small group of viruses or in vitro, to generate a more stable sequence for detailed laboratory analysis.

Centimorgan (cM): A unit of measure of recombination frequency. One centimorgan is equal to a 1% chance that a marker at one genetic locus will be separated from a marker at a second locus due to crossing over in a single generation. In human beings, 1 centimorgan is equivalent, on average, to 1 million base pairs.

Centromere: The chromosomal region where two sister chromatids are joined and to which spindle fibers attach during cell division.

Certificate of confidentiality: Letter issued by the DHHS to shield from all compelled inquiry.

Chimerism: Presence of a mixture of two genetically different cell lines throughout the body in an individual arising from the fusion of two fertilized ova.

Chromosomes: The self-replicating genetic structures in cells containing the cellular DNA that bears the linear array of genes in its nucleotide sequence.

Chromosome painting: Visual identification of individual chromosomes by tagging these with a fluorophore.

355

***Cis*-acting:** Relating to a molecular effect on the same chromosome; *trans*-acting relates to effects arising from different chromosomes.

Cistron: A genetic unit that encodes a single polypeptide; mRNA that encodes only one peptide is called *monocistronic*. This is the norm in eukaryotic cells.

Clade: Population group or group of species that includes the smallest number of shared ancestors.

CLIA: Clinical Laboratory Improvement Amendments regulate laboratories in the USA.

Clone: A group of cells derived from a single ancestor of the same phenotype.

Cloning: The process of asexually producing a group of genetically identical cells (clones).

Cloning vector: DNA molecule originating from a virus, a plasmid, or the cell of a higher organism into which another DNA fragment of appropriate size can be integrated without losing the vector's capacity for self-replication.

Codominant: Mode of inheritance where two traits can be apparent at the same time.

Complementary sequences: Nucleic acid base sequences that can form a double-stranded structure by matching base pairs; the complementary sequence to GTAC is CATG.

Concordance: Extent of agreement in phenotypic or genotypic expression of traits, often determined in twins.

Confirmatory testing: Independent replication of a test or investigation to strengthen reliability of a result.

Conserved sequence: A base sequence in a DNA molecule (or an amino acid sequence in a protein) that has remained essentially unchanged throughout evolution.

Contig map: A map depicting the relative order of a linked library of small overlapping clones representing a complete chromosomal segment.

Contigs: Groups of clones representing overlapping regions of a genome.

Copy number variants: Large duplications of DNA sequences, often extending over hundreds of thousands of bases.

Cosegregation: The tendency of traits or loci to be inherited together.

Cosmid: Artificially constructed cloning vector containing the *cos* gene of the phage lambda. Cosmids can be packaged in lambda phage particles for infection into *Escherichia coli*; this permits cloning of larger DNA fragments (up to 45 kb) than can be introduced into bacterial hosts in plasmid vectors.

Crossing over: The breaking during meiosis of one maternal and one paternal chromosome, leading to the exchange of corresponding sections of DNA, and the subsequent rejoining of the chromosomes. This process can result in an exchange of alleles between chromosomes.

Cytidine: Nucleoside containing the base cytosine and the sugar ribose.

Cytosine: Pyrimidine base; a constituent of nucleotides and the polynucleotides RNA and DNA. The cytosine (C) in DNA is usually paired with guanine (G) in the complementary strand.

Deoxynucleotides: The precursors for in vivo or in vitro DNA and RNA synthesis, containing a base (adenine [A], cytosine [C], guanine [G], thymine [T, or uracil [U]), deoxyribose, and phosphate.

Deoxyribonucleic acid (DNA): The double-stranded polynucleotide molecules that encode all genetic information in plants and animals.

Digenism: A condition that is dependent on the presence of gene variants at two distinct loci.

Diploid: Full genome with 23 chromosomes from one parent and 23 chromosomes from the other parent.

Diplotype: Combination of an individual's two haplotypes at a specific locus.

DNA replication: Duplication of a DNA strand, such as occurs in the nucleus during cell division.

DNA sequence: The specific arrangement of base pairs in a segment of DNA, for instance in a PCR product or an entire chromosome.

Domain: Distinct region of a large protein, often with specific functional properties, such as nucleotide or ion binding.

Dominant: Mode of inheritance where the trait always overrides the other (recessive) trait in the heterozygous state.

Double helix: The intertwined arrangement of two DNA strands that is stabilized by the pairing of complementary bases, i.e. adenine (A) with thymine (T), and guanosine (G) with cytosine (C).

Dietary Reference Intakes (DRI): Summarizes intake recommendations for healthy people.

Duplicon: A block of low-copy repeats (LCR); often located in pericentromeric chromosome regions, duplicons can be preferred targets of duplications or other gene fragments.

Estimated average requirement (EAR): The intake level for assessing adequacy of groups.

Endonuclease: An enzyme that cleaves its nucleic acid substrate at internal sites in the nucleotide sequence; restriction endonucleases recognize and cleave at specific sites with a 4–8 nucleotide sequence; homing endonucleases cleave double-stranded DNA at specific long (12–40 nucleotides) sequences.

Enzyme: Catalytic protein that decreases the activation energy required for a reaction and thus accelerates the reaction rate without changing the equilibrium of the reaction.

Epigenetic: Relating to chemical DNA modifications that influence gene expression.

Epistasis: Gene–gene interaction, meaning the impact of one gene or locus on the effect of another one.

Eukaryote: A classification comprising all organisms consisting of one or more cells with a cell membrane, nucleus, mitochondria, and other well-defined organelles. This definition includes all higher plants and animals and excludes archaea, blue-green algae, bacteria, and viruses.

Expressivity: Range of symptoms and outcomes associated with a particular genotype.

Exon: The portion of a gene that is translated into protein; most mammalian genes comprise several exons separated by introns (nontranscribed sequences).

Exonuclease: An enzyme that cleaves nucleotides sequentially from the free ends of a linear nucleic acid substrate.

FDCA: Federal Food, Drug, and Cosmetic Act (USA).

Fluorescence in situ hybridization (FISH): A physical mapping approach that uses fluorescent tags to detect hybridization of probes with metaphase chromosomes and with the less-condensed somatic interphase chromatin.

Founder effect: Population expansion from a few ancestors that can explain some common variants.

Frameshift mutation: Deletion or insertion that alters the pattern (reading frame) for assigning three sets of bases at a time to a codon; skipping or adding one or two bases generates mostly nonsense codon sequences that often include inappropriate stop codons.

Gamete: The reproductive cells (sperm or ovum) that carry a single set of chromosomes (23 for humans).

Gene: A DNA segment that encodes a protein or a functionally distinct RNA molecule.

Gene expression: Process that transcribes genomic DNA into messenger, ribosomal, and other types of RNA.

Gene family: Group of genes with a common heritage and related function.

Gene mapping: The process of measuring the relative positions of genes and the distances between them on a chromosome or plasmid.

Gene product: Relates to both mRNA and protein synthesized from a DNA template during gene expression.

Gene therapy: Transfer of DNA, usually via a virus or artificial construct, into somatic cells with the aim of balancing a harmful genetic variation.

Genetic drift: Shifting population frequency of a trait arising from random variation.

Genetic exceptionalism: The idea that genetic information should be treated differently from other kinds of medical data.

Genetics: The study of inherited traits, particularly of structural and functional effects and their transmission.

Genome: The sum of an organism's genetic and epigenetic information.

Genomics: The science and technology that investigates and uses information about the sequence and organizational principles of genomes.

Genotype: Pair of inherited alleles at a gene locus (except for X-chromosomal alleles in males).

Germline: Gametes and their direct precursors that provide the DNA for a new individual, i.e., ova, spermatocytes, sperm, and premorula embryonic cells.

Germline mutation: Modification of DNA in germline cells that affects future generations.

GINA: The Genetic Information Nondiscrimination Act of 2008 (USA).

Guanine (G):   A nitrogenous base constituent of DNA and RNA, which readily pairs with the base cytosine.

Guanosine: Nucleoside containing the base guanine and the sugar ribose.

Haplogroup: A group of people with shared ancestry, as shown by haplotype similarity.

Haploid: Referring to a cell or organism carrying only half the normal number of chromosome copies.

Haploinsufficiency: A condition in which physiological function is lost due to deletion of one allele despite retention of a normal allele.

Haplosufficiency: A condition in which the inactivation of an allele causes no detectable detrimental effects.

Haplotype: A series of alleles that are inherited together in a chromosome segment.

Hardy-Weinberg equilibrium: Correspondence of observed genotype frequencies in a population to numbers calculated from allele frequencies under the assumption of strict Mendelian inheritance. If a population is in perfect Hardy-Weinberg equilibrium with respect to alleles A and a at a particular locus, the probability (p) of an A/A homozygote is $(pA)^2$, the probability of heterozygotes (A/a) is $2.(pA \times pa)$, and the probability of an a/a homozygote is $(pa)^2$.

Haseman-Elston regression: Graph of marker locus position (Mb) vs. statistical probability ($-\log_{10}$ of p) of linkage.

Hemizygosity: Condition where function of half of the copies of a particular gene is lost, often due to partial or total loss of a chromosome.

Hemizygous: Carrying one allele when no second copy is present.

Heritability: Estimated percentage to which a trait is inherited.

Heterodisomy: Inheritance of two different versions of a chromosome from the same parent.

Heteroploid: Indicates that the number of chromosomes in a cell or organism is not an even multiple of the usual single (haploid) set of chromosomes.

Heterosis: Molecular heterosis refers to significantly higher or lower (negative heterosis) trait expression in carriers of a heterozygous genotype compared to carriers of the homozygous genotype of a particular polymorphism; a potential example may be the effect of the *PON1* L55M polymorphism on beta-cell function.

Heterozygote: Individual with nonidentical alleles at a particular gene locus; those with identical alleles are homozygotes.

Heterozygous: Carrying different alleles on a chromosome pair.

HIPAA: Health Insurance Portability and Accountability Act (USA).

Homeobox: DNA sequences that are extremely similar in many genes across species from fruit flies to humans, often controlling development of organs or limbs.

Hominins: Human species (*Homo sapiens*, *Homo ergaster*, and *Homo rudolfensis*) and their recent ancestors.

Hominoids: Hominins plus chimpanzees, gorillas, and orangutans.

Homolog: A gene whose sequence is similar to that of another gene in the same or another genome, possibly due to common ancestry (which would be an ortholog or paralog).

Homology: Indicates that a DNA, RNA, or protein sequence resembles that of another organism.

Homologous chromosomes: Pair of chromosomes that carry respective maternal and paternal DNA sequences, for example the two copies of chromosome 6 in somatic cells.

Homozygote: Individual with identical alleles at a particular locus; those with nonidentical alleles are heterozygotes.

Homozygous: Carrying the same variant forms (alleles) on a chromosome pair.

Hybridization: The process of joining two complementary strands of DNA or one each of DNA and RNA to form a double-stranded molecule.

Hypostatic trait: Genetic phenotype (such as blue eye color) that becomes only apparent when the effect of interacting (epistatic) genes is diminished or absent.

Imprinting: Methylation or other modification of germline DNA that influences expression of one or more genes.

Inborn error of metabolism: Rare inherited diseases with severe harmful effects on the biochemistry of the body.

In situ hybridization: Use of a DNA or RNA probe to detect the presence of the complementary DNA sequence in cloned bacterial or cultured eukaryotic cells.

Intein: Internal portion of a protein sequence that is excised to produce the mature mRNA template for protein synthesis.

Interphase: The period in the cell cycle when DNA is replicated in the nucleus; followed by mitosis.

Introns: The DNA base sequences that separate the protein-coding sequences (exons) of a gene; introns are transcribed into RNA but removed before the RNA is translated into protein.

Independent review boards (IRB): Their approval is needed for all medical research.

Isodisomy: Inheritance of two identical copies of a chromosome from the same parent.

Isotenic: Indicating that DNA sequences of different species are located in similar regions of homologous chromosomes.

Karyotype: Analysis of chromosomes by size and shape (microscopically) or on the basis of their DNA content (flow cytometry).

Kilobase (kb): Unit of length for DNA fragments, equal to 1000 nucleotides.

Knockout model: Strain of animal with targeted deletion of a gene.

Knock-in model: Strain of animal with targeted replacement of a genome sequence.

Linkage: An indirect measure of the distance between two loci on the same chromosome, derived from the probability of those two loci being inherited together; usually measured in centimorgan (cM, approximately 1 million bases in humans).

Linkage disequilibrium: A measure that describes how closely the alleles at two loci are statistically correlated.

Locus (pl. loci): The position of a particular DNA segment on a chromosome. The use of locus is sometimes restricted to mean regions of DNA that are expressed.

Locus heterogeneity: When a trait is determined by alternative gene loci. For example chylomicronemia may be caused by variants at the *LPL* (*lipoprotein lipase*) or *APOC2* loci.

LoD score: A measure of the statistical probability of linkage between markers or traits; LoD is an abbreviation of *logarithmic odds*; 3 indicates that linkage is 1000 times more likely than no linkage, -3 indicates that linkage is 1000 times less likely than no linkage.

Lyonization: Inactivation of one X-chromosome in women by imprinting. At some sites the selection of the suppressed gene depends on the gender of the originating parent.

Marker: Sequence at a known chromosome location.

Megabase (Mb): Unit of length for DNA fragments equal to 1 million nucleotides and roughly equal to 1 cM (centimorgan).

Megaphenic: Describing genetic or other factors that move the individual phenotype greatly outside the normal distribution.

Meiosis: The process of two consecutive cell divisions in the diploid progenitors of sex cells. Meiosis results in four rather than two daughter cells, each with a haploid set of chromosomes.

Memetic inheritance: Transmission of methods and behaviors from one person to another.

Mendelian disease: Human disease caused by mutations in a single gene.

Mendelian randomization: A study design where outcomes are compared by genotype.

Messenger RNA (mRNA): Molecule that encodes the information for polypeptides.

Metabolomics: The science and technology that investigates and uses information about the metabolic actions and products of genomes.

Metaphase: A stage in mitosis or meiosis during which the chromosomes are aligned along the equatorial plane of the cell.

Missense mutation: Single DNA variant in the coding region of genes that alters the sequence of the encoded protein.

Mitosis: The process of nuclear division in cells that produces daughter cells that are genetically identical to each other and to the parent cell.

Monocistronic: Refers to a gene sequence that encodes a single polypeptide.

Monomorphic: Sequence that lacks common variants.

Monozygotic twins: Siblings grown from the same fertilized egg.

Moonlighting enzyme: An enzyme that exhibits other than its standard activities, which may be responsible for an unexpected phenotype.

Mosaicism: Presence in tissues of two or more cell lines that differ genetically.

Multimodal distribution: Non-normal distribution generated by subgroups with different means.

Multiplexing: Performing the simultaneous amplification of several targets for variant identification or sequencing.

Mutation: Alteration to DNA sequence. Only mutations in germline cells (ova, sperm, and pre-morula embryonic cells) are heritable; this means that most acquired mutations (e.g., in cancer cells) are not transmitted to offspring.

Neural tube defects: Anencephaly, spina bifida, and related birth defects.

Nonsense mutation: DNA base change that introduces a termination codon in the coding sequence, which results in a truncated protein.

Northern blotting: A technique for the analysis of RNA by (electrophoretic) separation and transfer to a carrier (membrane) followed by detection with a labeled complementary DNA probe.

Nucleic acid: A large molecule composed of nucleotide subunits.

Nucleotides: The basic units for genetic transmission consisting of a nitrogen-containing base, a phosphate molecule, and a sugar molecule (deoxyribose in DNA and ribose in RNA).

Nucleus: Organelle in eukaryotes that contains the chromosomes and other molecular structures for RNA and protein synthesis.

Nutritope: An environment with a particular pattern of nutrient abundances and food toxins.

Oncogene: Mutated genes associated with excessive cell proliferation rates; they are common components of benign and malignant tumor genotypes.

Odds ratio: Compares the probability of a trait or event in the presence of a particular genetic factor to the probability of the trait or event in the absence of the factor.

Operator: A DNA sequence within a gene to which activators or repressors can bind and thereby influence gene expression.

Operon: A gene cluster that is transcribed together, usually under the control of shared untranslated regions.

Ortholog: A gene whose sequence is similar to that of another gene due to shared ancestry.

Overexpression: Abnormal production of a gene product (which may be a protein or an RNA molecule).

p: Indicates the short or small (*petit*) arm of a chromosome.

Paralog: A gene whose sequence is similar to that of another gene in the same genome, usually by duplication and subsequent divergent evolution.

Paucimorphism: Relating to a very rare genetic variant.

Penetrance: Probability of occurrence of a particular phenotype given the presence of a certain genotype. Specifically, penetrance may be the probability of someone with a certain genotype to get a certain disease or disease symptoms.

Phenotype: The physical or otherwise observable appearance of an individual with a particular genotype.

Phenocopy: Mimicking of the phenotype typically produced by a particular genotype of a gene. This may be exclusively due to nongenetic factors.

Physical map: A map of the locations of identifiable landmarks on DNA (e.g., restriction enzyme cutting sites, genes), regardless of inheritance. Distance is measured in base pairs. For the human genome, the lowest resolution physical map is the banding patterns on the 24 different chromosomes; the highest- resolution map would be the complete nucleotide sequence of the chromosomes.

Plasmid: Autonomously replicating, extrachromosomal circular DNA molecules, distinct from the normal bacterial genome and nonessential for cell survival under nonselective conditions.

Pleiotropy: Describing an effect that concerns diverse and unrelated body functions or different life stages.

Polymorphism: Commonly occurring difference in DNA sequence among individuals. Genetic variations occurring in more than 1% of a population are considered useful polymorphisms for genetic linkage analysis.

Positive predictive value: Indicates the probability that people with a particular test result will experience a specific outcome.

Private mutation: A variant that is particular to an individual or family.

Probe: Synthetic single-stranded DNA or RNA molecules for the amplification or detection of specific polynucleotide sequences by hybridization.

Proficiency testing: Activity organized by an independent certifying entity to assess whether a laboratory can perform a test correctly.

Prokaryotes: Cells or organisms, such as bacteria, without a separate nucleus and other subcellular compartments.

Promoter: A site to DNA to which RNA polymerase will bind to initiate transcription.

Proteomics: The science and technology that investigates and uses information about the sequence, structure, and organizational principles of protein products of genomes.

Proto-oncogene: A gene that promotes normal cell division in its native state, but can cause unrestrained proliferation upon mutation.

Pseudogene: Nonfunctional genomic sequence derived by descent from a functional gene.

Purines: Chemical structures that are characterized by an aromatic six-ring system consisting of one nitrogen atom and five carbons joined to a five-ring system with another two nitrogen atoms and one carbon atom. The most common examples of purines are adenine and guanine in DNA and RNA; the adenosine phosphates, adenosine triphosphate (ATP), adenosine diphosphate (ADP), adenosine monophosphate (AMP), and cyclic AMP; the guanosine phosphates, guanosine triphosphate (GTP) and guanosine diphosphate (GDP); the metabolites inosine, hypoxanthine, xanthine, and uric acid; and the presumably essential nutrient queuine.

Pyrimidines: Chemical structures characterized by a six-ring, basic compound that occurs in nucleic acids. DNA contains the pyrimidine bases cytosine (C) and thymine (T); RNA contains the pyrimidine base uracil (U) in place of thymine.

q: Indicates the larger arm of a chromosome.

QTL: Quantitative trait locus (plural loci).

Rare-cutter enzyme: Restriction enzyme that cuts at less than one in 10,000 base pairs.

Recommended dietary allowance (RDA): The intake level that meets the needs of most people.

Recessive: Mode of inheritance where the trait is always overridden by the other (dominant) trait in the heterozygous state.

Recombination: The process by which progeny derive a combination of genes different from that of either parent. In higher organisms, this can occur by crossing over.

Regulatory region: A DNA segment that influences gene expression.

Restriction enzymes (endonucleases): Bacterial enzymes that recognize and cut short palindromic nucleotide sequences with very high specificity.

Restriction fragment length polymorphism (RFLP): Variation between individuals in DNA fragment sizes cut by specific restriction enzymes.

Retrotransposon: Short DNA sequence that amplifies itself by being copied into RNA, then back into DNA that can then be reinserted into the genome. Retrotransposons are very common in both plants and animal, comprising nearly half of the human genome.

RNA: Polymeric ribonucleic acid, which usually contains a uridine moiety instead of the thymidine in DNA, encodes the messenger RNA (mRNA) templates for protein synthesis as well as a variety of functional and short regulatory sequences.

Ribosomal RNA (rRNA): The type of ribonucleic acid associated with the ribosomes, the cellular structures responsible for protein synthesis.

Ribosomes: The molecular structures, consisting of numerous protein components and specialized RNA, responsible for protein synthesis.

Satellite DNA sequence: Highly repetitive DNA sequences adjacent to the centromere of higher eukaryotes.

Segregation analysis: Investigation of the ratios of traits across generations.

Sensitivity: Probability that a test will give an affirmative result when the sample contains the target variant.

Sequence tagged site (STS): Short (200–500 bp) DNA sequence that has a single occurrence in the human genome and whose location and base sequence are known.

Sequence analysis: Deciphering or decoding of the sequential composition of polynucleotides (DNA or RNA) or proteins. In DNA or RNA, the sequence of nucleotides is determined; in proteins or peptides, the sequence of amino acids.

Sex chromosomes: The X and Y chromosomes in humans that determine usually the sex of an individual.

Shotgun method: Cloning of DNA fragments randomly generated from a genome.

Single gene disorder: Inherited disease caused by variation of a single gene (e.g., familial hypercholesterolemia, due to defective low-density lipoprotein receptor).

Short tandem repeat (STR): Repeats of two or more bases; STRs are commonly used for forensic identification.

Single-strand conformation polymorphism (SSCP): A method that uses differential electrophoretic migration of amplified DNA in denaturing gels to detect sequence variations.

Specificity: Probability that a test will give a negative result when the sample does not contain the target variant.

Somatic cells: Any cell in the body except gametes and their direct precursors.

Southern blotting: A method for the characterization of DNA that combines electrophoretic separation on a gel with detection with radiolabeled complementary probes.

Syntenic: Genes located on the equivalent chromosome in different species.

Tandem repeat sequences: Short repeating DNA sequences that may follow in the same direction (direct tandem repeats) or in opposite directions (inverted tandem repeats).

Telomere: Special DNA sequences at the ends of chromosomes that influence replication and stability of the chromosome.

Thymine (T): A pyrimidine base, pairs with adenine (A) in DNA and RNA.

Thymidine: Nucleoside consisting of the base thymine and the sugar ribose.

Trait: Observable property of an organism.

***Trans*-acting:** This describes a molecular effect on the same molecule; DNA recognition sequences for specific binding proteins such as the vitamin D receptor are *trans*-acting if they affect transcription of a remote DNA sequence; they are *cis*-acting when they affect nearby sequences.

Transactivation: Activation of gene transcription by a protein or RNA encoded at a distant locus.

Transcription: Generation of mRNA copies from nuclear DNA segments as the initial step of gene expression.

Transfer RNA (tRNA): Short segments of RNA with specific triplet nucleotide codons at one end and a segment that binds a particular amino acid with high specificity.

Transformation: Alteration of a cell's specific DNA sequences, often those affecting proliferation or differentiation.

Transition: Change of a pyrimidine base (cytosine [C], thymine [T], or uracil [U]) into another pyrimidine base, or of a purine base (adenine [A] or guanine [G]) into another purine base, in a nucleotide sequence, usually due to a mutation; a more common form of mutation than a transversion.

Translation: Ribosomal synthesis of protein with the use of mRNA templates.

Transposon: DNA segment that can move to another organism without having any homology to a sequence of the new host; may contain one or several genes and is an important mode of conferring resistance among bacteria.

Transrepression: Inhibition of gene transcription by a protein encoded at a distant locus.

Transversion: Change in a nucleotide sequence of a pyrimidine base (cytosine [C], thymine [T], or uracil [U]) into a purine base (adenine [A] or guanine [G]), or vice versa, in a nucleotide sequence, usually due to a mutation; a less common form of mutation than a transition.

Triploid: A cell or organism containing three sets of chromosomes; this may occur in cancer cells or in concepta where an ovum has been fertilized by two sperm.

Uracil (U): Pyrimidine base in RNA that pairs with adenine (A) during transcription.

Uridine: Nucleoside containing the base uracil and the sugar ribose.

Validation: Procedure of proving that a test works as intended and achieves the intended analytical result or biological outcome.

Verification: Confirmation with established credentialing procedures that specified performance or proficiency requirements have been fulfilled.

Virus: A noncellular infectious agent that uses the transcriptional machinery of infected host cells for replication.

Variable number tandem repeats (VNTR): Allelic variants defined by their different numbers of repeated oligonucleotide sequences in series.

Western blotting: Method for the characterization of proteins that combines electrophoretic separation with detection by antibody binding.

Wild type: Expression used to indicate the most common allelic form of a variant; use is problematic in humans because it has often been unclear in the past whether the allele is truly the most common one across the world. Neither the wild type nor the most common allele is necessarily the ancestral form.

Yeast artificial chromosome (YAC): A vector used to clone DNA fragments (up to 400 kb); it is constructed from the telomeric, centromeric, and replication origin sequences needed for replication in yeast cells.

Xenolog: A gene that has sequence similarity with a gene in another species because an ancestor has acquired the gene from the other species.

SCIENCE JOURNALS RELEVANT TO NUTRIGENETICS

American Journal of Clinical Nutrition, http://www.ajcn.org/
American Journal of Human Genetics, http://www.cell.com/AJHG/
British Journal of Nutrition, http://journals.cambridge.org/action/displayJournal?jid=bjn
Drug Metabolism and Predisposition, http://dmd.aspetjournals.org/
European Journal of Clinical Nutrition, http://www.nature.com/ejcn/index.html
Genes and Nutrition, http://www.springer.com/biomed/human+genetics/journal/12263
Journal of Nutrigenetics and Nutrigenomics, http://content.karger.com/ProdukteDB/produkte.
asp?Aktion=JournalHome&ProduktNr=232009
Journal of Nutrition, http://jn.nutrition.org/
Molecular Genetics and Metabolism, http://www.sciencedirect.com/science/journal/10967192
Molecular Nutrition & Food Research, http://onlinelibrary.wiley.com/journal/10.1002/%28ISSN
%291613—4133
Nature Genetics, http://www.nature.com/ng/journal/vaop/ncurrent/index.html
PLoS Genetics, http://www.plosgenetics.org/home.action
Public Health Genomics, http://content.karger.com/ProdukteDB/produkte.asp?issn=1662—8063
The Pharmacogenomics Journal, http://www.nature.com/tpj/index.html

INFORMATION ON INBORN ERRORS OF METABOLISM

Blau N, Duran M, Blaskovics ME, Gibson KM. Physician's Guide to the Laboratory Diagnosis of
Metabolic Diseases (2nd ed.). Springer, 2002, 716 pp. ISBN 978-3-540-42542-7.
Blau N, Hoffmann GF, Leonard J, Clarke JTR. Physician's Guide to the Treatment and Follow-up of
Metabolic Disease (1st ed.). Springer, 2006, 416 pp. ISBN 3-540-22954-X.
Scriver's Online Metabolic & Molecular Bases of Inherited Diseases, http://www.ommbid.com/

GENE INFORMATION

Disease descriptions, glossary, and lab directory, http://www.ncbi.nlm.nih.gov/sites/GeneTests/
National Library of Medicine, gene information, http://www.ncbi.nlm.nih.gov/sites/entrez?
db=gene
Online Mendelian Inheritance in Man (OMIM), comprehensive gene information, http://www.
omim.org/

DATABASES

1000 Genomes Project, http://www.1000genomes.org
Catalogue of GWAS, http://www.genome.gov/gwastudies/

Database of Genomic Variants, The Centre for Applied Genomics, Toronto, http://projects.tcag.ca/variation/

HapMap, http://www.hapmap.org/

Micronutrient-related pathways, http://micronutrients.wikipathways.org

NHLBI Exome Sequencing Project (ESP) Exome Variant Server (accessed June 2012), http://evs.gs.washington.edu/EVS/

OpenSNP, http://opensnp.org/ searchable by rs number, gives position on chromosome and in gene

SNPedia for allele frequencies, http://www.snpedia.com

DATA MINING AND DISPLAY TOOLS

dbSNP, http://www.ncbi.nlm.nih.gov/projects/SNP/

Display linkage disequilibrium and recombination, https://statgen.sph. umich.edu/locuszoom/genform.php?type=ourdata

Ensemble, http://www.ensembl.org

Entrez Gene, http://www.ncbi.nlm.nih.gov/sites/entrez?db=gene

GEN2PHEN Project, http://www.gen2phen.org/

GenBank, http://www.ncbi.nlm.nih.gov/Genbank/

Genetic Association Database, http://geneticassociationdb.nih.gov/

HuGE Literature Finder, http://www.hugenavigator.net/HuGENavigator/ startPagePubLit.do

HuGE Navigator GWAS Integrator, http://hugenavigator.net/ HuGENavigator/gWAHitStartPage.do

Human Gene Coexpression Database, http://www.geneticsofgeneexpression. org/network/

Integrative Genomics Viewer, tool for displaying associations, http://www. ncbi.nlm.nih.gov/projects/gapplusprev/sgap_plus.htm

National Center for Biotechnology Information (NCBI), http://www.ncbi. nlm.nih.gov/

Phenotype-Genotype Integrator, http://www.ncbi.nlm.nih.gov/gap/PheGenI

PhenX Toolkit (consensus measures of Phenotypes and eXposures), http:// www.phenxtoolkit.org

PolyPhen,tool for the exploration of the possible impact of amino acid variation on protein function, http://genetics.bwh.harvard.edu/pph/

PubMed, http://www.ncbi.nlm.nih.gov/sites/entrez/

Variant Name Mapper, http://www.hugenavigator.net/HuGENavigator/ startPageMapper.do

GENETIC SOFTWARE PACKAGES

HAPSTAT: Statistical analysis of haplotype-disease associations, http://www. bios.unc.edu/~dlin/hapstat/

ISHAPE and SHAPEIT: Accelerated and verified haplotype analysis programs [1], http://www.griv.org/ishape/ and http://www.shapeit.fr/

PEDCHECK software: Examine variants for Mendelian inconsistencies [2], http://watson.hgen.pitt.edu/register/

PREST software: Pedigree RElationship Statistical Test [3], http://fisher.outstat.toronto.edu/sun/Software/Prest

S.A.G.E.: Statistical Analysis for Genetic Epidemiology, open source software for the genetic analysis of family, pedigree and individual data, http://darwin.cwru.edu/sage/

SOLAR software: Sequential Oligogenic Linkage Analysis Routines, for estimating residual heritability [4], http://www.txbiomed.org/departments/genetics/genetics-detail?p=37

References

[1] Delaneau O, Coulonges C, Boelle PY, Nelson G, Spadoni JL, Zagury JF. ISHAPE: new rapid and accurate software for haplotyping. BMC Bioinformatics 2007;8:205.

[2] O'Connell JR, Weeks DE. PedCheck: a program for identification of genotype incompatibilities in linkage analysis. Am J Hum Genet 1998;63(1):259–66.

[3] McPeek MS, Sun L. Statistical tests for detection of misspecified relationships by use of genome-screen data. Am J Hum Genet 2000;66(3):1076–94.

[4] Almasy L, Blangero J. Multipoint quantitative-trait linkage analysis in general pedigrees. Am J Hum Genet 1998;62(5):1198–211.

Index

Note: Page numbers followed by "f", "t" and "b" indicate figures, tables and boxes respectively